Science, Style and the Study of Community Structure

An Example from the Central Mississippi River Valley

Carl Philipp Lipo

BAR International Series 918
2001

This title published by

Archaeopress
Publishers of British Archaeological Reports
PO Box 920
Oxford
OX2 7YH
England
www.archaeopress.com

BAR S918

Science, Style and the Study of Community Structure: An Example from the Central Mississippi River Valley

Printed in England by The Basingstoke Press

ISBN 1 84171 216 7

All BAR titles are available from:

Hadrian Books Ltd
122 Banbury Road
Oxford
OX2 7BP
England

The current BAR catalogue with details of all titles in print, prices and means of payment is available free from Hadrian Books

TABLE OF CONTENTS

LIST OF FIGURES

LIST OF TABLES

ACKNOWLEDGMENTS

The source of this volume was the author's dissertation research at the University of Washington. The University of Washington Department of Anthropology provided financial support, laboratory space and other resources needed to conduct this research. I also received financial support from the University of Washington Graduate School, Alice and Genevieve Niles Fellowship, and the George F. Dales Memorial Foundation.

I thank Jim Feathers, Terry Hunt, Robert Wenke, and Robert Dunnell for all of their advice and encouragement. John House of the Arkansas Archeological Survey was instrumental in the establishment of my field site and provided great guidance as I learned the practicalities of doing fieldwork in the Mississippi valley of Arkansas. I would also like to express my deep gratitude to the many kind residents of St. Francis and Crittenden counties, Arkansas who granted me access to land and information during my fieldwork. I thank Frank Barton, Ellie Beasley, Thomas Beene, Gordon Brent, Buster Briggs, Horace and Ed Cupples, Bill Felty, Rush Harris, Margaret Linn, Cuples, Lan Letner, J.J. Meals, Jeff Mitchem, Tim Mulvihill, Bert Pouncey, Jim Pugh, Larry Reynolds, Kenn Thompson, Gazzola Vaccaro, Christopher Wise, Amanda Wise, David Wallace, George Zanone, and many others. I was particularly fortunate to have received the hospitality and friendship of Ms. Ellie Beasley of Heth, Arkansas who provided me a 'base of operations' while I was in the field and aided in my research in innumerable ways. The staff and a large number of the 1996 University of Washington Field School students provided valuable assistance in generating my ceramic collections. In addition, the enthusiastic work of University of Washington Archaeology 299 and 499 students enabled me to finish the ceramic descriptions. Sarah Demb of the Peabody Museum was instrumental in helping me obtain copies of original PFG field photographs and site records. I am also thankful for the help of Lela Donat of the Arkansas Archeological Survey who helped me acquire a copy of the PFG field notes. Alejandro Lynch kindly provided the source for his Turbo Pascal program for calculating expected values for neutral memes. Mary Dunnell contributed substantial editorial assistance for which I am very grateful.

I extend much thanks to Dan Bush, Ethan Cochran, Steve Cole, George and Barbara Dales, John Dudgeon, Andy Grogan, Kim Kornbacher, Fran Hamilton, Sarah Sterling, and Kris Wilhelmsen for their longtime friendship, inspiration, and support.. I would like to give special thanks to Tim Hunt for his invaluable creative input during our early computer simulation days and beyond, Michael Pfeffer for his great friendship and inspiration, and Mark Madsen who continues to be a long-time contributor, friend, and compatriot in my research. A very special thanks goes to Deborah Schechter who, more than she knows, played a vital role in keeping my momentum moving.

Most of all, I thank Robert C. Dunnell for his intellectual contribution to this research and for his unwavering support and guidance throughout my graduate career. Through his work, Dr. Dunnell has provided me a model of genius and uncompromising commitment to intellectual standards that is necessary for conducting explorations along the frontiers of science.

DEDICATION

I dedicate this thesis to my parents, Christine and Thomas A. Lipo, who have long supported my never-ending inquires into the workings of things and have given me a deep sense of commitment and hard work.

CHAPTER 1: INTRODUCTION

Science is "the organized systematic enterprise that gathers knowledge about the world and condenses the knowledge into testable laws and principles" (Wilson 1998:53). It is also the greatest process for building knowledge ever devised (Hull 1988). The power of science is tremendous; there are no aspects of the physical world that science cannot potentially explain. Nonetheless, not all scientific endeavors enjoy the same degree of success. In some realms of research, such as nuclear science and aeronautics, scientists have built mature, robust and mathematically defined systems of explanations for the mechanisms involved in the workings of the atom and in the requirements for flight. In fields such as genetics and cosmology, scientists are just beginning to enjoy the fruits of scientific explanations. There are other arenas, however, that remain relatively poorly developed. Our ability to explain disease, for example, is just beginning to expand (*e.g.,* Ewald 1994; Nesse and Williams 1994; Ryan 1997) and only recently has scientific research started to produce functional models about the workings of the brain (*e.g.,* Deacon 1997; Minsky 1986; Pinker 1997).

Ironically, some of the most poorly understood parts of the empirical world are those that potentially will bring the greatest benefits. History, in particular, remains poorly understood and the development of historical science poses one of our greatest challenges (Wilson 1998). The explanation of history has long frustrated individuals interested in understanding how and why human events occur in the form and patterns they do. Attempts to explain and ultimately control history have a long tradition in human endeavors. Examples range from the rituals of scapulimancy that date to 5000 BP in China (Moore 1957), oracles, astrology and palm reading (Sagan 1996), to the writings of the philosophers of the Enlightenment like Condorcet (*e.g.,* Condorcet's 1794 *Sketch for a Historical Picture of the Progress of the Human Mind* cited in Wilson 1998: 16-21). Although often touted as *the* study of the past, archaeology is only one of the latest lines of inquiry in a series of marginally successful endeavors to build a science of history. Indeed, since the historical investigations of Nabonidus, the last king of the Babylon Empire in the 6th century B.C (Dougherty 1929), archaeologists have used physical remains to understand the rise and fall of empires, the spread of humans across the globe and the extensive alteration of the natural environment observed in our recorded history.

Yet, despite the centrality of the task and the attention of thousands of archaeologists working over centuries, social scientists have only begun to build an understanding of the workings of history. On the one hand, archaeology has become quite sophisticated and rigorous in terms of techniques and quantitative abilities. On the other hand, it is no nearer to becoming a science (Dunnell 1982). Progress in our discipline has been exceedingly slow. For more than a century, archaeologists have sought to build a scientific discipline. This goal was expressed by early Culture Historians (*e.g.,* see Dunnell 1992; Holmes 1886a, 1886b, 1892; Lyman *et al.* 1997; O'Brien and Lyman 1998; Wissler 1916a, 1916b, 1917), and later advocated more strongly by proponents of the New Archaeology (*e.g.,* Watson *et al.* 1971, 1984). Some archaeologists have recognized the enormity of building a scientific archaeology, but maintain the results will be worth the effort (*e.g.,* Dunnell 1980, 1992). As many have conceived, a scientific archaeology addresses the processes of cultural change and the origins of human cultural diversity (Binford 1962; Boas 1896, 1902, 1909; Dunnell 1980; Wissler 1917), thus fitting comfortably within the larger disciplinary goals of anthropology. Importantly, science would bring to archaeology more certain conclusions and, as a result, the growth of a cumulative body of knowledge about the past and the mechanisms of historical change.

Any number of reasons can be given for the failure of archaeology to emerge as a science, but it is generally agreed that the lack of a unifying theory has been the greatest hurdle to success. While better-developed sciences are built upon explanations derived from physics and chemistry, the sense-making system of archaeology is built on common sense rather than any particular body of theory. Common sense accounts are dominated by the use of empirical generalizations to make statements about the world; explanations built on common sense consist of summary statements about some specific set of observations (Binford 1968a, 1978, 1980; Dunnell 1971, 1980, 1982). Since empirical generalizations are modal summaries about the nature of the world, they are contingency bound (depending upon the set of observations that produced them) and cannot be true in an absolute or universal sense. Consequently, generalizations about one part of the archaeological record are not necessarily applicable to other parts of the archaeological record. Rather than being able to explain all realms of the archaeological record, the accuracy of empirical generalizations is a direct function of the temporal and spatial distance between the data set upon which the generalization is founded and the new observation or prediction (Dunnell 1982:17). An archaeology built on common sense is doomed to repeated failure.

Archaeologists should not take this failure as a sign of hopelessness. Prior to their success, other fields languished with problems resulting from the confusion of empirical generalizations with theory. Aeronautics, for example, had a long, floundering history before suddenly transforming itself into a scientific discipline of flight built upon rigorous theory. The famous myth of Deadalus and Icarus, kites in ancient China, medieval birdmen donning makeshift wings and leaping from towers and the renowned work of Leonardo Da Vinci are just a few examples of the long-standing human aspiration to emulate birds by observing the natural world. Early movie film images capture the multitude of attempts of individuals attempting to build flying machines based on this premise. These floppy, improbable, and now laughable machines all seek to capture the secrets of flight through variants of bird designs. Their creators built bird-shaped wings and tails to mimic the design of birds and rationalized their attempts at flight on the basis *ad hoc* generalizations about avian behavior and morphology. Despite centuries of often-fatal attempts and decades of research, little technical progress was made in aeronautics prior to the latter half of the nineteenth century. Yet, this scenario quickly changed. In just over one decade of work, two brothers from Ohio concluded the quest for heavier-than-air flying machines with their first successful airplane flight in 1903. What made the efforts of the Wright brothers different from all those who preceded them?

While the Wright brothers were unquestionably intelligent, their success was more a function about the way in which they approached their task than their mental aptitude. One of the main differences that set the Wright brothers apart from other researchers of the time is the scrutiny they paid to the progress other individuals made in their attempts at flight. In particular, the Wrights poured over the results of the German researcher Otto Lilienthal and the Frenchman Octave Chanute (Jakob 1990). Lilienthal, in particular, published tables of lift factors that were empirically generated by wings with different angles of attack as they were moved through the air. The Wrights expanded upon this and took Lilienthal's work a conceptual step further. Up until that point, testing conceptions of flight consisted of building full-scale model wings, attaching them to crude airframes and then launching them into the air with a human pilot. Success and failure in this enterprise was a slow, tedious and often deadly affair. Realizing that the *principles* of flight needed to be understood rather than merely the empirical manifestations of flight, the Wrights built a wind tunnel that accurately modeled the relationships between wing form and airflow (Jakob 1990). With this wind tunnel, the brothers evaluated the aerodynamics of tiny models of wings and made measurements of the lift generated by various orientations and shapes. To the brothers' surprise, the studies quickly rejected the work of Lilienthal. Lilienthal had made invalid assumptions about the relationship between airflow and lift by using a rotational model and had very poor precision in his estimates (Jakob 1990). The Wrights quickly developed new lift tables and checked the values under a variety of configurations and conditions. In this way, the Wrights began to build knowledge about the principles of flight. Thus, the key to the Wright's achievements is the *experiments* they used to test notions about the way they understood wings to work. Building knowledge about flight, subsequently, became a cumulative activity. This transition required, above all, moving from descriptions of the natural world to the study of the principles that form the framework within which the natural world can be understood to operate. In fact, after their wind-tunnel experiments, the Wrights built a full-scale model of a plane large enough to carry a human that finally flew. Scientific knowledge had been generated. Although the Wright flying machine bore little resemblance to a bird, it still flew as a bird because it took advantage of the same scientific *principles* that a bird uses to makes flight possible.

Although the process of scientific research is governed by the empirical world, the importance of separating the empirical world and the theoretical world cannot be over estimated (Dunnell 1971). In archaeological research, the struggle to separate these worlds continues (Dunnell 1982). Archaeological products remain empirical generalizations built on observations of the way we see human behavior take place in a day-to-day fashion. These efforts are entirely analogous to the improbable attempts at building flying machines without understanding the principles of flight. Like the early pioneers of aeronautic science, we are plagued with the use of an unanalyzed empirical world to guide our explanations. If we are to achieve the goal of building a scientific archaeology, it is necessary to focus on the principles behind history rather than the empirical details of history itself.

Dunnell (1971:26) recognized the difference between these two kinds of understandings with the concepts of the ideational and the phenomenological. The gap between the ideational notions that science constructs and the one that our senses perceive was aptly described by the English astrophysicist Sir Arthur Eddington in his account of "two tables" (Eddington 1953:xi). The first, the "familiar table," is a piece of wooden furniture, a flat surface mounted on four legs. It is an everyday object, made of "substance," that Eddington leans on to write. The second is the "scientific table," which is "mostly emptiness" with "numerous electric charges rushing around with great speed" (Eddington 1953:xi). To write, Eddington leans the emptiness of his scientific elbow on the emptiness of his scientific table. The representation that physics forms of an object is a theoretical construction resulting from hypotheses accumulated over the course of centuries. The world of science is a world of abstractions, a world of symbols. "Science aims at constructing a world which shall be symbolic of the world of the commonplace experience" (Eddington 1953:xv). That does not mean, however, that every symbol represents a precise fragment of the everyday world, or even something that we can explain in terms of sensory experience. Science resembles writing in the way that the marks of the written word symbolize the thing the word represents. Learning to move about in the scientific world, however, is not necessarily an easy task. As Eddington points out even in well-developed sciences such as physics, "we are always relapsing and mixing the symbols with incongruous conceptions taken from the world of consciousness. Untaught by long experience, we stretch a hand to grasp the shadow instead of accepting its shadowy nature" (Eddington 1953:xvii). Traditionally, the solution has been to confine discussion in mathematical symbolism, as it is "hard to avoid dressing our symbols in deceitful clothing" (Eddington 1953:xvii).

In the quest to build a science of history, it is as difficult to build a "scientific table" as it is important. We live in "familiar history" and, in everyday terms, understand that history is caused by a set of individuals, their actions and rationales. It is very hard to divorce ourselves from this understanding because to do so is entirely counterintuitive. Nevertheless, that is the point. Explanations built on the observations of a particular familiar history cannot account for changes in history across time and space anymore than notions of aeronautics built from common sense descriptions of birds can explain flight. Like the principles of flight, "scientific history" must be constructed. Like aeronautics, the study of "scientific history" is comprised of abstractions and not common sense summaries of the observations of historical change. These abstractions form a body of theory and are used to account for properties not immediately observable yet symbolic for the processes and patterns involved in "familiar history." This theory generates both our observations and explanations (Dunnell 1982). It is likely that this theory will move the study of "scientific history" into realms increasingly unlike that of "familiar history." Common sense, in this way, may prove of little use in the science of history. Nevertheless, with the loss of familiarity, we gain the power of understanding. A theory of history potentially gives us the means for understanding how and why historical change occurs and ultimately gives us the means for making better-than-random chance decisions in our stated goals (Dunnell 1982). A science of history is a powerful field indeed.

BUILDING A HISTORICAL SCIENCE

While many have agreed that building a scientific view of history is the major task facing archaeologists (*e.g.,* Binford 1962, 1968a; Nelson 1919; Spaulding 1960), there has been reoccurring confusion about the appropriate metaphysic for tackling this challenge. There are two ways to study the world. Dunnell (1982) has characterized these approaches as *time-like* and *space-like* views of temporal change. In a *space-like* view, time is perceived as discrete blocks. Periods are treated as if they are real entities and are used as parts of explanations. Relationships between these discrete blocks are perceived to remain constant during each period of time. Change between periods is transformational; one kind of period changes into another since changes in the relationships between things result in a change in the system as a whole. From Eddington's (1953) perspective, space-like views of time are equivalent to our "familiar table." Space-like views of time are consistent with the way humans intuitively deal with the world. They are commensurable with common sense.

Time-like views of history, on the other hand, are equivalent to the "scientific table." Here, we treat time as an extension of materialism into the temporal dimension and variation in both the spatial and temporal dimensions is seen as continuous. Types, species and all configurations of the empirical world are ephemeral. Relationships between things are constantly changing. In a time-like view of change, there are no Mesolithic and Neolithic periods but only changing frequencies of traits related to human and plant/animal interaction over time. Time-like views of change are historical and, as a consequence, critical to archaeology and other historical sciences. Such a view of change is non-intuitive because of the way in which humans tend to emphasize space but ignore time in their explanations. Time is not part of our common sense. Because of our mental machinery, we are not well equipped to examine the world in four dimensions simultaneously; ordinarily, we are limited to three. Consequently, as a focus of science our understanding of time is poorly developed. One of Charles Darwin's great contributions to science, for example, was to provide a theory with which it became possible to build explanations within a time-like view of change (Gould 1986; Mayr 1976, 1997:115-119).

Archaeologists are not entirely strangers to viewing the world in a time-like fashion. The early culture historians built chronologies with seriation, a method that is based on a time-like perspective of change albeit fortuitously. Seriation is a way of ordering frequencies of attributes and, as such, it is perfectly commensurable with a historical framework. In the absence of theory, however, culture historians turned to common sense to talk about their constructions (*e.g.,* Deetz and Dethlefsen 1965, 1971; Ford 1938a, 1949:35-37, 1962; Kroeber 1916; Phillips *et al.*1951:220-223; Spier 1917). Though continuous, and without any theoretical warrant to be otherwise, time/space sequences were turned into periods that were regarded as internally homogeneous, consistent with common sense understanding of culture. This was a process of *periodization* in which patterns are defined by sets of associated attributes that bear no necessary relationship to one another. Spurred on by anthropologists and changing interests in archaeology, behavior was reconstructed for these units using ethnography, common sense and analogy; attributes were defined in an *ad hoc* fashion and given meaning in a *post hoc* manner. In this way, culture historians moved from a time-like view of change (as embodied by seriation) to a space-like view of difference.

The new archaeologists of the 1960s expanded on culture historians' *ad hoc* transition from periods to behavior. Since their explicit goal was to be both scientific and anthropological, the attempts of the culture historians to be anthropological was (rightly so) regarded as unscientific and "normative" (Binford 1962; Flannery 1967). Consequently, the entire culture historical enterprise (unjustly so) was regarded as unscientific. Binford (1962) and others (*e.g.,* Butzer 1971; Clarke 1968; Flannery 1986) argued that archaeologists must view material remains as products of interactions between humans and their environment. Systems were seen as a rigorous and scientific means of addressing these anthropological interests. Systems were perfectly compatible with a space-like view of change. Common sense, ethnographic observations, and analogy could all be conducted within a space-like view of change. A systems approach, therefore, seemed to meet the scientific needs of the archaeologists. In fact, the first attempt at making anthropological analyses of archaeological material more scientific focused on the development of more rigorous analogies (*e.g.,* Ascher 1961; Binford 1967, 1968b, 1978, 1980; 1986; Gould 1968a, 1968b, 1969, 1978; Gould and Watson 1982; Klejn 1973; Longacre 1970b; Morwood 1975; M. Smith 1955; Stiles 1977; Thompson 1956; Tringham 1978).

However, while space-like views are appropriate for understanding questions relevant to how things work in a particular point in time, they are inappropriate for understanding change. Relations between things are not constant through time but change, as do the things themselves. Systems, therefore, are inappropriate for time-like change. Systems rely on the fact that relationships and things are constant and real. While they work in the present where the effects of time are minimal, they are perfectly useless for great temporal change. The archaeological record is not a "snap-shot" of time but rather an accretionary phenomenon (Binford 1978, 1981). The movement of archaeologists to adopt a systems approach was a misguided attempt to treat the failings of the culture historians -- the new archaeologists rejected the time-like component of culture history within which the strength of the paradigm lay.

History is appropriately viewed from a time-like perspective. In fact, most sciences that have traditionally studied the world as a "system" of phenomena have begun to turn to analysis of dynamic systems of continually changing phenomena and relationships (*e.g.,* Kaufmann 1993; Waldrop 1992). Ecologists have begun to conclude that such staid concepts such as "ecosystems" are nothing more than the historical trajectory of changing species and their interrelationships (*e.g.,* Davis 1976, 1985). Cultural phenomena should not be treated any differently. In a time-like view of time, there is no "past" system to reconstruct – only patterning of material caused by the historical interaction of phenomena through time.

Biologist Francois Jacob (1998:18) argues that :

> If it is difficult to predict the future, it is sometimes just as difficult to reconstruct the past. Scientific thought both refines and complicates the answer to the old question, where do we come from? The whole universe and the objects it contains, living or not, are like the products of an evolutionary process in which two kinds of factors are at play: on the one hand, the constraints that determine the rules of the game and demarcate the limits of the possible; on the other, the circumstances that direct the actual courses of events, the history of the game as it was actually played. The constraints can usually be formalized. Everything that depends on them can be predicted with high probability. The role of history, on the other hand, can be recognized and sometimes explained. But obviously the chain of events that will make history tomorrow cannot be predicted. This feature of the forces that shape our world is entirely contingent.

Thus, as archaeologists we face the challenge of building a scientific history that simultaneously treats the constraints of history that produce patterns and the circumstances that comprise the *actual* course of events. Unlike physics, the empirical domain of history is entirely contingent; the order of events matter and causation is derived from the order of those events (Gould 1986, 1989; Gould and Lewontin 1979; Mayr 1997). History matters. In addition, we must build models of the constraints that comprise the processes of historical change. Here, theory is needed to define both the units of measurement and their relationships (Dunnell 1971). Using models of these constraints we can, in turn, measure and understand the particular sequence of events that comprise history. It is the interplay between the formalization of the constraints and the study of the events that make up history that forms a science of the historical record.

Throughout the history of archaeological research, many (*e.g.,* Bettinger 1991; Binford 1977; Dunnell 1982) have argued that a scientific approach to explaining culture change requires theory. Scientific theory provides the essential component of all scientific explanations because it is a coherent system of propositions that stipulates construction of meaningful units, and specifies the nature of the relations between those units (Dunnell 1971). Theory articulates ideas with the empirical world. E.O. Wilson (1998:52) writes:

> Nothing in science – nothing in life, for that matter, makes sense without theory. It is our nature to put all knowledge into context in order to tell a story, and to re-create the world by this means. So let us visit the topic of theory for a moment. We are enchanted by the beauty of the natural world. Our eye is caught by the dazzling visual patterns of polar star trails, for example, and the choreography of chromosomes in dividing root tip cells of a plant. Both disclose processes that are also vital to our lives. In unprocessed form, however, without theoretical frameworks of heliocentric astronomy and Mendelian heredity, they are no more than beautiful patterns of light.

Although the concept carries with it a certain amount of weight, any statement about the world can be a theory in the colloquial sense. There are no special logical requirements for statements identified as theories. *Scientific* theories, on the other hand, do hold a special status. Scientific theories are constructed in such a way so that hypotheses (potential explanations) derived from them can be proved wrong. Lewontin (1974:6-12; also Dunnell 1982:7) has outlined three critical parameters by which theory (or an explanation generated by it) can be evaluated. *Dynamic sufficiency* is an assessment of the completeness of theory; are enough or the right variables being used to account for phenomena? *Empirical sufficiency* is an assessment of the measurability of the theory in the phenomenological world; can we evaluate hypotheses through empirical measurements? *Tolerance limits* are the standards of accuracy in a match between observations and hypothesis; how close is close enough? Since scientific theories are characterized by the logical requirement that hypotheses derived from them can be proven wrong empirically, it is the practice of scientists to make their "mistakes" as quickly as possible and begin again. It is this feature of scientific practice that gives science its iterative character. Thus, in light of the iterative process of theory construction, theories are the product of *informed* imagination and the best theories are those that generate the most fruitful hypotheses that can be "translated cleanly into questions that can be answered by observation and experiment" (Wilson 1998:53). Theories and their hypotheses compete with one another to explain the available data; the survivors become placed in the current set of "accepted wisdom" until new theories, generated from new imaginations that can better explain the data, become the victors. Lewontin's (1974) account and Dunnell's (1982) analysis for archaeology, emphasize the inventive aspects of constructing theory in concert with building the appropriate units. This effort stands in marked contrast to systematic empiricism (Willer and Willer 1973) and the analogical arguments of the New Archaeology (Dunnell 1982). Building a scientific archaeology requires the formal initiation of the iterative process. We need a theory that articulates principles of historical change with the phenomenological record of archaeology. Such a theory will link dynamic sufficiency with empirical sufficiency and tolerance limits. And, because the only scientific theory that can account for historical change and organic diversity is evolution, such a theory will be evolutionary (Dunnell 1980, 1989, 1992).

Archaeologists have recently begun to focus on evolutionary theory as a foundation for building a "scientific history." Although the details are still being worked out, there is growing consensus on the direction that archaeology *must* go in order to achieve an understanding of history (*e.g.,* Barton and Clarke 1997; Boone and Smith 1998; Broughton and O'Connell 1999; Dunnell 1980, 1992, 1995a; Lyman and O'Brien 1998; Neff 1992; Neiman 1995; O'Brien 1996a; O'Brien and Lyman 2000; Rindos 1984, 1989a, 1989b; Schiffer 1996; Teltser 1995b [and papers therein]). While some social scientists remain as the last bastion of the non-Darwinian thought in modern academia, there is a realization about the general applicability of scientific evolution based on arguments made decades ago (*e.g.,* Blute 1979; Dunnell 1978a, 1978b; 1980; Leonard and Jones 1987; O'Brien and Holland 1990; Rindos 1984). Evolutionary theory explains the origin and differential persistence of traits in living forms, *i.e.,* change. Evolutionary theory, therefore, is the only *scientific* theory that explains change (*why* rather than *how*). As such, a commitment to evolution arises from a commitment to an empirical epistemological standard (*i.e.,* falsifiability) and nothing more. Although there remain numerous critics of the application of evolutionary theory to the archaeological record (*e.g.,* Schiffer 1996), for the most part, the arguments that remain tend to be on the importance of the various constraints over the contingencies of history (*e.g.,* Ahouse 1998; Boone and Smith 1998; Dennett 1995; Gould and Lewontin 1979; Lipo and Madsen 1998; Lyman and O'Brien 1998; Smith and Boone 1998; Wilson 1998).

LEARNING FROM THE PAST: THE CULTURE HISTORIANS

Recognizing that evolutionary theory is the appropriate theoretical framework for explaining history is far from building a science of history based on evolutionary theory. Building a science requires the development of theory that can account for historical phenomena. The construction of science is not a single effort, however, but rather a process of back-and-forth evaluation of successes and failures. Scientific progress is iterative.

One of the reasons for the Wright brothers' achievements in developing the principles of flight, for example, came from their careful study of the successes and failures of others. Lilienthal, for example, built a monoplane glider in the 1890s that he was able to fly and control using his legs to shift his weight from side to side and fore and aft. This glider succeeded in flight until a strong gust of wind caused the craft to nose up, stall and crash from an altitude of fifty feet, killing Lilienthal. Octave Chanute and August Herring also built airplane prototypes that achieved various measures of success. The Wrights carefully studied these prototypes in their efforts to understand the principles of flight and these "failures" had much to do with their ultimate success. Much of their genius lay in their ability to "analyze critically what preceded them, reduce the remaining problems to their most basic elements, and design methods and devices for the practical resolution of these problems" (Jakob 1990:37).

In the construction of a scientific archaeology, we can work in the same way. Archaeology has not been entirely unsuccessful in explaining the past. In particular, in the quest for building a scientific archaeology the culture historians achieved limited success. Culture history emerged as the dominating paradigm in archaeology in the 1930s for two particular reasons. First, due to the work of the CCC and WPA, archaeologists developed a set of agreed upon field methods (*e.g.,* Griffin 1951; Heizer 1949, 1950; see O'Brien and Lyman 1998). This approach was eventually recorded in manuals of field methods (*e.g.,* Heizer 1949, 1950). Second, using distributions of stylistic attributes, archaeologists were able to order phenomena through time. It is this second development that allowed archaeologists to assess the veracity of conclusions. Until this time, archaeology had existed as part of a general and diverse interest in natural history (Dunnell 1985b). With the development of the chronological method seriation, archaeologists could make empirically testable statements about the archaeological record for the first time. These statements could be independently tested whenever other data on age were available. It was the first time that archaeologists could make statements and *be wrong.* Although culture historians could only offer intuitive rationale for their success (*e.g.,* Deetz and Dethlefsen 1965; Ford 1938a; 1949:35-37, 1962; Kroeber 1916; Phillips *et al.* 1951:220-223; Spier 1917), it became clear that the use of stylistic frequencies was a robust and empirically *successful* means of studying the past.

Archaeologists have experimented with frequencies of stylistic elements to measure change and for detecting spatial variation in cultural transmission among prehistoric populations since the early part of this century. Beginning with the work of the early culture historians (*e.g.,* Collins 1927a, 1927b; Kidder 1917; Kidder and Kidder 1917; Kroeber 1916, 1919; Petrie 1899; Rouse 1939; Spier 1917) and culminating in the work of Ford (Ford 1935, 1936a, 1936b, 1938a, 1938b, 1949; Ford and Quimby 1945; Ford and Willey 1941; Phillips *et al.* 1951) in the 1930s and 40s, archaeologists developed specific descriptions of the archaeological record, termed 'stylistic' elements, as useful measures of chronological change. Using variability in the occurrence and frequency of stylistic elements with methods such as seriation, culture historians were able to order parts of the archaeological record in chronological sequences. Further work with these stylistic elements (*e.g.,* Deetz and Dethlefsen 1965; Ford 1952) suggested that they were also useful indicators of spatial relationships. Subsequent work (*e.g.*, Griffin 1952; Phillips 1970; Williams 1954) extended the study of style to include spatial variability and used stylistic frequencies as measures of spatial structure among populations to create "phases." More recently, archaeologists have attributed a wide variety of behavioral functions to style, including social boundaries and identity (*e.g.*, Close 1978; Conkey 1990; DeBoer 1990; Fitzhugh 1987; Graves 1994a, 1994b; Longacre and Stark 1992; Sackett 1985; 1990; Sampson 1988; Sterner 1989; Weissner 1983, 1984, 1990).

Of all the examples of the use of stylistic frequencies to build knowledge of the archaeological record, the Lower Mississippi River Valley Survey by Phillips, Ford and Griffin (1951; hereafter PFG) stands as one of the most influential and certainly the most empirically sound. The PFG survey was a landmark both substantively and methodologically (Dunnell 1985a). From this study, archaeology gained an appreciation for regional scale analyses, quantitative sophistication in terms of sampling and seriation, ceramic types used for relating pottery, the chronology for the Southeast, and ultimately, phases – spatial units of prehistoric populations. The achievements of the PFG research were based entirely on frequencies of types of ceramics collected from 383 locations across the Mississippi river valley analyzed in a series of seriations constructed by James Ford. In a quest to build a scientific archaeology, therefore, the PFG study provides an appropriate starting point for investigating how the use of stylistic frequencies can potentially produce chronological and spatial knowledge about past interaction. An investigation into why PFG obtained their particular results, therefore, sheds light on this central issue.

MODEL BUILDING AND EXPERIMENTS

Careful and clever selection of the experiments to be performed is one of the greatest challenges to the scientist. Productive experiments allow scientists to distinguish between empirical expectations, resolve differences between proposed hypotheses, and reduce error terms for estimates. One of the remarkable accomplishments of the Wright brothers, for example, was to build wind tunnel experiments to test the wing configurations in simulated airflow. These experiments permitted the Wright brothers to reject earlier findings by workers such as Lilienthal and Smeaton. Indeed, the Wrights' wind tunnel experiments were so productive that after a single winter of tests, the brothers' subsequent attempt at building a full-size airplane was entirely successful. Heavier-than-air flight was achieved once a *model* was built to explain how flight operated in the real world.

Models play a critical role in relating empirical observations and theory. Models are the expression of theory in terms that account for a class of phenomena (*i.e.,* hypotheses). Theory is often qualitative; models are quantitative. As Michod (1999:16) points out in his discussion about models of natural selection:

> Models show what is possible, based on assumptions about how the world works. They cannot by themselves prove that a hypothesis is true. However, they can rule out poorly formulated or illogical hypotheses as well as suggest new hypotheses and fruitful lines of inquiry. By guiding experiment and observation models are an integral part of scientific discovery.

In the construction of a model about stylistic frequencies and their relationship to past behavior, two conditions must be considered: dynamic and empirical sufficiency (Lewontin 1974). Dynamic sufficiency can be evaluated through conceptual experiments that test interaction of units and their relations. In mature sciences, models are often represented entirely mathematically and consist of substituting various values into equations to see how the formulas behave in differing state-spaces. Although no such mathematics have been developed in archaeology, simulation and computer modeling potentially provide an interim means for evaluating the dynamic sufficiency of a model. Testability of a model confers empirical sufficiency. Together these aspects of models are used to develop empirical experiments. In empirical experiments, one tests the reaction of the real world in controlled cases where all variables except one are held constant and expected and observed results are compared. Information gleaned from these experiments is fed back into the model and the appropriate revisions made. This process forms the so-called "scientific method" (*sensu* Popper 1959) that is familiar to us now and that represents the core of the Wright brothers' success.

As Lewontin (1974:8) points out, because models can be evaluated dynamically and empirically, scientific work is an iterative process proceeding from descriptions of process with *ad hoc* variables to the creation of units with which to examine the model's performance in explaining particular empirical problems. Rather than a linear process, Lewontin stressed that the scientific process is a halting, back-and-forth enterprise, typically requiring many iterations through revision of units and descriptions of process before arriving at a robust model. The process, of course, never completely halts. Models are continually evaluated in terms of internal consistency and for empirical relevance through feeding newly generated data back into the model and adjustments made. As such, science continues on its bumpy, convoluted yet never-ceasing path.

THE CURRENT PROJECT: ONGOING EXPERIMENTS

PFG's work suggests that in the central Mississippi river valley, the frequencies of combinations of design elements on ceramics collected from the surface are useful indicators of chronological relationships within populations. Of particular concern to many evolutionary archaeologists has been the potential recognition of interacting populations within the PFG data. The frequencies of stylistic types suggest that there are spatial patterns within the ceramic data across the study area, though no explicit work was conducted at the time. Subsequent work with the PFG results has primarily emphasized these spatial aspects and ultimately resulted in the establishment of the "phase" as a unit of analysis, a unit that is taken to be equivalent to prehistoric groups (*e.g.,* Phillips 1970).[1]

Despite the potential of the PFG data for understanding chronology and spatial variability among past populations, little is known about the reasons for their results. We lack, for example, a dynamic and robust theory of the processes driving variability in stylistic frequencies. While stratigraphic excavations and radiocarbon dates have consistently supported the general *chronological* results of the PFG research (and usually at the level of "periods") and these results form the basis of our knowledge of the prehistory of the central Mississippi river valley, we have no independent means of assessing other conclusions (*i.e.,* time scale chronology and non-chronological conclusions) drawn from the data. We do not know, for example, the degree to which the PFG results are a function of the idiosyncratic way in which the samples were collected. Thus, although the PFG methods and their results served to help form the foundations of archaeological knowledge and practice and represent archaeology at its most scientific, we have little understanding as to why this should be so.

This dissertation conducts a series of experiments that show how variability in stylistic frequencies measures the structure of prehistoric communities. I demonstrate the steps needed for building a science of history within one realm of inquiry: building explanations of population structural changes through time. From my examination of the theory of transmission of stylistic traits, it is clear that culture historical methods such as seriation have the potential to detect population structure in time and space, but that crude units such as phases and their ethnographic interpretations are inappropriate for this purpose. These conclusions demand a closer look at the theoretical basis of stylistic frequencies and cultural transmission.

Chapter 2 explores the issue of building units that describe interacting populations and how archaeologists have previously made efforts in this regard. In particular, I explore the details of the PFG survey and how the *phase* (a unit that ultimately fails for studying past population structures) was derived from empirical observations of stylistic variability among ceramic types across space. Chapter 3 develops a theory of cultural transmission for explaining the distribution and variability in stylistic frequencies and a model that can account for sources of variability observed in archaeological data. Using this model, the traditional culture historical method of seriation (Dunnell 1970) is recast as a method for examining patterns of cultural transmission within stylistic frequencies. I use a modified frequency seriation method designed to examine the effects of space on the structure of homologous similarity.

[1] Ford's (*e.g.,* 1952) work stands out as an important exception as he did not pursue the space-like essentialism inherent in Phillips' (1970) phase scheme.

Historically, PFG survey data constitute the basis for whole cultural units (phases) in use in this region. In addition, these data were collected with the intent of studying homologous similarity. In Chapter 4, I reanalyze PFG's data for their St. Francis and Memphis local areas. My findings demonstrate the possibility of understanding interaction history using methods like deterministic frequency seriation, thus allowing the measurement of transmission patterns.

Since detailed information on how the PFG data were generated is not available and all the principals are dead, factors of sampling, counting and functional differences between deposits that potentially drive frequencies in decorated ceramics cannot be ruled out. Since seriations are constructed using class frequencies, measurement must be controlled in such a way that the methods used to gather data, the units, the size of the samples, comparability between classes, and other abundance evaluation procedures do not contribute variability unrelated to the target observations. New collections are necessary.

In Chapter 5, I specify the conditions necessary to generate these collections and evaluate the conclusions of the seriation analyses for the PFG data. These conditions include: controlling sources of variability driving stylistic frequencies due to transmission, deposition and taphonomic processes, measurement, sample size and sampling. With a well-defined research protocol, I conduct an empirical experiment of this model by recollecting ceramics from several of the original PFG locations in order to determine how much of the stylistic variability seen in the original PFG data set can be attributed to sampling, sample size or chance. These recollections involved surface collections in five locations (Beck, Belle Meade, Cranor Place, Nickel, and Rose Mound) as well as one location that was not examined by PFG (Holden Lake). In addition, I conducted limited excavations at Castile Landing. These investigations produced a substantial quantity of ceramics that I use to evaluate empirical results of the PFG work and my subsequent seriation analyses in Chapter 6. In Chapter 7, I examine the effects of classification on the seriation results and how changing measurement scale informs on the patterns of interaction. Finally, in Chapter 8, I summarize the results of the analyses and suggest new directions of research.

Through the detailed analyses of the PFG data and new collections, I show that a theoretically robust framework for studying the frequencies of decorated ceramics is sufficient for testing hypotheses about the patterns of interaction in the past. The stylistic frequencies I examine in the Mississippi valley that were initially collected by PFG are best explained as the product of locally interacting communities and *not* widespread communication across a single population. By carefully controlling sources of variability from collection methods, sherd descriptions, type assignments, sherd size, sherd form, and depositional distribution that potentially influence frequencies of culture-historical types, I am able to make a strong argument that past interaction between populations occurred within geographically distinct groups. These groups explain the distribution of traditionally defined phase designations but are theoretically equivalent to lineages since they have dimensions in both time and space.

Ultimately, these data provide an excellent *empirical* basis for building new hypotheses about the evolution of complexly organized social groups that are thought to have evolved in the Mississippi valley in late prehistory (House 1991; Morse and Morse 1983; Morse 1981, 1990; PFG 1951; Smith 1990).

The general purpose of this book, however, is not to generate empirical results about past populations, but rather to build and refine theory and methods for studying the past. In one sense, empirical results about the past are *by-products* of theory construction, part of the iterative process in which we evaluate how well our explanatory system can account for the material world. This thesis presents just one part in the overall sequences of studies that address the processes that produce variability in stylistic frequencies. It is through this kind of continued experimentation that we will enjoy the fruit of knowledge building: archaeology as an historical science.

CHAPTER 2: BACKGROUND

In recent years, a growing number of social scientists have argued that to understand human behavior within a scientific framework will require the application of evolutionary theory. Within archaeology, Robert C. Dunnell and others have led the charge to design an explicitly evolutionary approach to the past (*e.g.,* Barton and Clarke 1997; Boone and Smith 1998; Broughton and O'Connell 1999; Dunnell 1978a, 1980, 1989, 1992, 1995b; Leonard and Jones 1987; Lyman and O'Brien 1998; Neff 1992; Neiman 1995; O'Brien 1996a; O'Brien and Holland 1990; Rindos 1984, 1989a, 1989b; Teltser 1995a). From an evolutionary perspective, the principal mechanism explaining differential persistence of transmitted variation is natural selection. For selection to be operative, not only must variants be transmitted, but also at least some of the variants must interact with the environment and do so differentially (*i.e.,* result in fitness differences). Such variants are commonly referred to as functional or "adaptive," though the latter term is fraught with undesirable connotations in the human context. It is with this argument in mind, for example, that many writers talk about the use of evolutionary theory as taking a "selectionist" approach (*e.g.,* Boone and Smith 1998; Broughton and O'Connell 1999). It is entirely inappropriate, however inadvertent, to make such an equation. For decades, biologists have realized that not all variation results in differential interaction with the environment; viewed from a selectionist perspective, some variation is neutral (Crow and Kimura 1970; Kimura 1977, 1983; King and Jukes 1969). That does not mean, again as some writers supposed, such variation cannot be explained by evolution, only that it is not explained by selection. Such variation is explained by transmission processes in combination with sampling (*e.g.,* Gulick 1872, 1905; Wright 1931, 1932, 1940, 1948, 1949), processes that are Markovian in nature (Dunnell 1978a, 1981; Gould *et al.* 1977).

The extension of evolutionary theory to archaeological phenomena required two additional steps. The first, beginning in the 1950s, was the recognition that an organism's phenotype was not limited by physical characteristics, a view fostered by traditional "museum" studies of inanimate specimens rather than the field approach taken by biologists prior to this time. Behavior, just as much as the somatotype, constituted phenotypic elements. Indeed, it is difficult to imagine evolution of morphology without behavior – giraffes could never have acquired long necks if they did not try to eat tree leaves (Wcislo 1989).

The second step in the application of evolutionary theory to archaeology was the recognition of culture as an additional mechanism for trait transmission. This recognition was facilitated by the first change inasmuch as most morphological traits are transmitted genetically most of the time, while many behavioral traits are transmitted culturally, even in animals (Alejandro Lynch 1996; Payne 1996). Given the dominance of culture in the generation of the human phenotype, it is doubly embarrassing for anthropologists that biologists (*e.g.,* Bonner 1980) were the first to work out these relations. An additional mechanism for trait transmission does not imply a need for additional selective mechanisms (*e.g.,* intentionality guised as "cultural selection") as some have assumed (*e.g.,* Boone and Smith 1998; Boyd and Richersen 1985; Cavalli-Sforza and Feldman 1981; Durham 1976, 1979, 1990, 1991, 1992; Graves-Brown 1996; Soltis *et al.* 1995). Although from a common sense point of view intention is usually invoked in causation, evolutionary explanations of human behavior require nothing more than that which is required for explanations by natural selection. Positing mechanisms like "cultural selection" is simply generated by a confusion of folk explanation (our culture's "explanation" for change) with scientific causation.

The changes in archaeological theory and practice required to implement an evolutionary approach are substantial. No task is harder, however, than the development of a descriptive language that is needed to characterize and measure change in the quantitative terms required by evolution. In order for natural selection to operate, there are three conditions that a descriptive language must consider. Individuals must vary. This variation must confer differing probabilities of replication (*i.e.,* there must be fitness differences) and some of this variation must be transmitted with better than chance fidelity through successive replications (*i.e.,* variability must be heritable). The measurement of natural selection is a *post hoc* exercise where the frequency of a particular variant is compared to contemporary variants (so-called "relative" fitness). The tabulation of frequencies must be conducted within some bounded set, rather than on simple observations of single individuals. Thus, evolutionary explanations necessitate the definition of *populations.*

In addition, evolutionary explanations involve description of phenotypic variability within groups that represent *interacting* populations. This requirement is necessary for two reasons. First, interaction contributes to the fitness of individuals. From the perspective of the individual, the selective environment is composed not only of the physical surroundings, but also of other individuals and their behavior. Because fitness is relative and often frequency or density-dependent, Sewall Wright's (1932) metaphor of the adaptive "landscape" is best imagined as a rubber sheet, deforming up and down every time an individual is added or deleted from the population. The falsification of any particular hypothesis regarding mechanisms is predicated on our ability to associate the archaeological remains of phenotypes into groups representing interacting populations and evaluate their particular composition. Second, interaction generates heritability. The study of evolutionary history, therefore, must trace the connections of inheritance and transmission. In the archaeological record, there is no secure way to do this for individual phenotypes. Contrary to the goals of reconstructionist archaeology (Binford 1981; Schiffer 1995; Wyle 1995), it is difficult to imagine that individual phenotypes can be studied at all in the archaeological record except through the skeletal remains they leave behind. We must, therefore, derive information about heritability,from the frequencies of different artifacts within some segment of the record where these frequencies represent previously interacting populations. Both of these aspects of evolution demand that we generate frequencies for measuring variability and change among populations. Generating frequencies, however, requires some kind of counting unit and methods for identifying the units within which counts are tabulated.

Although archaeology is plagued by common sense and descriptions of modal tendencies, we have a long history of documenting variability and change. Culture historians, using arbitrary historical types as measurement units, and methods such as seriation for tracking change through time, have created potentially useful techniques and data sets for building evolutionary explanations (*e.g.,* Dunnell and Feathers 1991; Feathers 1990; O'Brien 1994). Despite these achievements, there remains no consensus on the nature of units within which variability is tabulated into phenotypic frequencies, a critical step in using evolutionary theory to create explanations. Definitions of appropriate phenotypic and population units are clear from modern evolutionary biology (Dunnell 1995a; Hull 1980; Madsen and Lipo 1993; Sober 1984). Archaeological methods for identifying functionally redundant counting units are available (Dunnell 1995a; Dunnell and Campbell 1977; Fuller 1981). However, we lack methods for identifying sets of phenotypes that formed interacting groups from archaeological evidence alone. Only then can meaningful frequencies be tabulated.

ARCHAEOLOGICAL UNITS

A variety of archaeological units could potentially be used to measure interacting populations. Indeed, the Americanist archaeological literature testifies to a long flirtation with the definition of "whole cultural" units comparable to "culture" or "society" as used by sociocultural anthropologists. Such units imply interaction among populations and are not far removed from their vernacular counterparts. Variations of these units include "ethnic groupings" (*e.g.,* Cordell and Yannie 1991; Holmes 1903), "cultures" (*e.g.,* Rouse 1939, 1955), "phases" (*e.g.,* Chapman 1989; Krause 1977; Lehmer 1966, 1971; Phillips 1970; Phillips and Willey 1953; Willey and Phillips 1955, 1958; Williams 1954, 1980), "foci" (e.g., Griffin 1943, 1946; McKern 1939), "provinces" (*e.g.,* Cordell and Plog 1979; F. Plog 1979; S. Plog 1983) and "polities" (*e.g.,* Hammond 1972; King and Freer 1995; Peregrine 1991, 1992, 1995; Upham 1982, 1983; Upham *et al.* 1981; Upham and Plog 1986). While anthropological interpretations dominate the *use* of these units, all are *defined by* and rest upon, similarities and differences between archaeological assemblages. All of these units assume that past communities create patterns in the archaeological record that are captured by these archaeological units.

The Midwestern Taxonomic Method (hereafter, MTM) was an early attempt to replace ethnographic-based cultural units with archaeological ones (McKern 1939). The MTM created a set of six hierarchical units (component, focus, aspect, phase, pattern and base) based on phenetic similarity (*e.g.,* Sokal and Sneath 1963) between components (more or less homogeneous assemblages) (*e.g.,* Dunnell 1971; Krause 1977; Trigger 1989). Relations between units were ahistorical in consequence of employing an *ad hoc* set of "traits" to assess similarity (McKern 1939). As chronological data became available, the MTM was replaced by the nonhierarchical and decidedly historical system of Phillips and Willey (Phillips and Willey 1953; Willey and Phillips 1955, 1958). Of the units proposed, Phillips and Willey's *phase* is the most widely employed cultural unit with both spatial and temporal components (Chapman 1989; Phillips 1970; Phillips and Willey 1953; Willey and Phillips 1955, 1958; Williams 1954). Since its introduction in the 1950s, the phase concept has become the primary means for describing spatially contiguous and similar archaeological material. Its definition, established by Willey and Phillips (1958:22), is an "archaeological unit possessing traits sufficiently characteristic to distinguish it from all other units similarly conceived, whether of the same or other cultures or civilizations, spatially limited to the order of magnitude of a locality or a region and chronologically limited to a brief span of time." Although rooted strictly in measures of artifact similarity, the notion of phase has come to have ethnographic meaning. Current use of phases in the Lower Mississippi river valley consists of treating these units as equivalent to integrated social systems and whose remains archaeologically mark the boundaries of interacting populations (*e.g.,* House 1991, 1993, 1995; D. Morse 1973, 1982, 1989, 1990; Morse and Morse 1983, 1996; P. Morse 1981, 1990; Smith 1990). Assumptions aside, understanding how phases might be related to interacting populations, however, requires understanding the origin of the phase notion and its introduction into archaeological practice.

PHASE: BACKGROUND

The phase concept has its roots in the archaeology of the Mississippi river valley. The massive scale of the mounds and the quantity of decorated pottery contained within late prehistoric sites of the Mississippi river valley have attracted attention since the earliest days of North American archaeology. In a letter to Thomas Jefferson dated 1813, for example, Henry Brackenridge (1818:154) reported that mounds:

> are to be found at the junction of all of the considerable rivers, in the most eligible positions for towns, and in the most extensive bodies of fertile land. Their number exceeds perhaps three thousand; the smallest not less than twenty feet in height, and one hundred in diameter at the base. Their great number, and astonishing size of some of them may be regarded as furnishing, with other circumstances, evidence of their antiquity. I have been sometimes induced to think at the period when those mounds were constructed, there existed on the Mississippi, a population as numerous as that which once animated the borders of the Nile, or of the Euphrates, or of Mexico and Peru.

As in other places across North America, most early investigators were concerned with mounds and their associated artifacts (*e.g.,* Brown 1926; Conant 1878, 1879; Dellinger and Dickinson 1940; Dickinson and Dellinger 1940; Evers 1880; Lemley and Dickinson 1937; Peabody 1904; Phillips 1939; Potter 1880; Swallow 1858, 1875). In the Mississippi river valley, the quest for burial artifacts has been particularly energetic and has occasionally reached a fever pitch. Over the years, literally hundreds of thousands of whole pottery vessels have been removed from the mounds by archaeologists and collectors. Indeed, the riches of these Mississippian sites can be found in the museums and homes of collectors throughout the region and large museums of the eastern United States and Europe.

One of the earliest descriptions of an excavation of a Mississippi valley deposit is attributed to Edward Curtis, who collected pottery along the St. Francis River for the Peabody Museum (Putnam 1881:18). Curtis gathered over 900 complete vessels from sites in the St. Francis area and claimed that 10,000 complete pieces of pottery could be obtained in six months with five good hands (Morse 1978:9). The next recorded individual to investigate the sites of the St. Francis was "Captain" C. W. Riggs who, with the help of his family, returned barrels of pottery back to his home in Cincinnati (Brose 1980; Riggs 1893). At the same time, "Captain" Wilfred Hall excavated throughout the St. Francis river area to deliver pottery to rich patrons in the east (Morse and Morse 1983).

The first systematic researchers of "Mound Archaeology," as the prehistoric record first came to be known, were from the Peabody Museums of Harvard and Yale, the Davenport Academy of Sciences and the Bureau of Ethnology (Griffin 1981). Thomas (1887, 1894) of the Bureau of Ethnology Division of Mound Exploration carried out a large-scale program of mapping and excavation in 23 states in an effort to determine the origins of the "Mound-builders." Holmes (1886a, 1903), in particular, brought a very high level of sophistication to early Mississippian archaeology and produced detailed descriptions and illustrations of ceramic vessels. Clarence B. Moore and his crew of excavators explored the waterways of the St. Francis in his own steamboat, *Gopher of Philadelphia*, stopping to dig at many of the large, prehistoric mound sites (Moore 1910, 1911). Although his primary goal was to recover whole pots, his publications, careful descriptions, and beautiful pottery illustrations brought the deposits of the Mississippi river valley to the attention of Americanist archaeological researchers.

PHILLIPS, FORD AND GRIFFIN (1951)

It was the work of PFG (1951) that firmly established the Mississippi river valley as one of the primary foci of American archaeology and from which the notion of phases as archaeological conceptions of past community structure is derived (Dunnell 1985a; Lyman *et al.*1997; O'Brien 1996a; O'Brien and Dunnell 1998). Building on Ford's start (1935, 1936a, 1939), Philip Phillips, James A. Ford and James B. Griffin made systematic ceramic collections across the Mississippi river valley (Figure 2.1). The primary goal of the PFG survey was chronological – to discover the origins of the "Middle Mississippian cultures" – the period populations who inhabited sites with large mounds and made elaborately decorated pottery. Between 1940 and 1948, crews spent a total of seven months in the field locating and mapping sites, making surface collections, and conducting test excavations. The amount of data collected and compiled for the final PFG report was, and is still today, truly impressive. PFG made more or less systematic ceramic collections from 383 different locations. These collections produced over 346,099 ceramic artifacts that were subsequently described and tabulated.

There are several features of the PFG study that are unique in terms of archaeological practice of the time and that have had a lasting impact on our knowledge of the archaeology of the Mississippi river valley and on current archaeological practice. First, the PFG survey was regional in scale, covering a vast area of the Mississippi river valley between the Red and Ohio Rivers that was, except for the early work of Thomas, unknown. The focus on sampling the archaeological record on a regional scale effort was stimulated by the earlier work of Ford (1936a) in Louisiana and Mississippi and Griffin (1938, 1939) in the Norris Basin, Tennessee and the Wheeler Basin, Alabama. Of their survey, PFG (1951:41) state that "the object was merely an adequate sampling of sites, sufficient to provide a safe coverage of the area and to insure against the omission of any significant cultural manifestation." PFG (1951:41) were well aware that their sample was biased toward "larger and more conspicuous sites" and against finding "early pre-ceramic manifestations." The researchers did what they could to minimize these problems through discussions with local farmers. Nevertheless, given its large-scale sampling design that attempts to build a representative picture of occupations in the valley, rather than just a collection of *ad hoc* site descriptions, the PFG survey stands as one of the most complete samples of the late prehistoric record ever generated. One of the benefits of the PFG sample is that it permits one to examine the history of large-scale *populations* and, potentially, how the region's communities formed through time. Such a wide-scale spatial and temporal view is not possible with limited site descriptions or excavations (Flannery 1976).

Second, following a tradition started in the Mississippi river valley by Ford, the PFG survey included surface collections in their analysis instead of focusing entirely on excavated materials. The use of surface collections resulted in ceramic assemblages with numbers large enough to provide *representative* sample of these deposits – something few other contemporary (and present day) archaeologists typically considered.

Third, Ford's earlier work in the Mississippi river valley (1936a) set high standards for the sampling methods used to generate the PFG collections. PFG (1951:43) state that "generally speaking, our only concern was to get as large a sample as possible and a reasonably honest one." Large samples were necessary for the seriation analyses in order to

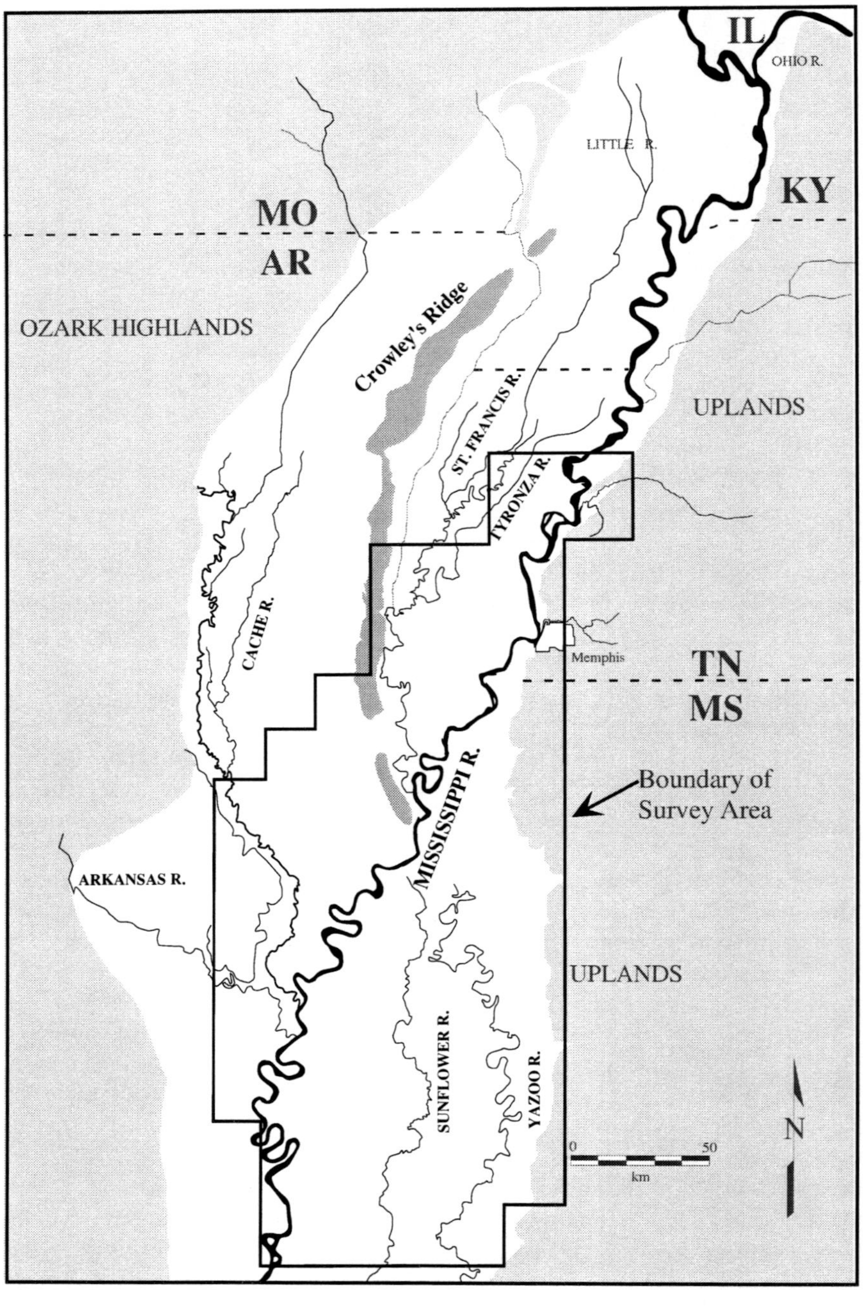

Figure 2.1. PFG Survey Area.

mitigate factors of chance in the frequency tabulations (PFG 1951:223, 232). Ford, in particular, was very concerned with the generation of adequate samples in the surface collections for the seriations (Ford 1936a:13-14; PFG 1951:223). In the PFG (1951:223) study, Ford states:

> We are also guessing that a random sample over fifty sherds is sufficient to indicate that proportionate type frequencies existing in the refuse from which the material was collected. A total of fifty is considered to be usable, but not particularly reliable. One hundred is much better and every sherd above one hundred is all to the good.

In addition, PFG were concerned that collections be made in such a way so as not to combine compositionally heterogeneous groups. If areas of sites showed evidence of having differing contents, these areas were separately collected. In addition, collectors were instructed to pick up every sherd they encountered so as not to bias the sample towards decorated or rim sherds. Local residents were hired to collect sherds. PFG (1951:43) reasons that "the only sure way to eliminate this difficulty [of biasing collections] is to hire local people to pick sherds up at so much per sack. You sometimes get brickbats and other extraneous material, but the sample is an honest one." As a result of this careful attention to sampling procedures, post-PFG work has shown that the PFG collections are remarkably reliable (House 1991; Morse 1981; Morse and Morse 1983; Phillips 1970; Pugh 1992; Williams and Brain 1983). Subsequent researchers in the Mississippi river valley have turned repeatedly to the PFG study with confidence that the results reflect the archaeological record rather than the way in which samples were collected.

TYPOLOGY

The PFG study was also significant in its rationale and methods for describing the pottery. PFG created the first comprehensive ceramic typological system for the central Mississippi river valley. Although there have been modifications over time (*e.g.,* Phillips 1970), this system still forms the basis of how sherds and vessels are classified and described in the region today. The PFG types were groundbreaking for a number of reasons. First, the classification scheme was designed to be applied to sherds as opposed to pots. Furthermore, PFG's typological system was designed to build a chronology of the Mississippi river valley. Second, PFG were very explicit about the creation of their types. These types were intensionally created (*sensu* Dunnell 1971) measurement tools for describing patterning in the archaeological record, not discoveries inherent in the world. Types were recognized by PFG as composites of several separate characters – paste, surface treatment and decoration – characters that PFG argued had separate histories. Type creation for PFG was an iterative process in which different combinations of characters were tested to see whether they were useful for building chronologies (*a la* Kreiger 1944).

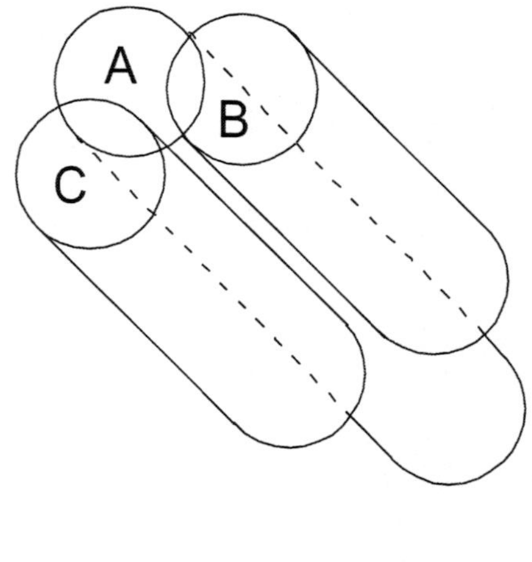

Figure 2.2. A Graphical Depiction of the PFG Perception of Types.

> The letters (i.e., A, B, C) represent types and the circles represents the limit of variability (often geographic) for each type. The diagonal lines form a cylinder that extends the circles into a third dimension, time. The length of the circle is the range of variability resulting from change through time. PFG point out that types overlap in both time and space. Thus, making any divisions in this continuous range of variability is entirely arbitrary.

PFG considered ceramic styles to be a continuum of variability in which individual ceramic vessels might vary within a potter's work, but tend to cluster around a norm (1951:62). In addition, the authors considered ceramic styles to be linked to population centers. In their model, the amount of communication between communities varied. Styles also moved between these centers and, consequently, vary roughly in proportion to the distances between the centers subject to ethnographic and geographic factors. Variability in style, therefore, can be envisioned as a continuous three-dimensional flow moving across space and through time (Figure 2.2). Given that variability in styles is continuous, it is sliced into a series of arbitrary chunks. These arbitrary slices of the ceramic continuum are pottery types. PFG (1951:63) note that "the most carefully defined types always overlap" and are, therefore, "created units of the ceramic continuum."

Out of this ceramic continuum, PFG described forty-five ceramic types for the Mississippi Valley, nineteen of which occur during the Mississippian period. These types are Parkin Punctated, Barton Incised, Vernon Paul Appliqué, Fortune Noded, Bell Plain, Neeley's Ferry Plain, Kent Incised, Rhodes Incised, Walls Engraved, Hull Engraved, Old Town Red, Carson Red-on-Buff, Nodena Red-and-White, Avenue Polychrome, Hollywood White-filmed, Wallace Incised, Owens Punctated, Leland Incised and Arcola Incised (PFG 1951:61-149). In general, these PFG types are defined by variability along two measurement dimensions: temper and surface treatment/decoration.

Temper

Following Griffin's (1938, 1939) work in the Wheeler and Norris Basins, the initial sorting of the PFG ceramics was by temper. This resulted in three loosely defined groups consisting of shell, clay, and sand. PFG types are, in effect, variants of these three groups defined by additional differences in surface finish and decoration. In general, the shell-tempered ceramics, identified by PFG as being from the Mississippian time period, are chronologically the latest and thus potentially contemporaneous. These ceramics are the focus of this study.

PFG continued to divide additional variability within the shell-tempered ceramics. As O'Brien and Fox (1994:31) point out, "the key to understanding the myriad of types and varieties proposed for ceramic material from the central Mississippi river valley lies in understanding the difference between ... fine paste and coarse paste." Fine paste refers to pastes that are composed of finely crushed, nearly powder-like fragments of shell. Undecorated sherds made from this kind of paste are described as Bell Plain; sherds that are composed of coarse paste are made from large, readily visible pieces of shell and are labeled Neeley's Ferry plain (Mississippian Plain of Phillips [1970]). These two kinds of pastes are the basis of the shell-tempered Mississippian ceramic types in this study.

Surface Treatment/Decoration

The greatest source of variability in the PFG ceramic typology is derived from decorative surface modification. This treatment can be, for example, incised patterns, punctated shapes or engraved figures. In other cases, the decoration involves extensive modification of large areas of vessel surfaces as seen in cord-marked, stamped, and slipped pots. Thus, the majority of PFG types are easily distinguished on the basis of decorative techniques (*i.e.,* incising, engraving, appliqué, noding, punctating, and slipping) that are visible on sherds. Since these types play such a large role in the archaeology of the Mississippi river valley and this study, I describe the major late prehistoric PFG types below.

Parkin Punctated

Parkin Punctated (Brown 1998:31; House 1991:359-360; O'Brien and Fox 1994:40; Phillips 1970:151; PFG 1951:110-114) was proposed by PFG to account for vessels with Neeley's Ferry paste that have their exteriors covered or nearly covered with punctations. For the most part, the punctations cover the entire vessel, but occasionally punctates occur in zones outlined by incised lines. These zone-punctated sherds are identified by PFG as Barton Incised rather than Parkin Punctated. Parkin Punctated includes an extraordinary amount of variability in terms of design and form. Punctate arrangements, for example, include random patterning, horizontal and vertical rows and overlapping lines. The form of the punctate varies from impressions of fingernails, sticks, hollow reeds, and pinching.

Barton Incised

PFG defined Barton Incised (Brain 1989; Brown 1998:12; House 1991:352; O'Brien and Fox 1994:36; Phillips 1970:44-45; PFG 1951:114-119; Williams and Brain 1983:126-133) as having a Neeley's Ferry plain paste, and groups of lines applied to a moist surface. The lines vary in width but are generally 0.5-mm to 3.0-mm. The lines are drawn parallel to one another and are oriented obliquely to the vessel rim, slanting downward from the lip of the vessel to the beginning of the shoulder area. The groups of lines often form alternating line-filled triangles or trapezoids. The incising on Barton Incised sherds occasionally includes cross-hatching of sets of parallel lines and even triangles filled with punctation (see Parkin Punctated).

Kent Incised

Kent Incised (Brown 1998:14; House 1991:352; O'Brien and Fox 1994:36; Phillips 1970:46; PFG 1951:126-129) was proposed by PFG (1951:126-127) to described sherds with "vertical incised lines extending from lower rim area to base, sometimes from the lip. The character of the line is similar to that for Barton Incised." The distance between lines averages around 1-cm apart, but varies from 4 to 17-mm. Occasionally, Kent Incised includes a herringbone effect between widely spaced lines or broken lines with punctates.

Wallace Incised

PFG defined Wallace Incised (Brain 1989; Brown 1998:14; O'Brien and Fox 1994:38; PFG 1951:134-136) as Neeley's Ferry plain sherds with incision by broad, round, or flat-ended instruments that produce a characteristic shallow U-shaped line, considerably broader than is usual for incised lines in this area. According to the PFG data, lines average 4-mm in depth. When lines meet, they tend not intersect with each other. Occasionally, brushing is employed to bring out the design by contrasting smooth and roughened areas. Rarely, punctations are used for the same purpose. Both recti- and curvilinear patterns occur with greater frequency on the body than on the rim. The most common designs on the rim area consist of parallel rows of oblique lines forming line-filled triangles, as in Barton Incised. Other rectilinear arrangements include groups of oblique lines slanting from the lip with a plain area between. Principle curvilinear rim design consists of broad shallow concentric festoons, or its reverse, concentric arches.

Rhodes Incised

Rhodes Incised (Brown 1998:20; House 1991:361; O'Brien and Fox 1994:39; Phillips 1970:157; PFG 1951:127) refers to Neeley's Ferry plain and, occasionally, Bell Plain sherds with "wide, deep generally U-shaped curvilinear incisions, closely spaced" (PFG 1951:127). The width of the incisions, according to the PFG description varies from ca. 2 to 3-mm. Designs include whorls, festoons, swastikas and triskeles.

Ranch Incised

Neeley's Ferry plain sherds with groups of thin, parallel, curved incised lines were referred by PFG as Ranch Incised (O'Brien and Fox 1994:37; Phillips 1970:156; PFG 1951:119-120). These curvilinear lines "often intersect one another to give the imbricated design somewhat like fish-scales" (PFG 1951:129).

Walls Engraved

Walls Engraved (Brain 1989; Brown 1998:22; Dye 1998; House 1991:362; O'Brien and Fox 1994:39; Phillips 1970:170; PFG 1951:127-129) refers to Bell Plain ceramics with fine-line engraving encompassing a wide variety of designs. The lines of Walls Engraved are called engraved, rather than incised, as they are executed on paste that has been dried or fired. The designs of Walls Engraved ceramics include triangular areas of cross-hatching, curvilinear bands of diamond-shaped cross-hatching contrasting with plain bands and, occasionally, "Southern Cult" motifs such as the long-nosed god, the feathered serpent and the weeping eye.

Hull Engraved

Hull Engraved (Brown 1998:22; Dye 1998; O'Brien and Fox 1994:39; PFG 1951:129) consists of the same defining characteristics as Walls Engraved except that it occurs on the insides of bowls rather than on bottles or jars. The designs of Hull Engraved vessels include patterns of parallel concentric arcs, "abutting against other groups to form a fish-scale-like imbricated pattern" similar to that of Ranch Incised (PFG 1951:129). The firing of Hull Engraved tends to an oxidized buff/red color rather than the reduced black of Walls engraved.

Fortune Noded

Fortune Noded (Brown 1998:47; House 1991:355; O'Brien and Fox 1994:41; Phillips 1970:83; PFG 1951:120-22) was proposed by PFG (1951:121) to describe Neeley's Ferry plain vessels with "hemispherical or conical nodes in rows, groups or as an all-over treatment of the vessel body." Characteristically, the nodes cover the entire body but occasionally the nodes are combined with incised and punctated decoration on the rim or shoulder area of the vessel.

Vernon Paul Appliqué

Neeley's Ferry plain vessels with wide, vertical ribs of clay applied to vessel exteriors usually below the neck area were identified by PFG as Vernon Paul Appliqué (Brain 1989; Brown 1998:47; House 1991:352; O'Brien and Fox 1994:42; Phillips 1970:167-168; PFG 1951:120).

Old Town Red

Old Town Red (Brain 1989; Brown 1998:44; House 1992:359; O'Brien and Fox 1994:34; Phillips 1970:145-146; PFG 1951:129-132; Williams and Brain 1983:191-193) was proposed by PFG to describe Neeley's Ferry *and* Bell Plain vessels with red slip applied to the interior and/or exterior of vessel walls. They (PFG 1951:131) note that Old Town Red includes both coarse and fine-tempered pastes and that "logically, there should be two divisions," but that the differentiation between red-slipped, Neeley's Ferry paste vessels from red-slipped Bell Plain paste vessels turns out to be difficult to sort. Later, Williams (1954:209-210) proposed Varney red-slipped as a type that was restricted to vessels of Old Town Red decoration but with peculiar forms. Varney red-slipped sherds were derived from "simple curved-sided bowls, shallow or deep... This is the shape often called the 'salt pan.' Another common shape is the large jar with a recurved rim" (Williams 1954:209). Later Phillips (1970:167) restricted the Varney red-slipped to "saltpans" asserting that saltpans "warrant special treatment." In general, however, most researchers have assigned red-slipped, shell-tempered sherds to Old Town Red (O'Brien and Fox 1994:35).

Carson Red-on-Buff

Carson Red-on-Buff (Brown 1998:44; House 1991:354; O'Brien and Fox 1994:35; Phillips 1970:63; PFG 1951:132) was used by PFG to describe Neeley's Ferry sherds that exhibit a buff-colored paste and broad bands of heavy red slip. The designs of the red-slipped lines include swastika-like spirals, vertical panels and stepped motifs. As PFG (1951:133-134) point out, many of the designs correspond to those of Rhodes Incised (swastika), Walls Engraved (swastikas and steps), and Kent Incised (vertical panels) (O'Brien and Fox 1994:35).

Nodena Red-on-White

Nodena Red-and-White (Brain 1989; Brown 1998:46; House 1991:358; O'Brien and Fox 1994:35; Phillips 1970:141-144; PFG 1951:133-134; Williams and Brain 1983:190) was used by PFG to describe Neeley's Ferry plain ceramics that exhibit a buff-colored paste and red and white slips to create a design on vessel exteriors. The designs of the red-and-white slipped areas are similar to those occurring on Carson Red-on-Buff.

Hollywood White-Slipped

PFG proposed Hollywood White-Slipped (Brown 1998:45; O'Brien and Fox 1994:36; Phillips 1970:90-91; PFG 1951:134) to describe Neeley's Ferry Plain and Bell Plain vessels with white slipping on their exteriors and, occasionally, interiors. PFG (1951:134) note that it is difficult to determine whether sherds with white slipping on exterior originate from the white portion of a Nodena Red-on-White vessel, or are indeed from a Hollywood White-Slipped vessel. The same is obviously true of Old Town Red.

SERIATIONS

Since chronology was the primary goal of the PFG study, seriations formed the central analytic effort. Ford was the primary architect of the seriations for the 383 assemblages. Little emphasis was placed on examining spatial variation in assemblage composition, and in order to meet the "same local area" condition, Ford divided the pottery collections into sets from five regions – St. Francis, Memphis, Upper Sunflower, Lower Arkansas, Lower Yazoo (Figure 2.3). Ford recognized that significant spatial variation existed in the collections and that no single seriation could be constructed using all of the assemblages. In examining how these seriations were made, it is important to note that these local areas were analytic contrivances, not archaeological discoveries; Ford just chopped the study area into chunks. As Ford (PFG 1951:224, emphasis mine) points out in the PFG report "we realized that the procedure we were adopting was fully arbitrary, and indeed was of the same kind of high-handed ruthlessness as were our decisions in regard to ceramic classification. We are again preparing to set up *artificial boundaries*, which this time are geographically defined, and draw the borderline cases back toward the selected concepts."

Ford (PFG 1951:224) defined the St. Francis area as a set of three 15' Mississippi River Commission quadrangles (PFG grid units 12-N, 11-N, and 11-O) in the region of the Lower St. Francis River that were already known to be similar in the archaeological literature of the time. He used several observations in making this definition. First, the sites were physically separated from the others to the east and south. Second, the survey did not extend to the north or west. Thus, Ford (PFG 1951:224) claimed that the division was made on the basis of "ignorance and a classical tradition; it couldn't be better." Earlier descriptions of the Walls Site by Calvin Brown (1926:288-319) and the distinctiveness of the sites in the area from the St. Francis group, helped Ford define a Memphis area. The three remaining areas, Sunflower, Lower Yazoo, and Lower Arkansas, were entirely arbitrary partitions of the assemblages across the survey area though PFG found a solution iteratively that separated the collections based on similarity (PFG 1951: 226).

Thus, five separate seriations were produced. Figure 2.4 displays the St Francis seriation. It should be noted that Ford took a materialist approach that generally contrasted starkly with the essentialist position of Griffin and Phillips. However, he capitulated to his co-authors on the matter of time, and broke the continuous sequence of events represented by seriation into chronological periods compatible with an essentialist view – Periods A, B, C, D. These arbitrary chronological "periods" were subsequently used for comparisons across the seriations and the order of the five sets of assemblages was concluded to be chronological. Indeed, subsequent excavations and radiocarbon dates have borne out these conclusions (*e.g.,* House 1991; Williams 1954; Williams and Brain 1983).

The Legacy of PFG: Phases

Williams (1954) was the first to apply the Willey and Phillips system anywhere and did so in the Mississippi valley. Phillips' study (1970), *Archaeological Survey in the Lower Yazoo Basin, Mississippi, 1949-1955*, however, has had the greatest impact on the area because it defined phases for the entire chronological sequence of the Mississippi river alluvial valley. Phillips' phases emerged directly from the results of the PFG analysis.

In particular, these phases are built on the "local area" blocks that Ford used to build his seriations (*i.e.,* St. Francis, Memphis, Upper Sunflower, Lower Arkansas, Lower Yazoo), and each of the separate time periods that chronologically break up the seriation sequences (*i.e.,* A, B, C, D). In his study, Phillips (1970:390) complained that "in the first LMS report (PFG 1951) even the faintest allusion to any such thing as a specific cultural unit smaller in scale than Baytown and Mississippian 'cultures' was scrupulously avoided. The majority view, not shared by Griffin, that any further 'slicing up' would be a wholly arbitrary business prevailed." Indeed, even before the publication of the PFG report Griffin began writing about two of the sub-areas (the Memphis and the Parkin areas) as the "St. Francis" and "Alpika" foci, treating them as distinct cultural units rather than analytic contrivances (Griffin 1946, Figure 7). Later, Griffin (1952:231-236) began referring to these units as "Parkin" and "Walls-Pecan Point." In his 1970 volume, Phillips codified these units and split the Period A assemblages of the PFG Memphis and St. Francis sub-areas into four distinct phases. The rationale behind creating the phases was entirely intuitive. Phillips noted that the composition of ceramic types in the assemblages in each group seemed more similar to each other than any other group. Using graphs of cumulative ceramic frequencies across assemblages in the four regions[1], Phillips argued that each set of assemblages should be lumped together (Figure 2.5). These sets became the Parkin, Nodena, Kent, and Walls phases (Figure 2.6).

Since its introduction in the 1950s, the phase concept has become the primary means for describing spatially contiguous and similar archaeological material. Recent archaeological work has been substantially based on and guided by the work of PFG with phases as the overall unit of analysis. P. Morse (1981, 1990), for example, worked for many years on the Parkin phase. D. Morse (1973, 1990) has centered much of his research on the Nodena phase. House (1991, 1995) has conducted extensive surveys and excavations of Mississippian sites in the Kent phase. Recently, Smith (1990) conducted a reanalysis of the Walls

[1] Phillips' (1970) graphs of cumulative frequencies is based on a model derived from Bordes (1952) analysis of Paleolithic tools.

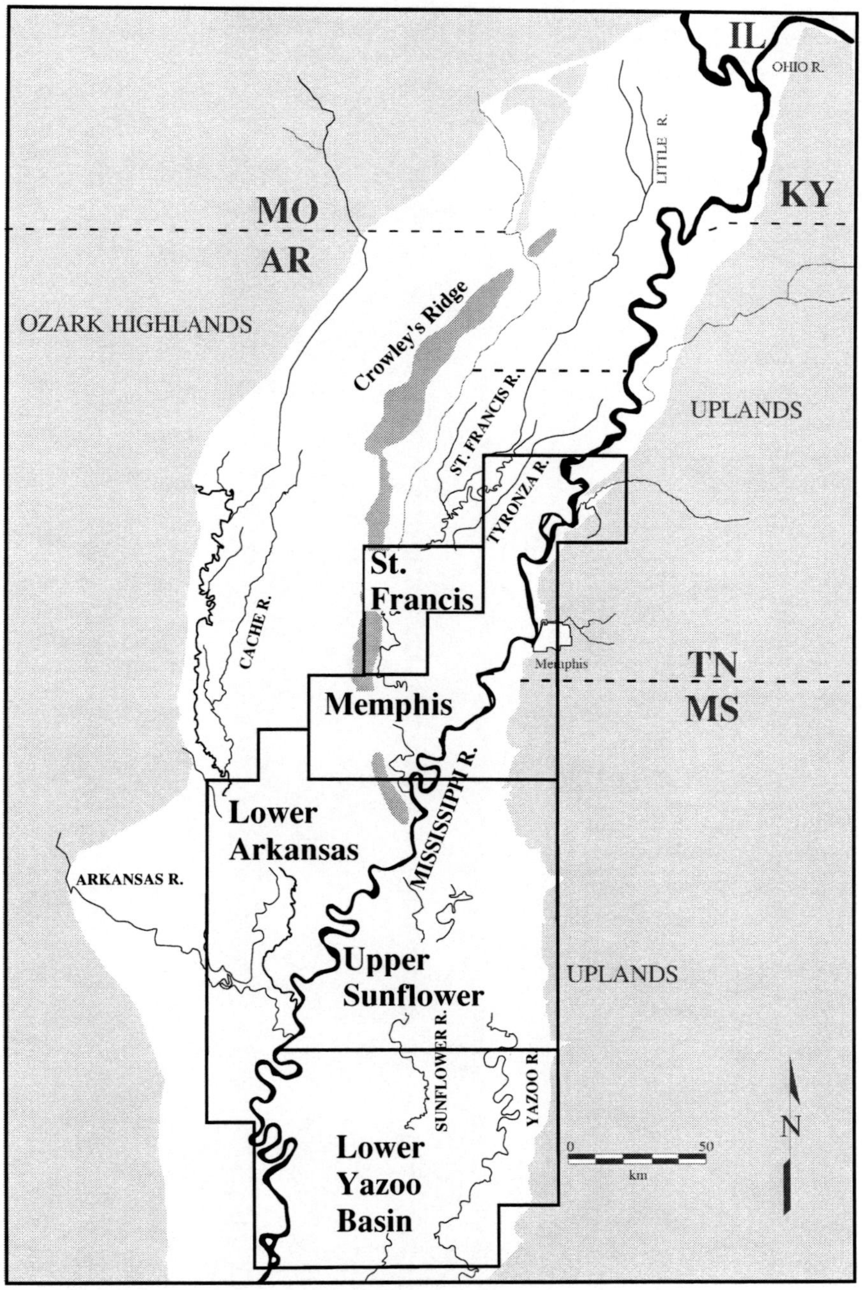

Figure 2.3. Subdivision of the PFG Survey Area into Analytic Units for Purposes of Seriation.

Figure from Philip Phillips, James A. Ford and James B. Griffin, *Archaeological Survey in the Lower Mississippi Alluvial Valley, 1940-1947*, Papers of the Peabody Museum of American Archaeology and Ethnology, vol. 25, 1951, p. 225. Reprinted courtesy of the Peabody Museum of Archaeology and Ethnology, Harvard University.

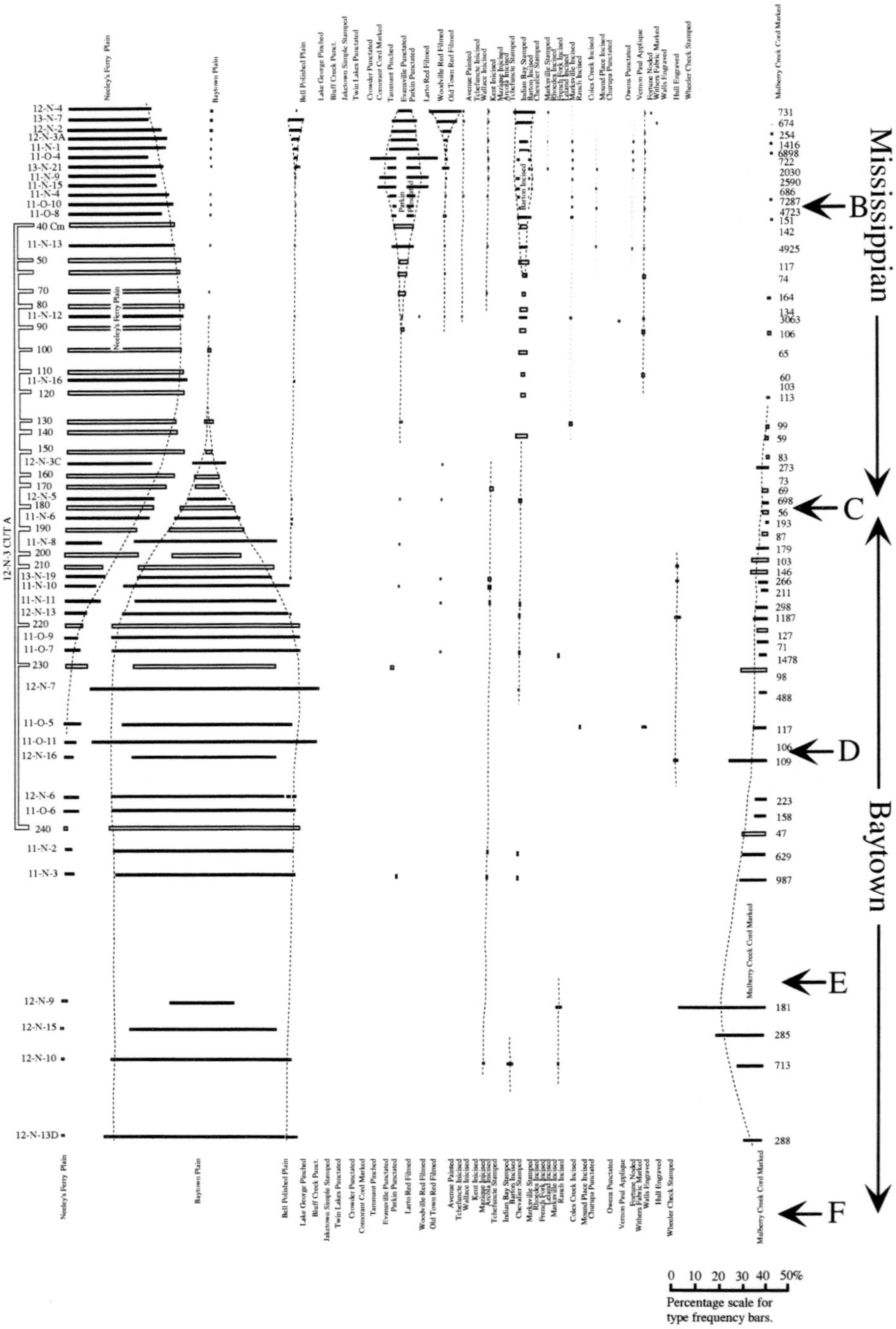

Figure 2.4. PFG Seriation for the St. Francis Area.

Figure based on Philip Phillips, James A. Ford and James B. Griffin, *Archaeological Survey in the Lower Mississippi Alluvial Valley, 1940-1947*, Papers of the Peabody Museum of American Archaeology and Ethnology, vol. 25, 1951, p. 237. Reprinted courtesy of the Peabody Museum of Archaeology and Ethnology, Harvard University.

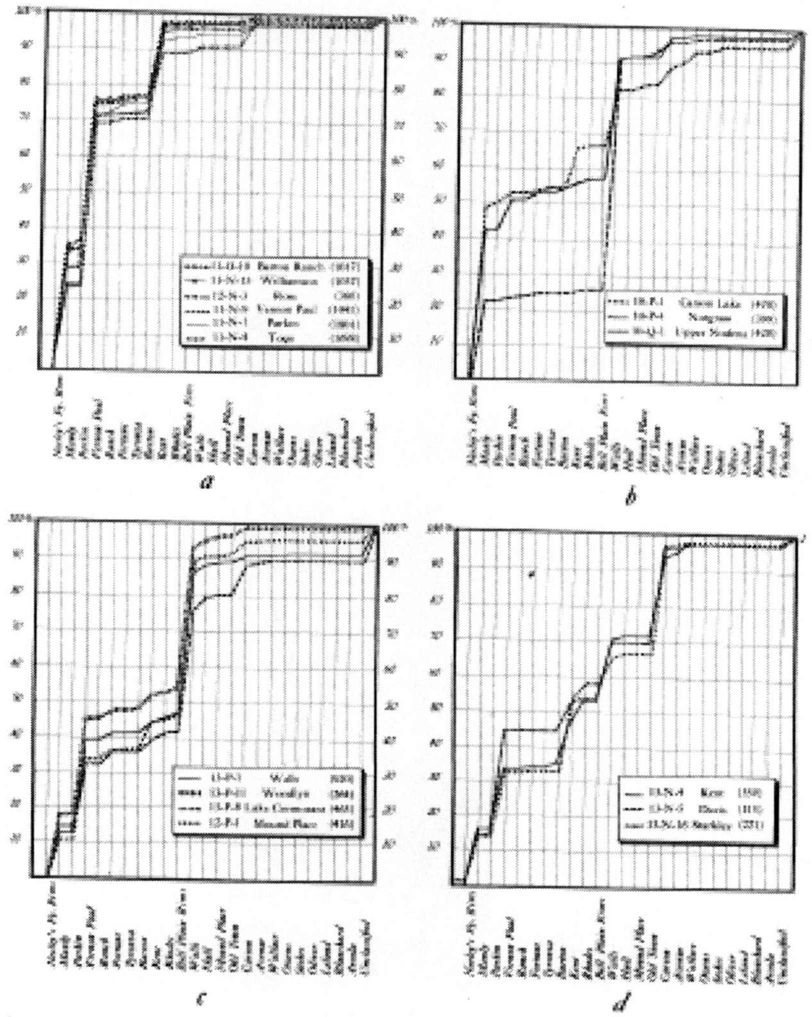

Figure 2.5. Cumulative Type Frequency Patterns of the Parkin, Nodena, Walls, and Kent Phases for the Mississippian Assemblages.

Figure from Philip Phillips, *Archaeological Survey in the Lower Yazoo Basin, Mississippi, 1949-1955*, Papers of the Peabody Museum of American Archaeology and Ethnology, vol. 60. Copyright 1970 by the President and Fellows of Harvard College, p. 931.

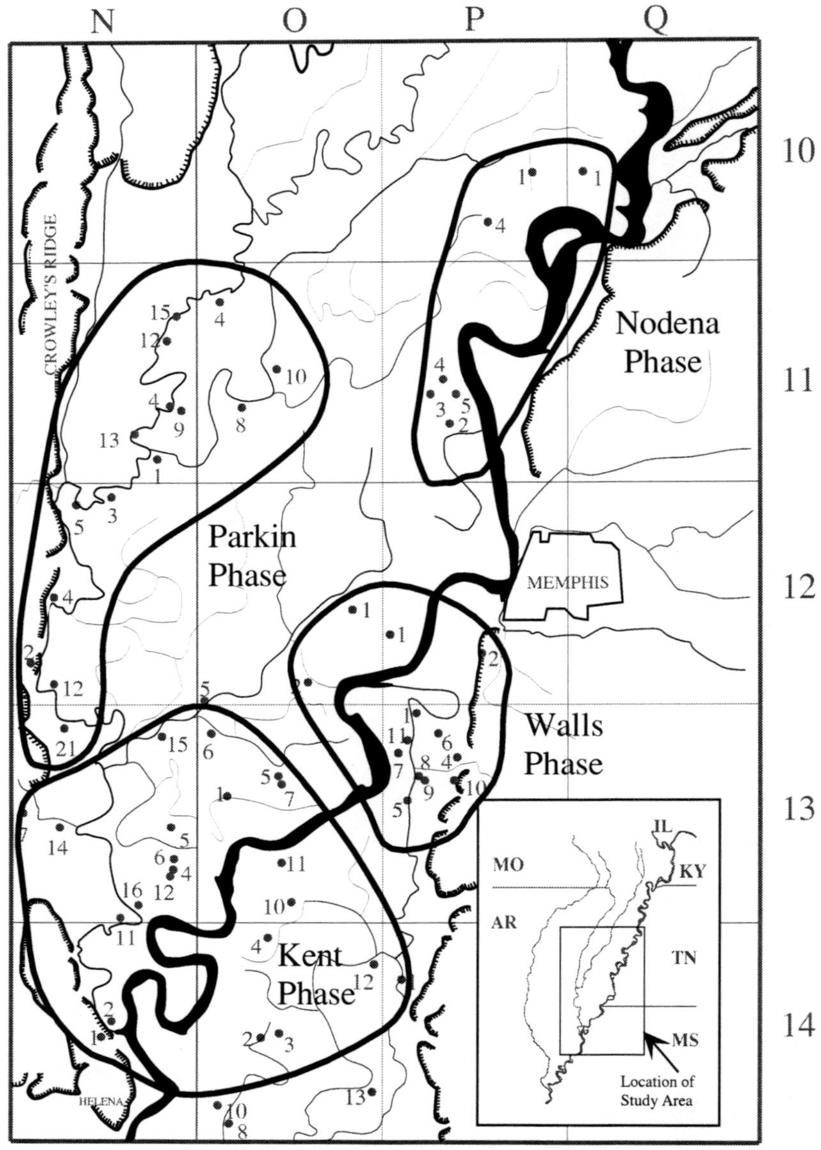

Figure 2.6. Phase Areas as Defined by Phillips (1970).

phase area, which resulted in the designation of a Horseshoe Lake phase and a slight reshuffling of the boundaries of phase areas. Given this history, it is not surprising that "phase" has been the dominant whole cultural unit used to describe the archaeology of the Mississippi river valley (*e.g.*, House 1991, 1993, 1995; D. Morse 1973, 1982, 1989, 1990; Morse and Morse 1983, 1996; P. Morse 1981, 1990; Smith 1990; Williams 1954).

Phases continue to be *the* primary analytic unit for studying spatial and temporal variability in the archaeological record in the Mississippi river valley. Recently, for example, Williams (1999) brought together archaeologists to review the status of phases in Southeast Missouri. Williams predicates his review of phases with the cautionary note that phases "can and should be re-worked or changed ... it's content may be changed, it can be thrown away, or even divided or partially re-named." The remainder of his discussion, however, is an examination of the empirical status of each phase used in the region. No consideration is given to the theoretical basis underlying the concept of phase. In addition, little attention is paid to the relationship between the attributes used in their definition and the nature of the variability these attributes are measuring in the archaeological record. Rather, phases are treated as empirical entities that undoubtedly exist but remain undiscovered and/or poorly known. The focus of research in this view is simply the discovery of the boundaries of the phases.

Reinterpreting and Reevaluating the Phase

Although it has been routinely suggested that "the equivalent of phase ... *ought* to be 'society'" (Willey and Phillips 1958:49; emphasis mine; see also Rouse [1955] for a similar view), early workers tended to be more cautious about the culture/phase equivalence (*e.g.*, Abbot 1972; Brain 1978; Phillips and Willey 1953) and realized that such equivalencies were accidental rather than structural. Pressure to be "anthropological," however, has tended to lead workers to ignore this wisdom in many areas, including the central Mississippi river valley. Here, many archaeologists treat phases as if they are the material manifestations of ethnic, political or linguistic units. It is not uncommon to hear archaeologists talk about phases as empirical entities. Archaeologists write, for example, that "it was necessary for the Parkin phase folk to live within fortified villages, safe from attack behind palisades" (Morse and Morse 1983:292), and that "the natural resource most important to Parkin phase people seems to have been arable land suitable to hoe agriculture" (Morse and Morse 1983:293). Central Mississippi river valley research has often focused on relating these phases to historical groups (*e.g.*, Brain 1978, 1985; Hudson 1985; Morse and Morse 1983, 1990, 1996; Phillips 1970; PFG 1951; Phillips and Willey 1953; Rouse 1965; *cf.* Abbott 1972).

Although ethnographic meaning has been attributed to material associated with phases such "meaning" is untestable save for the very brief contact period. We have yet to assess if phases are the result of structured interaction among populations. Are patterns of artifact similarity, noted in the archaeological record as "phases," the result of ethnic differences, or is similarity best explained by functional differences resulting from the evolution of integrated functional communities? Alternatively, it is possible that phases are convenient units that represent only a fraction of the archaeological record. If phases are inappropriate descriptions of the record, then how do we build aggregate units representing many related phenotypes? Similarly, what archaeological evidence can be collected that is capable of determining the degree of relatedness, or homologous similarity, between samples of the record? In the construction of a science of archaeology, the importance of the solution to this problem is not trivial.

Despite their centrality to the field, there is little discussion regarding the methods used for defining phases (*e.g.*, Eighmy and LaBelle 1996:56). Indeed, Phillips, one of the main proponents of the phase concept, candidly admits that the procedures for constructing phases are "regrettably non-objective" (1970:523-52). Still more recently, phase formulations and interpretations of central Mississippi river valley archaeology based on them have been criticized as too simplistic. It has been argued that neither ethnic, political, or linguistic units have straightforward analogs in artifact distributions (Fox 1992, 1998; House 1991; O'Brien and Fox 1994). Mainfort (1999:165), for example, has proposed that "some previously-defined late period phases represent relatively valid units (*i.e.*, statistically reproducible), while others are much less robust, and some lack an objective basis." Other reanalyses suggest that phases may have little empirical basis. Fox and others (Fox 1992, 1998; Mainfort 1995, 1999; O'Brien and Fox 1994) have demonstrated that phases such as those created by Williams (1954) have no defining characteristics and cannot be considered classes (*i.e.*, there are no necessary and sufficient conditions for membership). At the same time, cluster analysis of the ceramics frequencies used to place assemblages in particular phases shows that phases are not groups (*i.e.*, the members of one phase are not more similar to members of the same group than they are to members of others).

While particular formulations may be defective because of empirical mistakes (Fox 1992), much of the problem arises through the grouping legacy inherited from the MTM. Indeed, outside the Mississippi valley many phases are just renamed foci. Through Ford's work (1935, 1936a, 1936b, 1939, 1951; Ford and Quimby 1945; Ford and Willey 1940, 1941) the Mississippi alluvial valley already had a budding chronology by the mid-1930s, and thus the ahistorical MTM never gained a foothold in this region (Dunnell 1996a; Lyman and O'Brien 1998). To the extent that phases are actually groups (instead of classes), the generation of new data (ceramic assemblages) *requires* the reassessment of similarity between *all* assemblages, not just the new ones. This was never done; *ad hoc* extensional definition was used to assign new data to old names (Dunnell 1971). Thus, the "phases" of the Mississippi alluvial valley quickly lost their coherence and rationale, producing the mess documented by O'Brien and Fox (1994). The degree to which such units "worked" (*i.e.*, displayed time/space contiguity) is a function of the units used in assessing similarity (*i.e.*, ceramic types). So long as stylistic (*i.e.*, historical or neutral [Dunnell 1970]) traits dominated in the assessment of similarity, phases would necessarily display time/space contiguity. Because phases create boundaries regardless of the structure of the archaeological record, phases, their locations and distributions are inappropriate descriptions when our goal is to study historical interaction and detect boundaries if such exist. Phases in the central Mississippi river valley are probably variable mixtures of arbitrary dissections of continuous variability, sample variability (size, spacing), and underlying structure in stylistic class frequencies. More to the point, the use of phases, any phases, assumes a particular answer, a particular spatial configuration and compels all evidence to assume that form. While there is no reason to suppose that anything but an accidental association can exist between phases and ethnographic groups, it does not follow that community interaction patterns do not exist and have not played a major role in Mississippi river valley prehistory. It is only clear there is no warrant for believing that phases describe these patterns. More to the point, there is no way to test such correspondence when phases are used to describe the record. Thus, if we want to understand community structure in the archaeological record, we must develop new means to describe interaction between populations in space and time.

The variability upon which phases were initially based is not necessarily arbitrary or random (Phillips 1970). In part, phases are defined by Ford's partitioning of the PFG assemblages into five subsets for seriation. Stylistic frequencies of decorated ceramics produce both temporal and spatial variability. This variability is potentially reflected in phases. It is possible, therefore, that the culture historical methods, such as seriation used by James Ford, and phase definitions are telling us something about population interaction. Clearly, the effects of both *time and space* impact the results of seriation (Dunnell 1970) and the processes driving the stylistic frequencies upon which seriation results are based. It is likely then that patterns of stylistic frequencies and the seriation method can tell us something about past populations and their distributions. If so, the variability upon which phases were defined in their original formulation by Phillips and other culture historians potentially reflect the social landscape of past populations.

SUMMARY

The work of culture historians provides an excellent starting point for building methods for measuring changes in population interaction through time. In the Mississippi river valley, PFG used historical types defined by stylistic dimensions as measurement devices, and seriation as a method for ordering assemblages described using these types. Since these tools are sensitive to stylistic variability in both time and space, it is reasonable to expect that culture-historical methods might be employed with the archaeological record when our goal is to generate data for studying the form and distribution of community patterning. Although phases have their roots in the materialist work of early culture historians, their essentialist implementations in archaeological practice have produced anthropological and intuitive units that mask variability. Phases *assume* structure and create boundaries when applied to the archaeological record. Thus, researchers using phases have generated potentially erroneous (and unfalsifiable) interpretations of the spatial and temporal structure of the archaeological record. As traditionally defined and implemented, phases are inappropriate units for a scientific archaeology; phases *cannot* be used to explain *historical* phenomena. If we are to build units for studying population structure, we must examine the theoretical basis upon which seriation is built and how patterns in community interaction produce variability in frequencies in time and space. In the next chapter, therefore, I review the linkage between interaction, measures of homologous similarity and the effects of space and time transmission as a prelude to demonstrating the use of neutral traits to map patterns of interaction in the archaeological record.

> It is possible that gene-culture coevolution will lie dormant as a subject for many more years, awaiting the slow accretion of knowledge persuasive enough to attract scholars. I remain in any case convinced that its true nature is the central problem of the social sciences, and moreover one of the great unexplored domains of science generally; and I do not doubt for an instant that its time will come.
> – Wilson (1994:353)

The success of culture historical methods in building chronologies is remarkable. The seriations of PFG, for example, have withstood nearly 50 years of excavation and evaluation; the chronological conclusions of PFG have yet to be falsified. These achievements are extraordinary considering that until relatively recently, (*e.g.,* Dunnell 1970, 1981; Lipo and Madsen 1996, 1997; Lipo *et al.*1997; Neiman 1990, 1992, 1993, 1995; Teltser 1995a), there has been little theoretical development explaining why this might be so. The culture historians worked literally in a theoretical vacuum. Although adept at defining and utilizing types for studying the past, culture historians were never able to provide theoretical justification for why culture-historical types are sensitive to time and why seriation chronologically orders assemblages when described by these types.

The lack of theory in culture historical explanations did not go entirely unnoticed; PFG were painfully aware of their limited ability to *explain* why cultural phenomena behave as they do. In their discussion of how they were able to obtain chronological claims about the frequency of decorated ceramics, PFG (1951:220) confessed that "we have not discovered a natural law operating independently of our own humble efforts." Thus, although the methods *worked* for building chronologies, PFG recognized that they had no theoretical warrant for their success.

In the absence of this warrant, PFG justified their results using a series of explicit *empirical* claims about the way in which their measurement units, types, detect differences in cultural interaction. PFG (1951:220) consequently assumed that types are "more or less sensitive instruments for measuring cultural change with the passage of time and distribution over space." In light of this assumption, PFG argued that assemblages could be ordered chronologically when they are composed of objects described with these chronologically sensitive types. PFG (1951:220) claimed that:

> if pottery types are successful measuring units for a continuous stream of cultural ideas, it follows that when the relative popularity of these types is graphed through time, a more or less long, single-peaked curve will usually result. Put in another way, a type will first appear in very small percentages, will gradually increase to its maximum popularity, and then, as it is replaced by its succeeding type, will gradually decrease and disappear.

Interestingly, these empirical assumptions are not the same sort of assumptions that led the new archaeology of the 1960s into circular traps of empirical generalizations as described by Dunnell (1982). PFG were rather explicit that their observations about the behavior of types must themselves be explained and do not provide the basis of a falsifiable account. They (PFG 1951:219, emphasis PFG) explained, for example, that "we consider these assumptions as a set of probabilities which lead to conclusions that are our best guesses. Not that we intend to apologize for this admission. This we think is the real method of science. We are trying to expose our limitations and are not setting out to *prove* anything beyond all doubt." Thus, although PFG lacked an ability to explain why their methods produced the results that they did, they fully understood that their conclusions were a function of the way in which their units and methods measured the world.

In this way, culture historians rationalized their success. Empirical assumptions provided justification for the behavior of measurements of the archaeological record made with *types*. "Successful" types for building chronologies are those that show unimodal popularity curves over time. These types are known to have passed the "test of historical significance" (Kreiger 1944; see also Dunnell 1986:170-171; Lyman *et al.* 1997:6; Teltser 1995a:54). Type construction, therefore, is a major part of the rationalization of culture history. PFG (1951:220-221) write, for example, that "this peculiar character of type popularity distribution through time is something we have helped to bring about through our own conceptualization of the pottery types that manifest said behavior. How the curves come out is partly controlled by how the types are defined." These claims about "popularity" distributions by culture historians stand as empirical excuses in lieu of a theory to explain why types act the way they do.

The lack of explicit criteria for creating types is seen in an analogy that PFG provide for the popularity principle and the properties of seriation (PFG 1951:221-222). As an illustration about the assumption of the behavior of pottery types, PFG describe how the process of measuring stylistic change in pottery is like observing changes in the frequency of modes of transportation in Ohio. They present fictitious data to demonstrate that if transportation types have characteristic peaks of "popularity," the compositions of popularity modes can be used to pinpoint a sample in time. The example sheds light on the *process* behind building chronologies using the unique combination of frequencies at any point in time. The behavior of the fictitious transportation types perfectly demonstrates how combinations of types, when their distributions display a unimodal, "popularity" curve, form unique combinations in time.

However, the goal of PFG's example is to provide a basis for an assumption about the behavior of decorative ceramic *types*. Although the example is only illustrative, it does point to a central confusion about types and the reasons for the behavior of objects measured with the types. Change in transportation systems is not equivalent to change in pottery decorations. Commonsensically, changes in decorative patterns can be attributed to observation and communication between potters, while the changes in the popularity of transportation are most likely functional (*i.e.*, frequencies shaped by selection). This suggests that PFG rationale for choosing the peculiar kinds of types that can be used to build chronologies is *ad hoc*. Since "popularity" distributions were treated as an empirical property of types, *all* proper types were considered to have chronological properties (e.g., Krieger 1944). In this view, it would not matter whether an example was given for ceramic decoration or transportation. This indifferent treatment of variability by PFG was not an isolated case. In general, all culture historians suffered from an inability to specify the kinds of attributes that could be used to build types, and to account for why decorative attributes on ceramics were particularly useful for building chronologies. This confusion resulted in the creation of culture-historical types, including PFG's, that contained a mixture of decoration and other kinds of attributes such as thickness or hardness. Without theory, it was not possible to systematically manipulate the construction of types with different kinds of attributes in order to control the behavior of phenomena described with these units. Only the *post hoc* test of historical significance – time and space contiguity – required of proper type distributions (*sensu* Krieger 1944) prevented complete collapse of the system.

One concept that has often been singled out as central for explaining the success of culture historical methods is the notion of "style" (Dunnell 1978a, 1995b). Beginning in the work of Holmes (1886b) and emerging in the early work of Kroeber and other culture historians (e.g., Kroeber 1919; Spier 1917), archaeologists have suggested that "stylistic" attributes are useful descriptions for measuring cultural change in the archaeological record. In his classic work on Zuñi pottery, Kroeber (1919:242) claimed that the variation he found useful for making chronological distinctions between pottery samples was "purely stylistic" because it did "not vary in purpose" (Kroeber 1919:239). However, Kroeber provided no further warrant for this proclamation.

Without criteria for judging whether an attribute is "stylistic" or not, use in archaeology research has been inconsistent. In subsequent culture-historical research, style has been usually treated as synonymous with decoration, though its implementation has varied from researcher to researcher. PFG types, for example, are comprised of a mix of decorative, shape and temper attributes. Other culture historians routinely and indiscriminately mixed a wide variety of attributes in the construction of their types (*e.g.*, Griffin 1938, 1939; Phillips 1970).

Although "style" appears to be a promising concept for explaining types, without a theory to guide our understanding it is not possible to consistently identify "stylistic" traits. Consequently, culture historians lacked criteria for building class definitions. Moreover, they had no ability to explain how "stylistic" attributes produce characteristic patterns in time and space. Nor did culture historians have an understanding of the mechanisms that drive frequency change in ceramics attributes. What are stylistic measurement units measuring? In general, these questions have remained unanswered until the last twenty years; "explanations" have largely consisted of common sense accounts of the popularity principle (Lyman *et al.*1997:118).

The general lack of theory explaining culture-historical types and seriation is more than just an annoyance: it limited our ability to move forward and develop archaeology as a science. Without a theoretical framework, we have no means to extend methods to address new questions. Are there aspects of type-frequency data, for example, that have been routinely ignored by researchers but that could be exploited to gain new knowledge about past populations? One interesting avenue of study in this regard is spatial variation in type frequencies. This variability has been traditionally interpreted by culture historians as *noise*, or distortions that confound chronological information. With only common sense accounts about popularity to explain stylistic variability, spatial variability can only be interpreted as noise at the worst, or simply equivalent to temporal variability at best (*e.g.,* Deetz and Dethlefsen 1965, 1971).

Significant to this discussion, our inability to explain *why* patterns of spatial variability coexist with chronological variability means that phases are simple statements of artifact similarity. Phases reduce to "assemblages that are similar to one another once temporal differences are controlled." While this definition certainly forms empirical groups, without theory it is not possible to explain why non-temporal similarity varies between assemblages since there is no theoretical relation between phase units and the attributes used to build them. Groups formed at the scale of phases provide *potentially* useful information about a number of processes that drive artifact similarity. These processes include measurement, observation, sampling, and behavior of prehistoric populations. Without theory to guide us, we have no means to specify the components involved in producing artifact similarity much less sort them out in terms of priority and precedence.

Finally, the inability to provide an explicit explanatory framework has historically led to general confusion and reoccurring arguments between archaeologists. The most famous of these arguments has come to be known as the Ford-Spaulding Debate (Dunnell 1986; Lyman *et al.*1997:147-149). The "debate" consisted of a series of papers in which James Ford and Albert Spaulding argued over the nature of types. Ford maintained that types were extensionally-defined (*sensu* Dunnell 1971) combinations of attributes, while Spaulding insisted that types were "real" reflections of cultural norms and discoveries of the analyst (Ford 1954a, 1954b; Spaulding 1953a, 1953b, 1954). The debate was never fully resolved mostly due to a lack of means for evaluating each author's claims. As Lyman *et al.* (1997:118) point out, "the absence of theory, and thus of theoretical units, typically result in awkward trial-and-error efforts to sort specimens into meaningful groups capable of getting some analytical job done, such as measuring time. Such efforts are ripe for all sorts of counterclaims and arguments over the meaning of groups. In the absence of theory, there is no way to resolve the resulting debates."

The development and implementation of theory, therefore, is a central concern for explaining why culture-historical methods behave the way they do; theoretical units avoid the problems of empirical generalizations since theory *explicitly* specifies the necessary and sufficient conditions for group membership and defines the relation between units. In this way, theory enables us to move beyond simple descriptions so that we can make explanations of how and why processes occurred in the past. Theory is what allows inquiries to become *scientific* investigations.

Over the past 22 years, Robert C. Dunnell and others have proposed using evolutionary theory to explain the past (e.g., Barton and Clarke 1997; Broughton and O'Connell 1999; Dunnell 1978a, 1980, 1989, 1992, 1995a; Leonard and Jones 1987; Neff 1992; Neiman 1995; O'Brien and Holland 1990; O'Brien and Lyman 2000; Rindos 1984; Teltser 1995a). From an evolutionary perspective, the principal mechanism explaining differential persistence of transmitted variation is natural selection. Biologists, however, have realized that not all variation results in differential interaction with the environment, or in other words that some variation is selectively neutral (Crow and Kimura 1970; Kimura 1977, 1983; King and Jukes 1969). As in other aspects of an evolutionary approach to archaeology, Dunnell (1978a, 1980) first made the link between selective neutrality and the archaeological concept needed to operationalize it: the concept of style.

While at first glance style may seem a strange concept to situate in an evolutionary context, artifact classes based on a notion of style have played a pivotal role in archaeology and thus require explanation. Artifact styles were the cornerstone upon which culture historical chronologies were built and served to define the spatial and temporal groupings we still use today. Later, early "New Archaeologists" used stylistic classes in an attempt to examine social organization (e.g., Deetz 1965; Hill 1968; Longacre 1964, 1970a). While these uses of style may appear difficult to reconcile, Dunnell (1978) pointed out that if one considered stylistic attributes to be selectively neutral, variation in their representation over time and space would reflect transmission and sampling alone, rather than the interaction of people with their physical environment. Over regional scales, patterning in stylistic classes would reflect the continuity and position of cultural lineages on a landscape; local-scale patterning in stylistic classes would likely reflect community organization.

Dunnell's notion of style as selectively neutral variation provides the guiding concept necessary to build a science of cultural transmission for archaeology to use with past populations. The concept alone is not sufficient, however, to allow us to actually study and map transmission in the record. Methods grounded in evolutionary concepts while still realistic for archaeological samples are required. In this chapter, I review the connections between the archaeological concepts of style, "neutrality" as it is used in evolutionary theory, and allied notions such as "drift." These concepts are part of a theoretical framework that provides a basis for explaining why PFG obtained the results they did and why culture-historical types are sensitive for studying time. In addition, this framework provides a means for explaining variability in pottery-type frequencies detected by phase units and directing research in the study of spatial patterns. In particular, I recast seriation using evolutionary theory to be a tool for examining interaction among populations and assessing observations in the archaeological record such as phases. I then review methods available for assessing neutrality in real data sets.

STYLE AND NEUTRALITY IN ARCHAEOLOGY

The first formal use of the word style in reference to the choice of attributes for describing artifacts comes from Holmes (1886b:451). Holmes occasionally employs style in a way that suggests he uses the term to designate decoration in contrast to functional properties (*e.g.,* shape) of artifacts. Holmes' use of the term implies that style is a description of phenomena, rather than the phenomena themselves but he is not consistent in his usage. During the early part of the 20th century, the Holmesian use of the term style led to a major breakthrough in archaeology when "style" was further developed in the work of culture historians like Kidder (1915), Kroeber (1916, 1919), Nelson (1919) and Spier (1917). In particular, southwestern archaeologists connected the style notion to chronological change observed through stratigraphy. These workers demonstrated that if artifacts were described with what Kidder (1915:453) termed "stylistic" classes, the frequency of these classes display characteristic "monotonic" distributions. As described earlier, it is this "monotonic" distribution of stylistic descriptions that could be used to date otherwise undatable assemblages. Culture history was founded on this principle.

There was, however, no attempt to explain why stylistic classes exhibited monotonic distributions. Seriation was accepted to "work" and types used in the seriations were generally agreed upon to be built from "stylistic" attributes. In the view of culture historians, style was considered to be anything that worked in the construction of seriations and evaluation procedures were trial and error (*e.g.,* Krieger 1944). Debates involving the culture-historical type or stylistic classification focused on ontological issues (*e.g.,* Ford 1954a, 1954b; Spaulding 1953b, 1954). Theoretical matters, such as criteria for defining classes, were ignored altogether. The relation between the properties identified as style and the distributional characteristics of stylistic classes thus remained obscure and the stylistic or historical type remained intuitive.

The advent of the new archaeology brought little clarity to the explanation of style. Instead, the situation has become considerably worse. Archaeologists such as Longacre (*e.g.,* 1964, 1970a), Hill (*e.g.,* 1966, 1968) and Deetz (*e.g.,* 1965) used the same notion of style as did the culture historians (*e.g.,* Ford 1935, 1936a; Kidder 1915, 1917; Nelson 1919), though they made a point of claiming to do otherwise. To make matters worse, they used the same attributes (*e.g.,* Spaulding 1954) as their predecessors. Rather than construct new classifications that would describe the record in a manner appropriate to the mechanisms they wished to use in explanations, the new archaeologists simply held to the classes already in use (Dunnell 1995b, 1996b). Built on this premise, the product of the new archaeology was primarily a series of new interpretations of old measurements (Dunnell 1986). However, since there were no changes in the relation between attributes and mechanisms, the new interpretations were generated without warrant (Dunnell 1996b).

The situation has steadily deteriorated through time to a point where style has a variety of meanings and the choice is left to the individual researcher. In their recent survey of "style," for example, Conkey and Hastorf (1990:1) report that:

> What we think of as style is pervasive in human society, no matter how we define it and style is involved in all archaeological analysis, whether it is covertly or overtly discussed. It is style that creates and defines types, culture types, and even types of evolutionary trajectories. Style is at work and can be studied at the level of the individual, a group, or a society.

Sackett (1990:33), in fact, goes so far as to praise the degree of ambiguity inherent in the current use of "style." Ironically, despite or perhaps *because* of this confusion, it is generally agreed that "the study of style and its place in research and interpretation in archaeology is central and determining" (Conkey and Hastorf 1990:1).

STYLE AS A THEORETICAL CONCEPT

In 1978, Dunnell introduced a theoretical basis by which the concept of style could be understood and proposed a means by which the behavior of "stylistic" traits could be explained. Using a framework grounded in the principles of Darwinian evolution, Dunnell (1978a:199) defined style as "those forms that do not have detectable selective values." Stylistic traits, he argued, are best explained as *neutral* traits. Stylistic similarity, then, is not impacted by natural selection but rather is the direct result of cultural transmission. One of the primary characteristics that marks stylistic traits is *equivalence,* since one of the central attributes of neutrality is functional redundancy. Meltzer (1981:314) suggests that "in many instances, the choice between certain kinds of design elements on ceramics is not a functional consideration, but rather is historically determined and selectively 'neutral,' because there is no inherent advantage between one element and the next. The actual presence of the design, however, has a selective value because that particular design serves to mark a certain individual or group boundary (or whatever other function it may serve)." O'Brien and Holland (1990) suggest that engineering studies are a way of distinguishing stylistic from functional traits (*e.g.,* Gould 1977). Engineering studies involve the analysis of performance for different alternatives. The smaller the degree of performance differences for any particular function, the more likely the traits will have neutral distributions. [1]

The explanation of "style" as neutrality has a relatively profound impact on the potential for evolutionary theory to account for human behavior seen in the archaeological record. Dunnell (1978a:198) argued that not only must we explain the "fixation of a particular form as a consequence of the increased Darwinian fitness that its presence confers on its transmitters," but we must also recognize that "not all elements can be assigned unambiguous positive or negative selective values." As in evolutionary biology, elements that confer relatively equivalent fitness contributions to their bearers within a given environment are defined as neutral traits. The recognition that some, if not many traits behave as if they are adaptively neutral, is an essential component of the evolutionary archaeology envisioned by Dunnell (1978a, 1981, 1992, 1995a). Importantly, the need for concepts such as neutrality is particularly acute in archaeological research since Dunnell (1978a:198) suggested that "if anything, cultural transmission should act to increase the capacity for ... neutral traits."

The effect of connecting the "style" concept to a larger theoretical framework was to resuscitate archaeological investigations within realm of science. Dunnell gave archaeologists a direction for building methods that are useful for exploring why particular cultural attributes have distinctive distributions in time and space. In this way, we are able to explain the success of the culture historians and to develop new tools in the study of the archaeological record.

Drift

A related concept in any discussion of neutrality is *drift.* Drift occurs as a result of change in frequencies of traits that is caused by sampling error (Wright 1931). Neutral traits in small populations are particularly susceptible to drift. Dunnell (1980) first linked drift among neutral traits and the properties of "style" when he noted that stylistic elements are characterized by their "unimodal" behavior in time and space and that in evolution, changes due to drift have a Markovian structure imparted by inheritance. These two observations led Dunnell (1980) to argue that cultural phenomenon noted as style display Markovian temporal structures indicative of drift and a lack of selection. This relation makes methods such as seriation, which rely on Markovian distributions to function, central for building explanations of historical phenomena

For the most part, interest by archaeologists in neutrality has focused less on neutrality *per se* and more on the Markovian

[1] However, understanding function requires control over the selective environment. The forms of ceramic vessels might have large performance differences in terms of their ability to transport materials, but could well be neutral traits if the ceramics were never involved in carrying. Ulimately, it is how traits behave across space and through time that determine whether they are stylistic or functional.

properties that neutral traits exhibit in time and space. In particular, Dunnell (1978a, 1980) and others (*e.g.,* Lyman *et al.* 1997; Teltser 1995a) have argued that culture historians succeeded in building chronologies with stylistic (*i.e.,* neutral) traits because they were independent of external conditions. Consequently, it was possible to define purely historical, non-repetitive classes with frequencies that could be ordered chronologically.

Although there has been significant discussion about the association between style, neutrality, and drift since Dunnell's original argument in 1978 (*e.g.,* Meltzer 1981; O'Brien and Holland 1990; Teltser 1995a), no specific arguments have been made to link style, the conditions under which drift occurs, and neutral theory. For the most part, stylistic traits are assumed to be neutral based on their "drift-like" distributions (*e.g.*, Allen 1996; Beck 1995; Kornbacher 2000; Pfeffer 1995; Wilhelmsen 2000). Nevertheless, since drift is a function of sampling error that occurs in populations of neutral and non-neutral traits, the link between drift and neutrality is simple. Style (*sensu* Dunnell 1978a) denotes archaeological classes that are manifestations of the theoretical concept of neutral traits. Drift-like characteristics in trait distributions, however, do not necessarily indicate that these traits are neutral since drift does not equal neutrality. The occurrence of drift depends on small effective population size since it amounts to a sampling effect in mating, whether panmictic within local groups or structured across a larger population. Neutrality, therefore, does not necessarily result in strong drift effects in populations of attributes. In addition, we can certainly imagine drift in traits that are *non-neutral* but population sizes have to be small for this effect to be substantial. Conversely, there is every reason to suspect that neutrality is common in large populations, where drift can have little effect. In other words, in large populations, the frequency changes in neutral traits are governed strictly by transmission and the introduction of novelty (*i.e.,* mutation). In such situations, it is unlikely that new variants will increase in frequency.

Clearly, simply stating that the concept of neutrality is capable of explaining stylistic traits is insufficient for understanding the conditions under which particular sets of attributes will or will not have unimodal distributions. Fortunately, biology has a long and rich tradition in studying neutral traits from which we can draw. Thus, to expand our understanding of neutrality and its relation to style, I turn now to the origins of neutrality and examine its current use in biological research.

NEUTRAL THEORY: BIOLOGY TO CULTURE

The notion that there is heritable variability not related to individual adaptation dates back to Darwin. Darwin (1859:374) argued that "on the view of characters being of real importance for classification, only in so far as they reveal descent, we can clearly understand why analogical or adaptive characters, although of the utmost importance for the welfare of the being, are almost valueless to the systematist." In addition, Darwin (1859:366-367) claims that "in formerly discussing certain morphological characters which are not functionally important, we have seen that they are often of the highest service in classification."

Despite Darwin's recognition that there are differences in descriptive traits, however, the notion that some traits have no adaptive value remained controversial among biologists who held that natural selection remained supreme in driving variability among organisms (*e.g.,* Wallace 1889). In 1872, for example, Gulick argued that natural selection could not account for the local differentiation of Acatinellid snails in the Hawaiian Islands (Gulick 1872, also 1888). The publication of Gulick's paper stirred up tremendous controversy among biologists such as Wallace who rejected the notion that differentiation among species could be produced by anything but the action of natural selection (Wallace 1889:148, 150). The debate over the role of natural selection in producing all observed patterns of variability in organisms was not quickly resolved and continued over the next sixty years (Provine 1986:216-220).

The existence of non-adaptive traits and their effects in evolution was finally quantified and codified with the development of the "neutral theory" in the 1950s and early 1960s.[2] The theory came about in the late 1960s, as protein sequence data were beginning to be available for comparative studies. At the time, Motoo Kimura, a population geneticist and staunch neo-Darwinist, was working to integrate population genetics with molecular data. In 1967, he compared the amino acid sequences of hemoglobin and cytochrome *c* from several mammalian species. By extrapolating the number of sequences to estimate the total genome, he was surprised to find that the number of mutant substitutions was too large to be tolerated by a strict interpretation of the theory of natural selection as held at the time (Ohta 1996). Based on this discrepancy, he proposed the neutral mutation-random drift hypothesis. This hypothesis states that most mutant substitutions at the molecular level are caused by random genetic drift rather than natural selection. Kimura (1968) published his findings in 1968 in *Nature*. Further evidence was published shortly after Kimura by J.L. King and T.H. Jukes (1969) who considered additional biochemical evidence. Together, these papers caused quite a controversy among evolutionary biologists (Ohta 1996).

Since Kimura's initial neutral theory proposal, there have been numerous modifications that proved necessary for its application to biochemical evolution. In the 1970s, biochemists such Ohta (1976) proposed that the 'nearly neutral theory,' in which mutations are very slightly deleterious, better accounts for the data on protein evolution and polymorphisms that was being collected at the time (Ohta 1996). In the original neutral theory, new mutations

[2] The development of the "neutral theory" occurred despite the fact that many scientists thought that the genetic mechanisms of natural selection were mostly solved and neo-Darwinism was at its peak (Ohta 1996:673).

were assumed either selected or neutral. In the new theory, the focus of research was on the interaction between random genetic drift and selection. The neutral theory, consequently, became quite complicated. In the 1980s, comparison of DNA sequences became possible and it was noted that great chunks of DNA such as synonymous sites or pseudogenes exhibited rapid evolution (*e.g.,* Kimura 1983:159). Consequently, the stricter neutral theory that was more in favor of Kimura's original formulation came back into favor (Ohta 1996). More recently in the 1990s, new DNA sequence data have shown that patterns of substitutions at selectively important sites differ from patterns of substitution at unimportant sites. Change in the unimportant sites, it seems, can be better accounted for by neutral theory, whereas important sites are more heavily influenced by natural selection (Gillespie 1991). The differences in substitutions, therefore, provide an opportunity to detect selection. Many scientists began to argue once again that strictly neutral theory is not entirely satisfactory (Gillespie 1991; Kreitman 1996; Ohta 1996).

The current nearly-neutral formulation is as follows: random genetic drift and selection both influence the behavior of very weakly selected mutations with drift predominating in small populations, and selection in large populations (Gillespie 1991; Ohta 1996). Most new mutations are deleterious and most mutations with very small effects are likely to be slightly deleterious. Such mutations are selected against in large populations, but behave as if neutral in small populations. Researchers (*e.g.,* Gillespie 1991) have developed complicated models for testing these weak selection scenarios.

The rejection of the strictly neutral theory, however, does not have a substantial impact on the practical usefulness of the neutral theory in building evolutionary explanations of variability in the archaeological record. As Mayr (1963:212-21) has pointed out, there is no compelling theoretical reason why neutral traits cannot exist or that, within our ability to detect them, alternative forms of traits could not have equivalent costs and benefits to an organism (Dunnell 1980). Neutral theory remains at the core of modern evolutionary biology. Neutral theory addresses relative differences in fitness between alternative traits, not absolute fitness. Since the degree to which traits are "equivalent" depends both on the engineering relation between the traits themselves and the strength of sampling effects given population size. In small populations, small fitness differences, for example, may be effectively neutral due to sampling effects where the same difference would be significant in a larger population.

The first sentence of Kimura's (1983:1) *The Neutral Theory of Molecular Evolution* provides a basis for method development using the neutral theory: "The neutral theory asserts that the great majority of evolutionary changes at the molecular level, as revealed by comparative studies of protein and DNA sequences are caused not by Darwinian selection but by random drift of selectively neutral or *nearly neutral mutants.*" However, what is a nearly neutral mutant? Kimura (1983:48) proposes the following standard: "If a mutant has a selection coefficient much smaller in absolute value than $1/(2N_e)$, it behaves like a neutral mutant" where N_e indicates the effective population size. Thus, from genetics we can build the basis of an empirical standard – traits with fitness differences less than $1/(2N_e)$ – by which it is possible to evaluate the observed distribution of traits against the expectations of the neutral model.

Consequently, far from being a model about the way all traits work, neutral theory is more commonly used today as a means for generating *null models*, against which we can detect the action of drift, selection, and other evolutionary mechanisms (Wayne and Simonsen 1998). That is, despite uncertainties about details, the neutral model yields explicit empirical expectations that provide a conceptual framework for explaining variation and studying evolution. The neutral model is, therefore, a "barometer for detecting selection acting" on the variability of parts of the phenotype (Kreitman 1996:682). From an empirical point of view, incorporating neutrality into methodology and explanations is critical to the success of evolutionary programs for geneticists, biologists and archaeologists alike. As a null model in archaeological research, our challenge is to build methods for evaluating hypotheses of neutrality in the context of cultural transmission even as we reject them.

NEUTRALITY AND CULTURAL TRANSMISSION

Evolutionary concepts such as neutrality cannot be applied directly to cultural phenomena. The notion of neutrality was constructed to account for patterns of variability seen in amino acid sequences. It is necessary to ask, therefore, whether variability in the distribution of cultural traits can be explained using the neutral model. As a *theoretical* concept, the neutral model is true *by definition* and determining whether it applies in any particular case is an empirical question (Mayr 1963). As noted above, there is no *a priori* reason that the neutral model will not hold for cultural transmission (Dunnell 1978a, 1980; Lipo and Madsen 2000; Teltser 1995a).

There may be, however, empirical restrictions on any particular genetic model that render it inappropriate for studying cultural phenomena. Case in point: genetic transmission operates through reproduction and genetic material is vertically transmitted from parent to offspring (Cavalli-Sforza and Feldman 1981). Unlike genetic transmission, cultural transmission is free to operate in directions other than vertical. It is common, therefore, for researchers to divide transmission into modes: vertical, horizontal and oblique (*e.g.,* Boyd and Richersen 1985; Cavalli-Sforza and Feldman 1981; Sober 1991). Horizontal transmission applies to transmission within members of the same generation. Similarly, oblique transmission describes transmission from an individual to a member of a different generation (Cavalli-Sforza and Feldman 1981).

In each of these scenarios, cultural transmission is conceived as distinct but analogous to genetic transmission in which discrete packets of information are passed between individuals in meaningfully defined generation groups. It is not clear, however, if a genetic model is appropriate in the case of cultural transmission. Using a model that assumes generations requires transmission events that are discrete. While the units and events of transmission are prominent in genetic transmission, the same is not true for cultural transmission. Cultural transmission is continuous through both time and space and hence there are no clear, empirical, temporal and spatial "packages" of transmission. Slicing the continuum of transmitted information involves defining measurement units.

It is necessary therefore, to insure that these measurement units are applicable to the cultural case. The first question to be asked is: what is the unit of transmission? In genetics, the answer is well developed. Genetic transmission occurs in a packet of information that is spatially and temporally discrete. Transmission of genetic material occurs in the process of nucleotide provisioning in DNA and of amino acid specification in proteins. Cultural transmission, however, has no such discernible units. Cavalli-Sforza and Feldman (1981:73) define the cultural unit, for example, to be "the result of any cultural action (by transmission from other individuals) that can be clearly observed or measured on a discontinuous or continuous scale."

The clarification of two points is necessary for this definition to be of use. First, although the definition is similar to the way in which culture-historical types are often conceived, types are not units of transmission. Lacking any inherent scale constraints, cultural transmission may vary in scale. Characteristically, Ford was clear on this point. His comparison of southeastern seriations centered on the examination of *attributes* such as "rocker-stamping" and "paddle-stamping" across time and space even though types were used to construct his seriations (Ford 1952). Deetz and Dethlefsen (1965), on the other hand, typify common-sense perspectives prevalent in archaeological literature when they assumed erroneously that gravestone *types* were transmitted in time and space. The assumption is, of course, false: types are constructions of archaeologists.

Second, attributes themselves are ideational slices of dimensions that are chosen by the analyst to measure the world in a particular fashion for a particular purpose. Entities transmit between one another along an infinite number of dimensions. Ideational filters break down the continuous nature of cultural transmission into internally identifiable chunks. The division of the cultural streams occurs *post hoc* by the engaged entities and in subsequent analyses. During the process of transmission, humans break the stream of information into things that we can identify in terms of words, sentences, expressions, emotions, gestures,and ideas (among other things). Since cultural transmission has no embedded scale or content, it is the responsibility of the researcher to develop a theoretically meaningful set of dimensions to divide the empirical products (Dunnell 1971). As Osgood (1951) pointed out, we must be careful to distinguish empirical and non-empirical components of culture so that the operation of different evolutionary processes may be measured (Dunnell 1971). Cultural transmission *occurs* within the ideational realm. Transmission, though, must be *identified* by observing empirical phenomena.

One of the important initial steps for developing a theory of cultural transmission using a model other than genetics is pinpointing an equivalent unit of transmission. Genetics, in many ways, is well understood because the molecular mechanisms involved in information transference are well studied. Here, the relation between theoretical units and empirical units are used interchangeably in model building. Cultural transmission, on the other hand, is poorly understood. As Hallpike (1986:46) states "theories of basic units of culture do not rest on any evidence or on any sociological theory at all, but are simply proposed because one is trying to explain culture on the basis of a neo-Darwinian theory of natural selection."

Consequently, there is little agreement about the appropriate unit of analysis for cultural transmission. On the one hand, Cavalli-Sforza and Feldman (1981) take an epidemiological approach and study the distribution of cultural traits across populations in which traits are passed either from parents to children (vertical transmission) and from child to child or unrelated adult (horizontal transmission). They describe cultural characters as "second order organisms" but define them very loosely. Alternatively, Boyd and Richerson (1985) and Durham (1991) offer studies of transmission that make use of particulate models of cultural traits while simultaneously arguing that particles are not a necessary part of their theory. These theories are presented from the point of view of individuals that emphasize how cultural traits affect individual fitness and biological reproduction. Finally, Dawkins (1976) coined the term *meme* to refer to any cultural trait capable of being transmitted. In his "memetic" approach to the study of cultural transmission, analytic focus is on the packet of information (memes) that is replicated through human communication and within human brains. The memetic perspective has seen substantial support in the past several years, occasionally as merely a metaphor but increasingly as a body of theory (*e.g.,* Best 1997; Best and Pocklington 1999; Blackmore 1999; Gabora 1996; Aaron Lynch 1996; Alejandro Lynch 1996; Lynch and Baker 1993, 1994; Lynch *et al.* 1989; Marsden 1998; Payne 1996; Payne *et al.* 1981, 1988; Pocklington and Best 1997).

Although there are no agreed-upon empirical units of cultural transmission, it is possible to build units of transmission that are theoretically equivalent to genes. Williams (1966) defined the gene as the unit that segregates and recombines with appreciable frequency. Using Williams' definition as a starting point, Pocklington and Best (1997:81) define cultural transmission units as "the largest units of socially transmitted information that reliably and repeatedly withstand transmission." Importantly, cultural transmission units are measurements of the *effect* of transmission on variability. In this way, culture-historical types are entirely compatible with this definition of the unit of cultural transmission.

Since units of transmission are ideational and not "directly" observable, it is necessary to devise means by which the effects of transmission can be identified. Well understood by the early cultural historians (*e.g.,* Colton and Hargrave 1937; Ford 1936a, 1938a, 1938b; Ford and Griffin 1938:3; Kidder 1915, 1917; Kroeber 1916:44; Nelson 1919; Spier 1917) and pointed out by PFG (1951:220), the task of measuring cultural transmission is accomplished through the use of historical types (Dunnell 1970, 1986; Lyman *et al.* 1997). Historical types are ideational units composed of combinations of traits. Observations are made along an infinite number of intensionally defined dimensions. The historical character of these types is derived from the fact that they, due to the peculiarities of transmission, are represented by samples of empirical units that have unique and contingent sets of traits. Thus, it is the intersection of these unique combinations of traits with the set of dimensions used to measure attributes that creates historically significant types with bounded temporal and spatial distributions. Importantly, temporal and spatial boundedness arises because of the interaction of the measurement process and the history of transmission that occurs between entities that inhabit the empirical world when the traits are stylistic or neutral.

Instead of viewing transmission solely from a genetic perspective (and cultural transmission as merely an analog to genetics), it is more useful to employ a materialist ontology and view the genetic model as a special case limited to DNA transmitters. There are two analyzable components in a continuous model: temporal and spatial. The temporal component arises because transmission events, as all physical processes, require time. Since interaction is from-to and empirical entities have spatial coordinates, interaction events have a spatial component. Interaction through space, however, is not entirely different from that through time (Deetz and Dethlefsen 1965, 1971; Dunnell 1970). The effects of spatial transmission on trait frequencies are qualitatively the same as, yet quantitatively different from, the effects of time on trait frequencies.

BIRD-SONG RESEARCH AND NEUTRALITY

Before proceeding further, it is useful to examine the application of neutral theory in evolutionary biology. If the neutral model assumes discontinuous transmission of attributes, for example, empirical predictions may not be consistent with observations. In addition, the model may not be empirically sufficient to account for the cultural case. Evidence that neutral theory is, in fact, relevant to empirical research is exemplified by studies of song distributions amongst bird populations. What makes this body of work even more compelling as a stimulant to archaeological research is that many bird species transmit songs and song elements culturally rather than genetically (*e.g.,* Lang 1995; Lang and Barlow 1997; Lynch and Baker 1993, 1994; Lynch *et al.* 1989; Payne *et al.* 1981, 1988; Payne 1996). Bird songs are simple examples of cultural transmission: transmission is continuous, non-genetic and most songs are learned from other birds. In addition, depending on the species, some variability in song form, composition and style appears to be related to mating and defense, but some does not appear to have any specific function (Payne 1996). Thus, application of neutral theory to the study of bird songs provides a useful guide for how the model can be applied in studies of cultural transmission generally.

In the literature on bird communication, songs are treated as if they are composed of a number of song elements (memes) each of which is transmitted independently between birds. There is a consensus among bird-song researchers that within acoustical constraints of the environment, song elements are functionally equivalent (Lynch *et al.* 1989). Payne *et al.* (1981), for example, concluded that differences in persistence of song traditions in indigo buntings (*Passerina cyanea*) were a matter of chance. Slater *et al.* (1980) showed that the distribution of song types of chaffinches in a British population conformed to the neutral model and suggested that the differences in the frequencies of different song types do not arise from preferential copying. Lang (1995) has shown in his studies of the Eurasian Tree Sparrow (*Passer montanus*) that bird songs meet the expectations of the neutral model. Significantly, Slater *et al.* (1984) found no significant differences in chaffinch song types between several British populations, some of which occupied very different environments. In other cases, however, song differences can potentially be attributed to the effects of natural selection. In the songs of British and New Zealand chaffinch populations, for example, Jenkins and Baker (1984) found that differences in song attributes can be attributed to the prevalence of chaffinches in dense pine forests in New Zealand, which have different acoustic properties from the broad–leafed woods of Great Britain. What is significant here is that the explicit theoretical framework provided by the neutral theory permitted these researchers to evaluate hypotheses about the distribution of cultural types across populations.

METHODS FOR EXAMINING NEUTRALITY

One important aspect of the investigation of the neutrality of cultural traits is the recognition that neutrality is a concept to account for the distribution of phenomena and not a phenomenological property of organisms. Consequently, types, memes, attributes, or traits are not, by themselves, neutral or non-neutral. Rather, any particular type or set of attributes can have neutral or non-neutral distributions at a various times and places (Dunnell and Feathers 1991). Attribute distributions change through time and can shift back-and-forth from distributions consistent with neutrality to ones characterized by mechanisms involved in selection. In the case of bird-songs, researchers have demonstrated how one cultural variant will meet the expectations of the neutral model in one particular time and setting but be under selection in another (*e.g.,* Jenkins and Baker 1984). Methods are needed, therefore, to evaluate whether particular attributes and their attendant empirical distributions meet the expectations of the neutral model. We need *null models* that can be used to evaluate existing empirical distributions. Using the general theory of neutrality, it is possible to derive two separate null models for neutrality; one based on functional equivalence and one based on stochastic change within closed arrays caused by drift.

Equivalency

The theory of neutral mutation predicts that variability in a population is a balance between the origin of new forms due to mutation (and transmission from external populations) and the extinction of old forms due to drift (Lynch and Baker 1993). In this case, there are forms that exist in a population that are functionally equivalent. Using Kimura and Crow's (1964) infinite alleles model in which memes belong to a single locus with infinite alleles, a standard model of neutral mutations can be summarized as follows: there is a population of N memes chosen from an infinite series of types with no selective differences among them. Successive generations are formed by a process in which the N genes, or memes are obtained by repeated sampling with replacement from the N genes or memes of the present generation. Each time a gene or meme is sampled, there is a probability v that a unique mutation occurs or that a new form enters the population by immigration. At equilibrium, the diversity of forms in a population is related to mutational and migrational input and to extinction by random drift by the equation:

Equation 3.1. $$s_e \approx 2Nv + 1$$

where N is the population size and v represents the combined effects of mutation (μ) and migration rates (m). Ewens (1972) has shown that the total number of distinct alleles or memes (s) in a sample and the sample size (n) are sufficient to provide an expected distribution of meme frequencies. This expected distribution is used as a null hypothesis for evaluating the distribution of attributes for neutrality. From the expected and observed distributions, a maximum likelihood method can be used to determine if the observed sample corresponds with the expected values of the model. If the distribution of types of memes is not strongly impacted by selection, it is expected that one or another combination should not be strongly favored. Types of memes, therefore, will be evenly distributed in an ideal population (*i.e.*, one entirely composed of neutral traits). In a process similar to resampling that uses the original distribution as a means for generating the expected distribution, Ewens' method (1972) calculates the likelihood that an observed distribution was generated from an even distribution of the same composition of variant combinations. The observed distribution of memes is first calculated as a meme identity (I) statistic. Watterson (1978) showed that I is a good test statistic for examining the expected and observed distributions. I can be calculated as follows: if there are S different memes in a population of size N, and if the frequency of the *kth* meme is p_k, then we can define *meme identity* as:

Equation 3.2. $$I = \sum_{k=1}^{S} p_k^2$$

The expected value of I can then be obtained by using the equation:

Equation 3.3. $$E(I) = \frac{1}{(\theta + 1)}$$

In Equation 3.3, θ is a population parameter that is derived by Equation 3.4:

Equation 3.4. $$\theta = 2N_e v$$

In Equation 3.4, N_e is the effective population size of memes, v is the combined mutation and immigration rates. Since N_e and v are not directly measurable, θ is estimated using a maximum likelihood method derived by Ewens (1972) and Equation 3.5:

Equation 3.5. $$E(k) = \sum_{i=0}^{k-1} \frac{\theta}{\theta + i}$$

where k is the number of variants. Although Equation 3.5 has no analytic solution, it can estimated by iteratively changing the value of θ until $E(k)$ equals the observed number of variants. The expected value of the meme identity has also been obtained by simulation using algorithms based on Ewens method and developed by Stewart (1977; see also Lang 1995; Lynch and Baker 1993).

The method, therefore, provides a means for deriving a null distribution of neutral traits, assuming that only the processes of mutation and sampling drive variability in trait frequencies. Using resampling on an empirically observed distribution of traits, the test generates an expected distribution, and iteratively calculates the likelihood that the original assemblage was derived from a multinomial distribution of the same composition, driven by stochastic processes. Deviations from the predicted distribution indicate that factors other than immigration, mutation, and the elimination of types due to random extinction shape trait frequencies. These factors include selection, sampling, and measurement errors.

Ewens' method has been profitably used in studies of cultural transmission outside of archaeology (*e.g.*, Lang 1995; Lynch and Baker 1993). In an analysis of bird songs, for example, syllable variants that comprise each bird song within a population are treated as memes. Ewens' method can then be used to evaluate whether the distribution of song types meets the expectations of the neutral model. Lang (1995), for example, demonstrated that within-sample variability was primarily neutral for Illinois and German populations of Eurasian Tree Sparrows. Furthermore, the distribution of bird-song attributes resulted from immigration and mutation, and the subsequent elimination of memes was due to random "extinction."

Markovian Distributions

Among archaeologists, the existence of Markovian distributions of traits through time has been used as an indication of their neutrality (Dunnell 1980; Neiman 1995; Teltser 1995a). With only a few exceptions (*e.g.*, Gould *et al.* 1977), this aspect of neutrality is unexplored by other evolutionary disciplines. As archaeologists, we may be uniquely situated to exploit this feature of neutral traits in the study of cultural transmission.

Markovian distributions are generated in any finite population. In a culturally interacting, finite-sized population, frequencies of traits are dependent upon the composition of the trait population from which those traits were derived. Over time and in an environment in which selection has a small or no effect, the frequency of any particular trait is determined by the previous generation. Free to vary, yet contingent on the past, these frequencies will have a probability of increasing and decreasing at any point in time determined by the previous frequencies *and* random processes. This distribution is said to be Markovian because of dependence upon abundance in the immediately preceding population configuration (Dunnell 1978a; Neiman 1990; Teltser 1995a:60).

Using a general model of cultural transmission in which individuals learn from each other, Neiman's (1990, 1995) and my own simulations (Lipo *et al.* 1997) show that these stochastic processes can produce "monotonic" distributions through time. Primarily, this is a function of measuring change within a population of stochastically changing variants of sufficient number using a closed array of frequencies. As Neiman (1990) notes, erratic change increases dramatically at small sample sizes; the "smoothness" of the monotonic distribution is, in part, a product of sample size. Sample size is a function of both collection effort and formation processes that produced the populations sampled.

Although Markovian distributions have often been used as an indication of the neutrality of traits, there has been little statistical work to evaluate this model against archaeological samples. Rather than simply make claims about neutrality, it should be possible to determine the degree to which a particular set of observations meets the expectations of the neutral model. To meet this need, I developed a statistical test to evaluate how closely an empirical distribution matches a neutral null model. The neutral model, considering the effect of Markovian change in closed array frequencies, states that while there may be change within the frequencies of classes, this change should not be strongly biased in one direction or another. This test, based on a random walk analysis, examines whether the sum of differences in frequencies between pairs of assemblages, given a particular order, is statistically different from zero. Thus, this test evaluates whether the frequency of any particular type meets the ideal "monotonic" model, while recognizing the effects of sampling, and that the distribution may or may not meet the ideal "battleship" shape.

Since the value of the differences between type frequencies among assemblages is a product of sample size, I followed a standard bootstrapping routine (*e.g.,* Efron and Tibrishani 1993). This procedure involved generating 1000 pairs of bootstrapped assemblages from the original sample size using the original frequency values as a null distribution. I then calculated the sum of differences between the frequencies of types in pairs of bootstrapped assemblages. This procedure allowed me to produce a normal distribution of difference values for each pair of assemblages from which a mean and standard deviation were calculated. Using the 95% confidence intervals, I then evaluated whether the mean of the summed differences between pairs of assemblages was statistically distinguishable from zero. Types with distributions that are statistically distinguishable from zero do not meet the expectations of the neutral model since in the case of neutral traits we expect frequencies that are not strongly increasing or decreasing. Conversely, those types with distributions across assemblages that have summed differences that are statistically indistinguishable from zero cannot falsify the hypothesis that these types meet the expectations of the neutral model. In this case, frequencies have a statistically-significant trend that is not consistent with the model.

A Hypothetical Example

As an example of this method, I generated a set of sample assemblages with types whose frequencies changed both monotonically and directionally across assemblages (Table 3.1). I then conducted the random walk test for these sets of assemblages. The results are shown in Table 3.2.

Table 3.1. Example Set of Ten Hypothetical Assemblages with Five Types.[a]

Assemblage	Type 1	Type 2	Type 3	Type 4	Type 5
A	10	40	0	5	45
B	20	30	0	15	35
C	25	20	0	25	30
D	30	5	10	30	25
E	30	0	25	35	10
F	20	0	30	40	10
G	10	0	45	40	5
H	5	0	40	50	5
I	5	0	35	55	5

[a]The numbers for types 1-5 were created to sum to 100 across each assemblage. The distribution of Type 1 was set to have a monotonic distribution; the remainders have directional changes across the assemblages.

Table 3.2. Results of the Random Walk Test Conducted for Table 3.1 Sample Assemblages.

Type	Mean Difference between Assemblages	Standard Deviation	Sample Size (Sherds)	Number of Assemblages with Type	C.I. Lower 95%	C.I. Upper 95%	Includes-Zero?[a]
1	0.008	0.090	155	9	-0.052	0.068	YES
2	0.115	0.054	95	4	0.061	0.169	NO
3	-0.053	0.092	185	6	-0.127	0.022	YES
4	-0.072	0.063	295	9	-0.113	-0.030	NO
5	0.046	0.059	170	9	0.007	0.086	NO

[a] 'YES' and 'NO' values refer to whether the mean differences between assemblages for the type fall within the 95% confidence intervals. Only Types 1 and 3 in this example meet this criterion. We would be, therefore, unable to falsify the neutral hypothesis in these two cases.

The results of the random walk test demonstrate how it is possible to test a null hypothesis of monotonic distributions for series of frequencies. In the test, I evaluated whether I was able to distinguish the sum of differences between pairs of assemblages for percentages of a single type from a total value of zero at a 95% level of confidence. Inability to falsify the null hypothesis suggests that the distribution of frequencies may meet the expectations of the neutral model. Falsifying the null hypothesis, on the other hand, indicates that either sample size or sampling is impacting class frequencies or that the type is not composed of attributes that are compatible with the neutral model. Types with distributions in the assemblages that are statistically distinguishable from zero are indicated by a 'NO' in the rightmost column of Table 3.2. This latter result might lead one to questionwhether the attributes being measured are under selection.

One consideration in the use of the random walk test for evaluating the neutrality hypothesis is the requirement it places on the number of assemblages used. Increasing the number of assemblages helps one evaluate whether types produce descriptions of material that meet the expectations of the neutral model or not. The larger the number of assemblages, the smaller the error term, since the error term is a function of the number of assemblages used to compare the sum of differences against zero. Larger numbers of assemblages reduces the chance of making a Type II error .

SUMMARY: ARCHAEOLOGICAL EXPLANATIONS AND NEUTRALITY

Dunnell's (1978a) explanation of stylistic variability as neutral variability was an important milestone in the development of a scientific archaeology (Rakita and Hurt 2000). With the introduction of this theoretical framework, it became possible to derive a dynamically sufficient account (*sensu* Lewontin 1974) for the properties of historical types long used by culture historians for building chronologies (Dunnell 1978a; Lipo *et al.*1997; Lyman *et al.*1997; Teltser 1995a). The concept of neutrality provides a means for explaining existing archaeological notions such as style and the properties of archaeological methods such as seriation. Investigation into the properties of neutral traits reveals aspects of attribute distributions that are not a function of selection but of past population structure, population size and transmission systems. Ewens' statistics and the random walk test are two methods for examining null hypotheses about expected distributions under the neutral model. These methods use neutrality as a null hypothesis for trait frequencies, a statistical background against which it is possible to examine the interplay of transmission and selection on culturally transmitted information.

CULTURE HISTORY AND CULTURAL TRANSMISSION

Since the connection between neutral variation and the archaeological notion of historical types was made (Dunnell 1970a, 1978, 1981), it has become possible to explain why seriation worked as a dating method and why spatial variation was a confounding factor in its early use by culture historians (Dunnell 1970; Telster 1995). If stylistic variation were neutral, then the temporal distribution of stylistic (historical) types would be unimodal because of the probabilistic nature of transmission (Dunnell 1982). This supposition, originally based on Dunnell's proposed model (1978a; personal communication) of Gould *et al.*'s (1977) descriptions of the shape of evolutionary change in paleobiology, has been confirmed and extended greatly by Neiman (1990, 1995).

Two other components are critical. Dunnell (1970) analyzed Ford's (PFG 1951:219-223) assumptions for the use of seriation, all of which were phrased as conditions that had to be met by the assemblages to be seriated. Dunnell showed that Ford's "local area" criterion, that all of the assemblages had to come from the same "local area," was incorrect. With the local area notion, Ford was attempting to exclude space from influencing assemblage placement in a seriation. Drawing boundaries around local areas was not an appropriate solution because culture trait transmission necessarily has both spatial and temporal components.

While one set of stylistic classes applied to a given area might not detect spatial variation, increasing the amount of space, or using another more detailed set of classes, would include spatial variability even though each analysis used the same set of potsherds. Seriation was not conceived as a theory to explain neutral variations by culture historians like Ford, but as a dating method. As a theory, it was, and is, dynamically insufficient.

A second element must also be understood from this historical perspective. Ford (PFG 1951) saw that assemblages to be seriated had to be drawn from the same tradition (*sensu* Phillips and Willey 1953:618-628). This was correctly attributed to a condition that the assemblages had to meet. The reasons this should be so are significant. As Dunnell (1970) noted, the ability to create a seriation was the test for determining whether a group of assemblages constituted a single tradition. What Ford (PFG 1951:223) meant by "tradition" was cultural continuity or the idea that "the culture of the area in the main probably changed gradually rather than by means of mass migration from other areas." This condition requires that the assemblages being seriated have to be drawn from assemblages that were historically related. In evolutionary terms, this requirement meant that the assemblages must be samples from the same lineage.

Thus, why seriation works (and does not work) is explained by evolutionary theory (the neutral theory in particular) and the appropriateness of seriation as a tool to explore interaction is clearly warranted. Carefully expanded, the method is more than a means for chronologically ordering material described by stylistic types. Rather, we can *refocus* our understanding of the seriation methods to be a general means for explaining the distribution of neutral traits. This refocusing is my task here.

BUILDING A MODEL OF TRANSMISSION

Since cultural transmission is continuous through time and space and cannot be viewed directly, models for studying cultural interaction within populations can become exceedingly complicated. To proceed further in the development of a model of cultural transmission, we must begin with a simple set of assumptions about the nature of transmission. Neiman's (1995) recent formulation of neutral trait transmission provides a defensible point of departure because it rests on a set of basic assumptions. In Neiman's model, all individuals in the population are assumed to have equal probability of interaction with each other. Thus, trait frequencies should behave stochastically due to drift. If this model was translated into a spatial context, it would mean that space imposed no energy costs and that all individuals would, over the long run, interact with each other. In panmictic[1] populations, I would expect that overall similarity between samples of individuals from across the space would be similar at any point in time. Such a model is simplistic, but as Neiman (1990, 1995:Figure 2) showed, when viewed as relative frequencies, transmitted attribute frequencies form unimodal curves of the kind seen in frequency seriation. As Dunnell (1970) notes, this behavior is due to the effects of panmictic transmission between individuals in the population without any restrictions on interaction caused by spatial costs, geography or social factors.

This effect can be seen clearly by simulating interaction using Neiman's assumptions, allowing individuals to interact over great distances, and accumulating many time slices of such transmission events. A spatial transect through the space and time "cube" (since individuals interact over a 2-dimensional landscape through time) is illustrated schematically in Figure 3.1. Time is displayed on the vertical axis, with the bottom of the figure being the start of the simulation, and space along the horizontal axis. The "threads" traveling from bottom to top are individuals. Each individual has a letter value that denotes an attribute currently held along one of several dimensions. Intersections between threads are, when viewed at a moment in time, interactions between individuals as they move around the space and encounter one another. The result of an interaction can be (1) a switch in traits for both individuals, (2) one individual copies the other's trait, or (3) neither individual changes their original phenotype. Transmission events are depicted, then, by a symbol change in a thread exiting one of these interaction "intersections." Using Neiman's assumptions of global interaction (Figure 3.1), I ran a simulation and tabulated the frequencies of three variants for six time durations, as indicated between the fine horizontal lines. As seen at the right side of the diagram, with no spatial biases on interaction, the frequencies form unimodal series.

In many real world cases, however, interaction is constrained by the spatial or social structure of the population. Individuals are not free to move or interact at random with all members of the population. They may be constrained by distance (as when the members of a population are patchily distributed) or they may be socially constrained (as when individuals' opportunities to interact with other members of the population are limited by network size or time constraints). In my experiments, therefore, I relaxed the assumption of panmixis both homogeneously (in the form of a simple distance-decay function [Hagerstrand 1952; Renfrew 1977]) and inhomogeneously (in the form of clumping in addition to distance). Figure 3.2 depicts a homogenous population with a strong bias towards interaction close to "home," along with an attribute (A) diffusing through the population from left to right over an appreciable period of time. Again, frequencies are sampled at five points in time marked by horizontal lines. The global frequencies are shown at the bottom, while frequencies in three regions of the transect are shown across the top. It is notable that the regional results show different frequency histories, and different histories of class richness. While all are unimodal (or nearly so, given the effects of sampling), they display peaks and valleys offset from one another, representing the unique history of transmission in a regional

[1] In panmictic transmission, individuals have an equal chance of communicating with one another; there are no spatial biases.

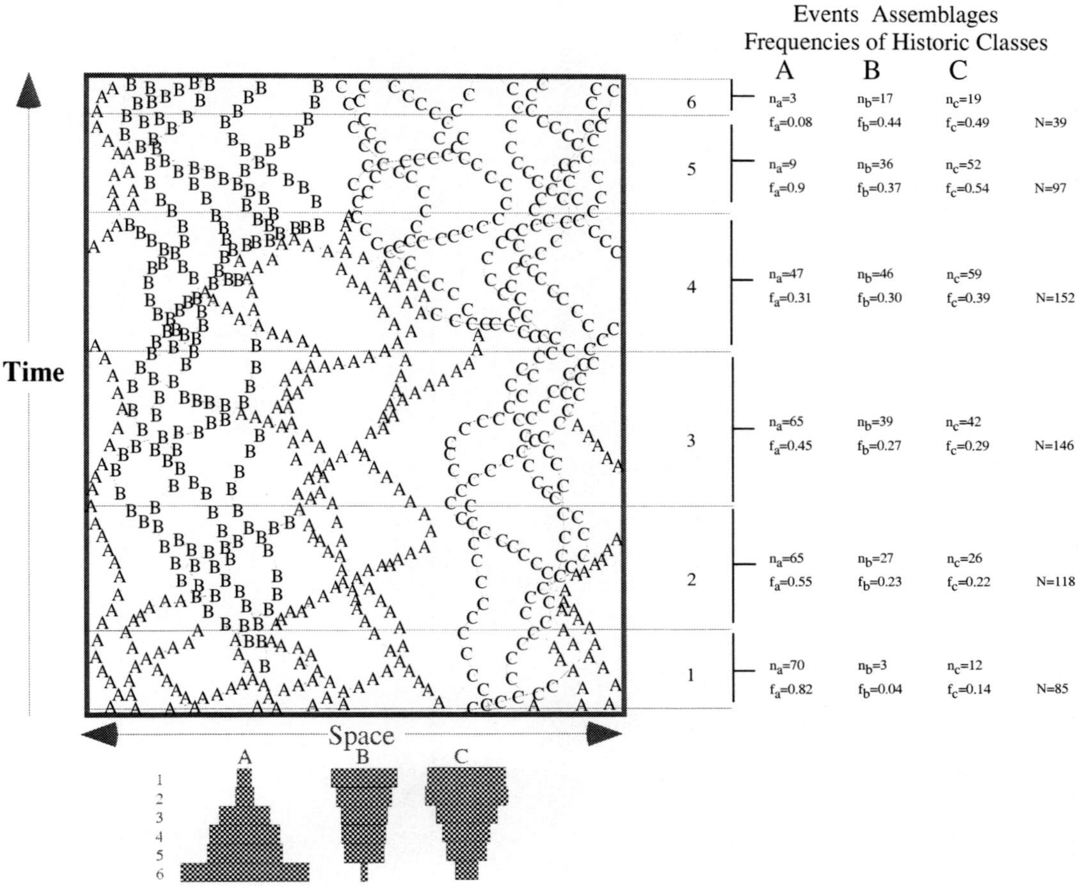

Figure 3.1. Transmission Among Individuals with Different Phenotypic Traits Labeled A, B, And C, Depicted within a Uniform Population with No Spatial Restriction on Interaction Frequency.

When individuals encounter each other there is a probability, not predicted by the form of the trait(s), that the individual takes the trait of the other individual. Although in this simulation the frequency of taking a trait was set at 50%, this frequency can also vary in relation to the frequency of the traits (*i.e.*, density-dependent). Individual letters (A, B, C) represent discard events that are aggregated over a period of time into depositional events ("assemblages" 1-6, represented areas between the horizontal lines). Traits frequencies (f_a, f_b, f_c) are tabulated for each each type (n_a, n_b, n_c) in each assemblage of differing sample sizes (N) along the right-hand side. A seriation using these trait frequencies is located at the bottom of the figure. Note that in real cases errors in sampling will yield numbers that do not match the model. Consequently, error terms must be used.

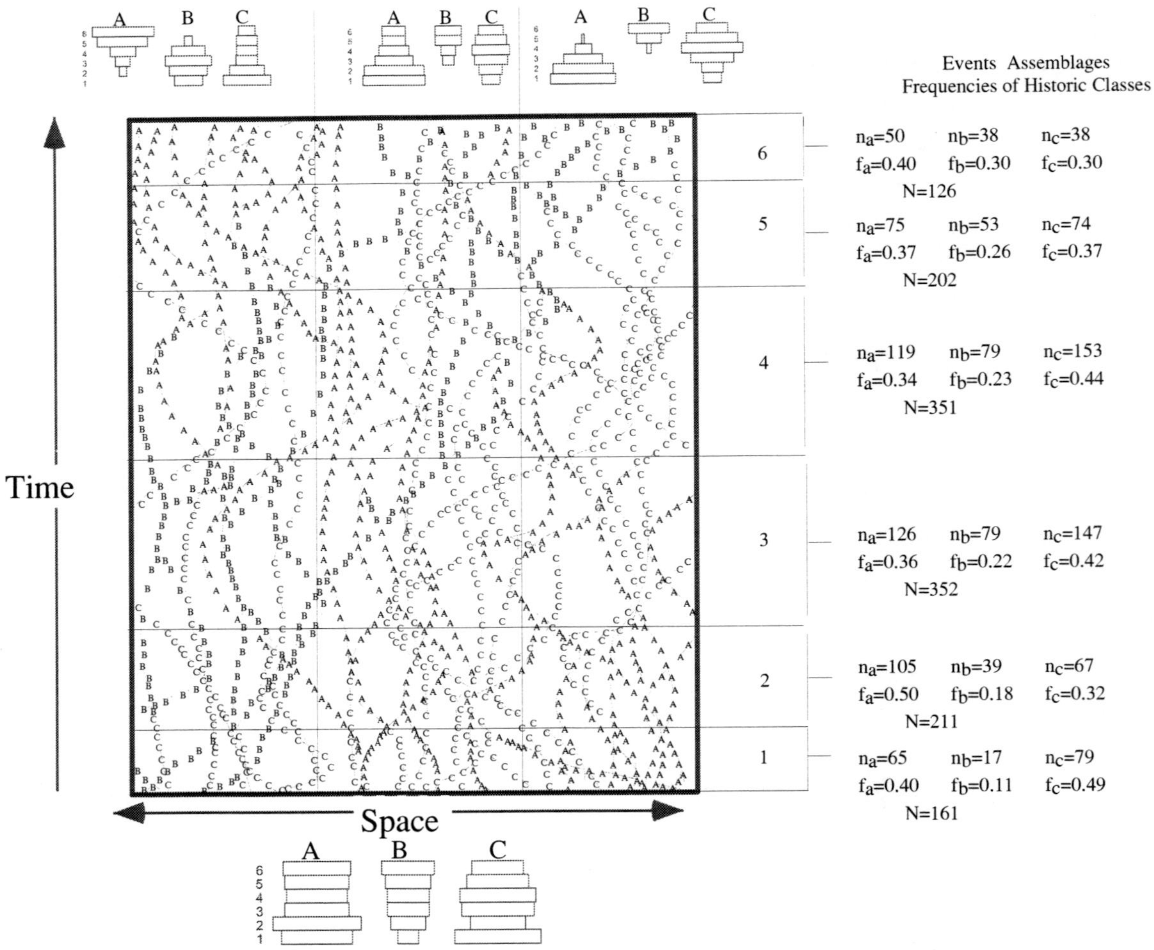

Figure 3.2 Transmission Among Individuals within a Uniform Population where the Probability of Interaction is Strongly Affected by Distance.

Individual letters (A, B, C) represent discard events that are aggregated over a period of time into depositional events ("assemblages" 1-6, represented areas between the horizontal lines). Traits frequencies (f_a, f_b, f_c) are tabulated for each each type (n_a, n_b, n_c) in each assemblage of differing sample sizes (N) along the right-hand side. In this simulation, individuals interact only with local neighbors and trait B moves from left to right through the population. At the bottom of the figure, a seriation that considers the entire space does not meet the unimodal expectation of the model. Seriations calculated at the top for spatially limited samples indicated by the vertical lines, however, meet the expectations of the model and form perfect unimodal curves As mentioned in Figure 3.1, real cases require the use of error terms due to the effect of sampling.

population. Now, the effectiveness of Ford's "local area" solution is explained.

As shown in Figure 3.3, when populations tend to interact within a short radius relative to the entire region, real differences in the rates of interaction between areas of the space and in trait frequencies appear at any moment in time. Although there is a constant density of individuals in each panel, frequency of encounter varies. Thus, even though individuals are uniformly distributed throughout the space, variants will diffuse more quickly through the left and right panels because the interaction frequency is higher, and will take longer to spread through and across the central panel. The same results are seen in Figure 3.4 where individuals are unevenly distributed in space and interaction occurs infrequently between the two "clusters." Despite the similarity of patterns of interaction, these two scenarios are archaeologically distinguishable by studies of the spatial distribution of assemblages. Simulations of Neiman's (1990, 1995) transmission model, with the additions of limited interaction radii and inhomogeneous population distribution, yields the kind of behavior we see in real archaeological data sets: areas within which data form unimodal curves, and larger regions of space over which unimodality breaks down because several locally interactive groups have been artificially lumped together.

The simulation work of Lowen and Dunbar (1996; see also Brauchli *et al.* 1999) shows that there are quantitatively different dynamics than those predicted by more conventional panmixis or mean-field models of cultural transmission such as those proposed by Boyd and Richerson (1985). Monte Carlo simulations produce results that show, depending on the details of the transmission process (*e.g.*, number of cultural parents and their importance in the transmission process), new variants will appear globally in the population much more infrequently than deterministic panmictic models predict. In addition, the rate of change in the frequency of variants in a panmictic model is much less than the structured model. In general, Lowen and Dunbar (1996) found panmictic models are a limiting case of all structured population models. Specifically, panmictic models become equivalent to structured models only if each individual in a structured population interacts directly with every other individual in the population. It is not surprising, therefore, that the rates of cultural transmission processes are consistently greater in panmictic models than those of structured population models.

Panmictic models can incorporate some of the aspects of structured population models in special configurations of interaction, but they cannot generate the detailed dynamics of structured populations through time (Brauchli *et al.* 1999; Lowen and Dunbar 1996). In addition, there are stochastic components to structured populations that the simple panmictic models do not encompass, including discrete population units and finite population size. Finally, panmictic models do not account for extinction processes and cannot address the local dynamics of transmission processes. As shown in simulations presented here, these configurations have significant quantitative and qualitative impacts on the results of transmission.

HERITABILITY AND CLASS FREQUENCIES

During my simulation experiments, I noted that panmictic populations whose members were free to interact equally over the entire space produced nearly perfect seriations, whereas any restrictions on the radius of interaction would destroy our ability to include the entire test population in one seriation. Frequencies of traits in one area of the simulated population were out of "sync" with frequencies in other regions, preventing any single ordering from yielding a perfect solution. In the absence of independent temporal control over a set of assemblages, failure of a set of assemblages to seriate might be exploited to test hypotheses about the history of interaction in a region.

EFFECT OF CLASSIFICATION ON MEASURES OF TRANSMISSION

Frequencies of stylistic types in assemblages are, however, more than simply the results of transmission in past populations. They derive some of their particular character from additional sources: counting units, duration, classification, and sampling (Dunnell 1970, 1981; Ford 1949; PFG 1951). In the work of PFG, Ford was cognizant of the duration problem and efforts were made in the field to eliminate duration effects (PFG 1951:219). On the basis of notes on their type tabulation sheets, it is also clear that further minimization of the effect of differences in duration was accomplished by dropping assemblages that were "mixed" or anomalously rich as determined by the seriation analyses (PFG 1951:219; *cf.* Dunnell 1970).

Ford (1949, 1952; PFG 1951) accounted for variability in the frequencies of stylistic types by noting that the unimodal character of historical classes was derived from filtering a continuous stream of transmitted information through a classification so that variants could be counted. Ford's explanation, however, is incomplete. Unimodality is created both by the stochastic or Markovian character of neutral trait transmission and by describing the frequencies of those traits with a set of classes treated as a closed array. If there are many classes forming the closed array, as in a classification with many dimensions and/or attribute classes per dimension, fine temporal and spatial discriminations can be made (Dunnell 1971). If there are few classes forming the closed array, the resulting discriminations will be coarse in scale. This elementary fact suggests that there is no single seriation solution for identifying groups of assemblages that comprise a lineage. Rather, seriation can be used to map quantitative changes in the history of regional and local interaction. The method I employ, therefore, creates seriations using a nested set of classes at variable levels in order to see how interaction changes with level.[1] If interaction is conditioned solely by distance (assuming the types used are strictly stylistic) then varying classification level should produce differences in how assemblages

[1] This method was inspired by the work of R.C. Dunnell (1970).

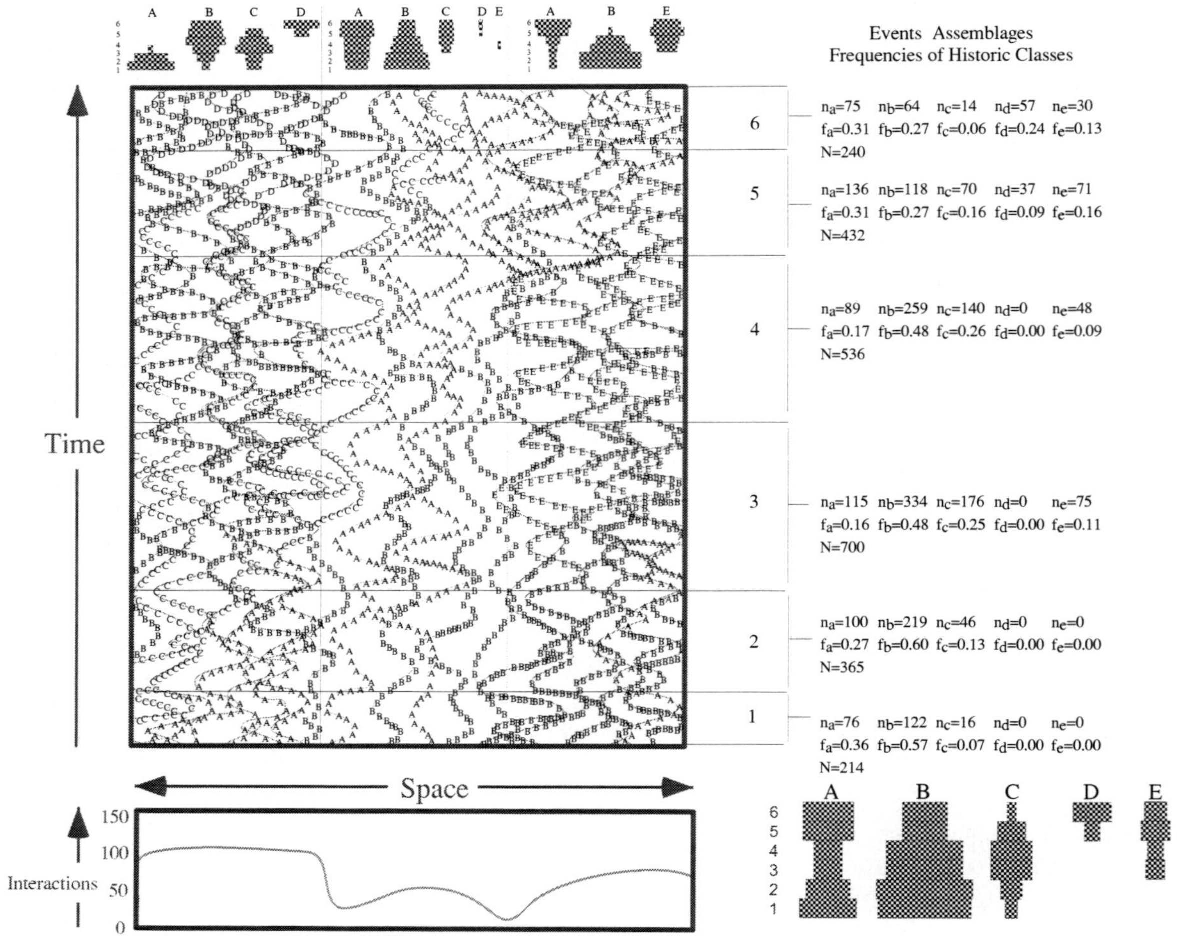

Figure 3.3. Transmission Depicted within a Population Varying in Interaction Frequency across Space.

Individual letters (A, B, C) represent discard events that are aggregated over a period of time into depositional events ("assemblages" 1-6, represented areas between the horizontal lines). Traits frequencies (f_a, f_b, f_c) are tabulated for each each type (n_a, n_b, n_c) in each assemblage of differing sample sizes (N) along the right-hand side. Two "populations" of individuals are apparent in this simulation separated by an area of less frequent interaction. A graph of the average frequency of interaction through time is shown at the bottom.

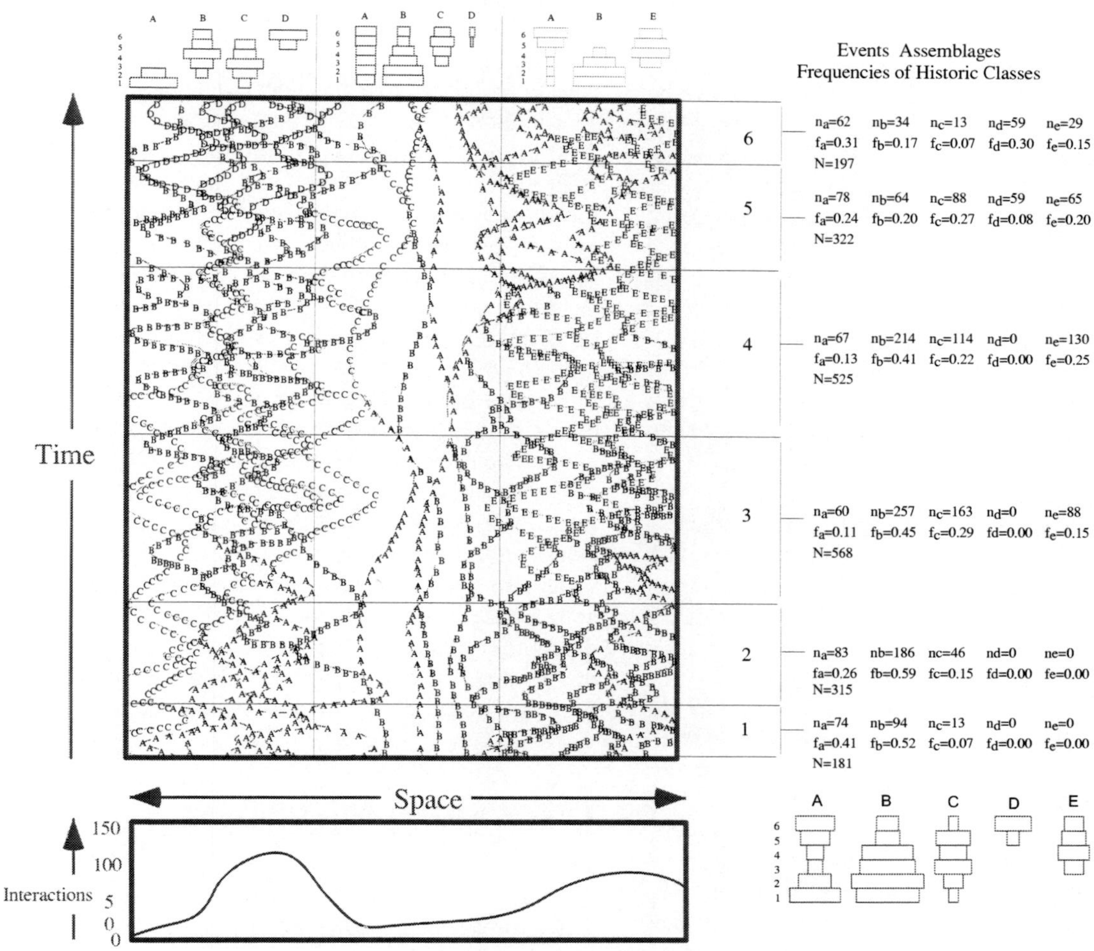

Figure 3.4. Transmission Depicted within a Population of Varying Density across Space.

Individual letters (A, B, C) represent discard events that are aggregated over a period of time into depositional events ("assemblages" 1-6, represented areas between the horizontal lines). Traits frequencies (f_a, f_b, f_c) are tabulated for each each type (n_a, n_b, n_c) in each assemblage of differing sample sizes (*N*) along the right-hand side. A graph of the average frequency of interaction through time is shown at the bottom. Note that the frequency of interaction is the same as in Figure 3.3. Archaeologically these scenarios (shown in Figures 3.3 and 3.4) can be distinguished through the study of spatial distributions.

correspond to changes in spatial "boundaries." If, on the other hand, interactions are structured by barriers (physical or social), spatial boundaries will be reflected in inflections, seemingly unrelated to distance at some classification levels while related to distance at others. The effect of classification is presented in more detail in Figure 3.5. Figure 3.5 shows two sets (A and B) of assemblages distributed in space.

In hypothetical example A (Figure 3.5b), interaction between locations is based solely on distance; individuals in locations that are closer together interact more often than those in locations farther away. Consequently, similarity between assemblages is a function of distance. Since groups defined based on similarity are a function of the classifications used to describe the assemblage, changing the level of classification by increasing or decreasing the number of dimensions used in descriptions changes the inclusiveness of the assemblages in any group. Given that distance drives interaction and thus neutral trait similarity, group inclusiveness should also be a function of spatial distance. In example A (Figure 3.5a), interaction (and thus similarity) is biased only by distance, such that the number of dimensions in the classification is decreased from four to one (indicated by the thickness of the lines in the figure), the spatial scales of the resulting groups increase uniformly and are correlated with distance. In hypothetical example B (Figure 3.5b), interaction between locations is biased by space *as well as* physical or social boundaries. For example, the middle two assemblages in the figure interact more with the assemblages on the right hand side of the figure than is simply predicted by the distance separating them. In this case, as the number of dimensions is changed, the groups formed do not change smoothly in spatial inclusiveness. Using a classification of two or three dimensions consistently produces two spatially coherent groups of assemblages. The disjunction in the measurement of these two assemblage sets (*i.e.,* the group of assemblages on the left and the right of the figure) cannot be simply explained by distance. Once sampling and differences in deposition are considered, one would have to evaluate whether this pattern is produced by interaction structured by physical or social boundaries. By varying classification level, the effects of classification on spatial structure can be disentangled from those of social and/or physical barriers.

IMPORTANCE OF NEUTRALITY IN THE STUDY OF CULTURAL TRANSMISSION

An important point, not easily captured in simulations or mathematical models of transmission, is the effect of selection on the eventual distribution of traits. Most phenotypic attributes have the potential of informing on interaction (Beck 1995), but the degree to which any particular trait is useful for studying interaction at a particular spatial and temporal scale will vary depending upon the strength of evolutionary processes (other than transmission) acting upon the trait. If selection or other evolutionary mechanisms do not act to change trait frequencies, the geographic distribution of the trait through time is determined by the spatial structure of individuals in the environment and the frequency with which individuals encounter each other (which may vary among communities). In other words, under null conditions, the frequency of interaction is equivalent to the spatial structure of the population. As discussed earlier, attributes measured along a dimension of phenotypic variability that do not affect fitness and thus whose frequencies are unaffected by selection, are neutral or stylistic.

Attributes that make either a positive or negative contribution to individual fitness (functional attributes) will be structured geographically, not only by the distribution and connectivity of individuals from lineage to lineage, but by the characteristics of the selective environment. The effects of a change in the selective environment on stylistic and functional attributes are depicted in Figure 3.6. In the right-hand panel, the distribution of two neutral traits, with respect to an environmental gradient (solid line), is random. These traits have distributions that are capable of informing on interaction between transmitters. In the left-hand panel, the distribution of alternatives is conditioned strongly by the environmental gradient. If we were to use these functional attributes, we might hastily conclude that there is little interaction between the two regions. Thus, in contrast to simulations, in real archaeological cases we must be careful which traits we use to delimit lineages. Therefore, building appropriate measurement units requires evaluating class distributions of types across space and through time against the neutral model. The two methods presented earlier provide a means of making this evaluation.

ANALYTIC METHOD: SERIATION

The seriation method presented here in which frequencies and error terms are compared in an iterative fashion using spatial hypotheses to build statistically valid solutions is distinct from the means by which many contemporary archaeologists order assemblages in "seriations." The common approach to seriation today is a probabilistic technique. In this type of technique, some form of a matrix ordering algorithm is used to shuffle assemblages until the best fit to a unimodal curve is achieved, with the resulting order accepted as chronological (*e.g.,* Johnson and Nelson 1990; Love 1993; Marquardt 1978). The probabilistic approach is inadequate, however, as it offers no means of differentiating violations of seriation principles from sampling and measurement error.

Probabilistic methods uniformly reduce type frequencies to a set of similarity measures, which are scaled according to some algorithm. Probabilistic techniques always produce an order, even with data of poor quality because information concerning sampling error is not used and there is no null model against which frequency patterns can be evaluated (Dunnell 1981). My research is particularly sensitive to issues of sampling because I am interested in using seriation not as a dating technique *per se* but as a means of explaining frequency variation in stylistic classes generally. I reject the probabilistic approach and, consequently, use a deterministic approach to seriation. This method accepts only those

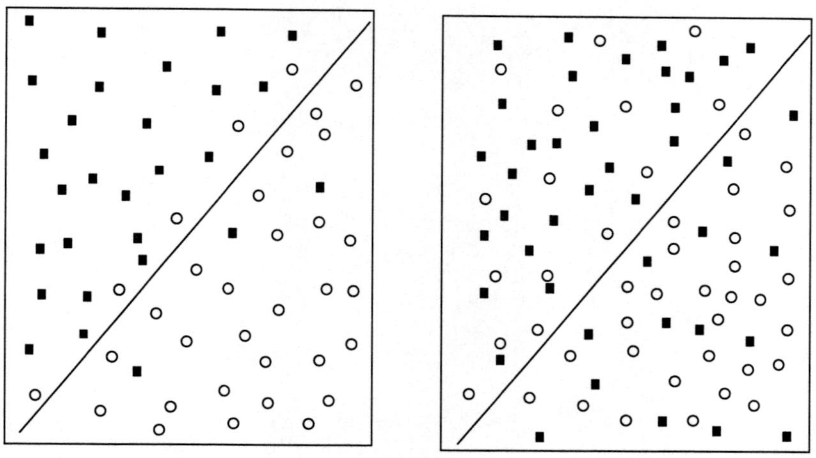

Figure 3.6. The Effects of a Change in Selective Environment on Stylistic and Functional Attributes.

In the right-hand panel, the distribution of two neutral traits (depicted by ■ and O symbols) with respect to an environmental gradient (solid line) is random; these traits have distributions that are capable of informing on interaction between transmitters. In the left panel, the distribution of alternatives is conditioned strongly by the environmental gradient. Failure to distinguish functional attributes from neutral attributes in the case of the left hand panel indicates that there is little interaction between to two regions. Care must be given when choosing traits to be used to delimit lineages.

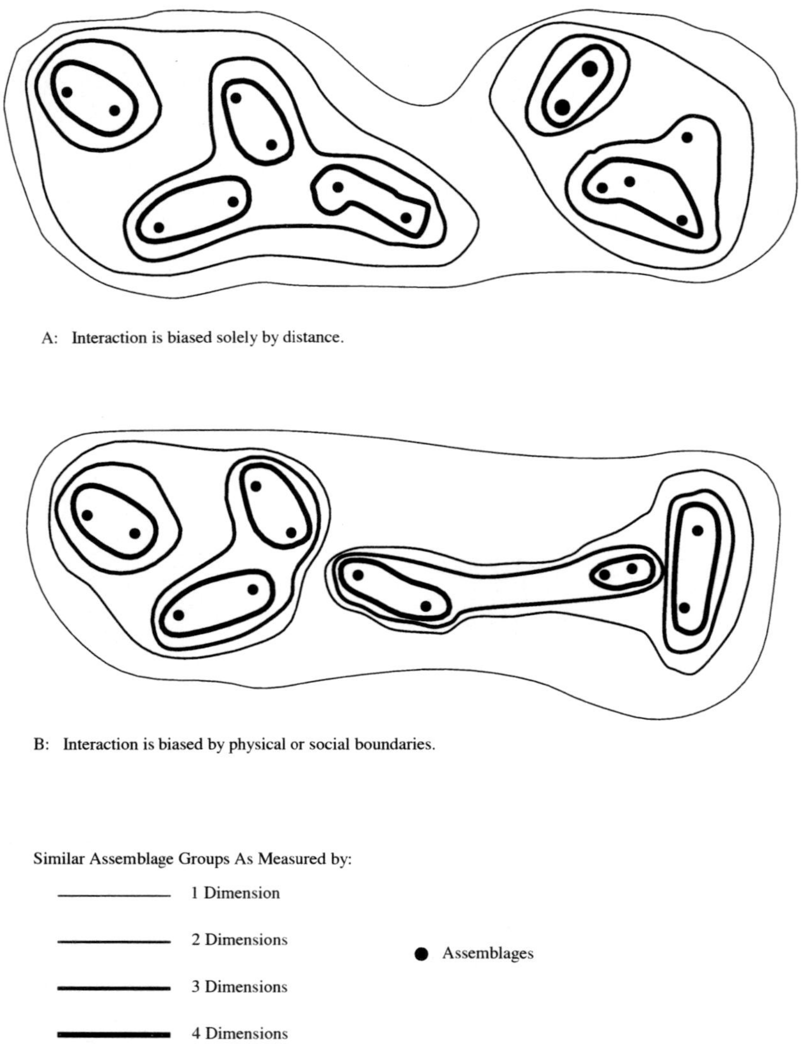

Figure 3.5. Nested Set of Stylistic Classifications.

Describing assemblages with a nested set of stylistic classes at variable classification levels permits the examination of how interaction changes with degree of specificity in class definitions. See the text for discussion about this figure assemblages into a particular seriation that fit the unimodal model formed by assemblages already in the seriation, with degrees of freedom in assignment dictated by sampling (Dunnell 1970, 1981, 2000).

SAMPLE SIZES

Usually assemblages are samples of much larger deposits. Because of this, comparisons between assemblages must be treated as hypotheses. Error terms are calculated based on sample size. Two approaches for evaluating sampling error were used in this analysis. First, because seriation is sensitive to the number of classes present in any assemblage (richness), and it is well known that in "small" assemblages richness may be a function of sample size (Grayson 1989), I examined the effects of assemblage size on the number of classes seen in PFG assemblages from the St. Francis and Memphis regions of their survey area. These data are useful in assessing whether assemblages are too small to be reasonably included in seriations (*i.e.,* differentiating "absences" due to sample size rather than population differences). Figure 3.7 shows plots of assemblage richness versus assemblage size for all assemblages in my study area. This figure demonstrates that while there is a familiar trend of richness increasing as sample size increases, there is also considerable variation in the relation between the size of a sample and the number of types observed. Usual practice accepts any assemblage that falls onto the asymptotic portion of the graph as large enough to have reasonable samples of all classes. However, this practice assumes that all differences in richness are due to the effects of sample size, and excludes possible variation in richness through time or across space. I did not want to exclude these possibilities by concluding that all assemblages with smaller sample sizes and low richness were poor samples, so I used bootstrap methods to examine richness and sample size for each assemblage.

Bootstrapping is a method for evaluating the degree to which a test statistic calculated for a particular sample is an estimator of the unknown, underlying population (Berry *et al.* 1980, 1983; Efron 1982; Efron and Tibrishani 1993; Fisher 1936; Good 1994; Mooney and Duval 1993; Noreen 1989; Simon 1990). The method relies on the creation of a distribution of values for the test statistic created by resampling original data repeatedly. In my evaluation of richness and sample size, I created a series of richness distributions constructed of randomly drawn resamples n which the resamples systematically increased in size (Appendix A). For example, with an assemblage of original sample size of 300, I drew 10 sherds randomly with replacement from the original sample and calculated richness. This procedure was repeated 1000 times and a richness distribution was calculated. I then resampled 20 sherds from the original sample and calculated a richness distribution. This process was continued by increasing resample sizes until the original sample size was reached. Significant changes in the richness distribution as sample size increased were caused by differences in sample evenness and hence require greater sample sizes to confidently assess richness.

By plotting the means and variances of the richness distributions, I was able to evaluate the degree to which assemblages had a sufficient sample size to confidently estimate population richnesses. On the basis of this assessment, I divided the assemblages into three groups (Figure 3.7). The first group ('A' assemblages) contained those assemblages that had richness mean and variance values that reached an asymptote with the original sample richness prior to the final sample size. The second group ('B' assemblages) included assemblages that had mean resample richnesses or variances (but not both) that did not approach an asymptote as sample size was increased. Finally, the third group ('C' assemblages) consisted of assemblages that had neither asymptotic mean richness nor variance, but suggested that increased sampling from the archaeological record would result in significant increases in the number of artifact classes represented.

Examining the distribution of the bootstrap-assessed types in a richness and sample size plot (Figure 3.7) demonstrates the limitations of simple sample size versus richness plots (*sensu* Grayson 1989). In Figure 3.8, even large sample sizes have a mix of 'A' (indicated by the diamond [♦] symbols) and 'B' assemblages (indicated by the square [■] symbols) that do not have asymptotic mean and variance values. In addition, the figure shows that while many small assemblages fail to have either asymptotic means or variances ('C' assemblages, indicated by the triangle [▲]), in certain conditions, small assemblages can meet one or both of the assymptotic criteria. There is no simple relation between richness and sample size requirements and, consequently, no one "global" sample size across all assemblages will result in a reliable set. Minimum sample sizes, therefore, must be determined for each sample. Ford's claim that fifty sherds are sufficient for creating samples representative of deposits, while recognizing the impact of sampling, is clearly inadequate.

THE ITERATIVE SERIATION METHOD

Unlike similarity methods, the seriation method proposed here simultaneously considers the effects of sample size during the evaluation of orders between assemblages. When making each seriation attempt, assemblages were added in an iterative fashion, and various orderings treated as a hypothesis test concerning fit to unimodality. Testing the fit involved pairwise comparison of assemblage frequencies using confidence intervals for class frequencies calculated on the basis of sample size following Beals *et al.* (1945) method. Here, a normal approximation to a confidence interval for a binomial variable was calculated using a type I error rate of 0.05 (modified from Cochran 1975:57). Assemblages that fit the increasing or decreasing expectations of the unimodal curve model were accepted into a given seriation group. Assemblages that did not fit, within calculated confidence limits, were rejected and placed in different groups. Obviously, this technique is not a mechanical algorithm in the spirit of multidimensional scaling but an iterative, operator intensive technique closer to Ford's (1949, 1962; PFG 1951) approach. With Timothy D. Hunt of Emergent Media, Inc., I built a visual interface in

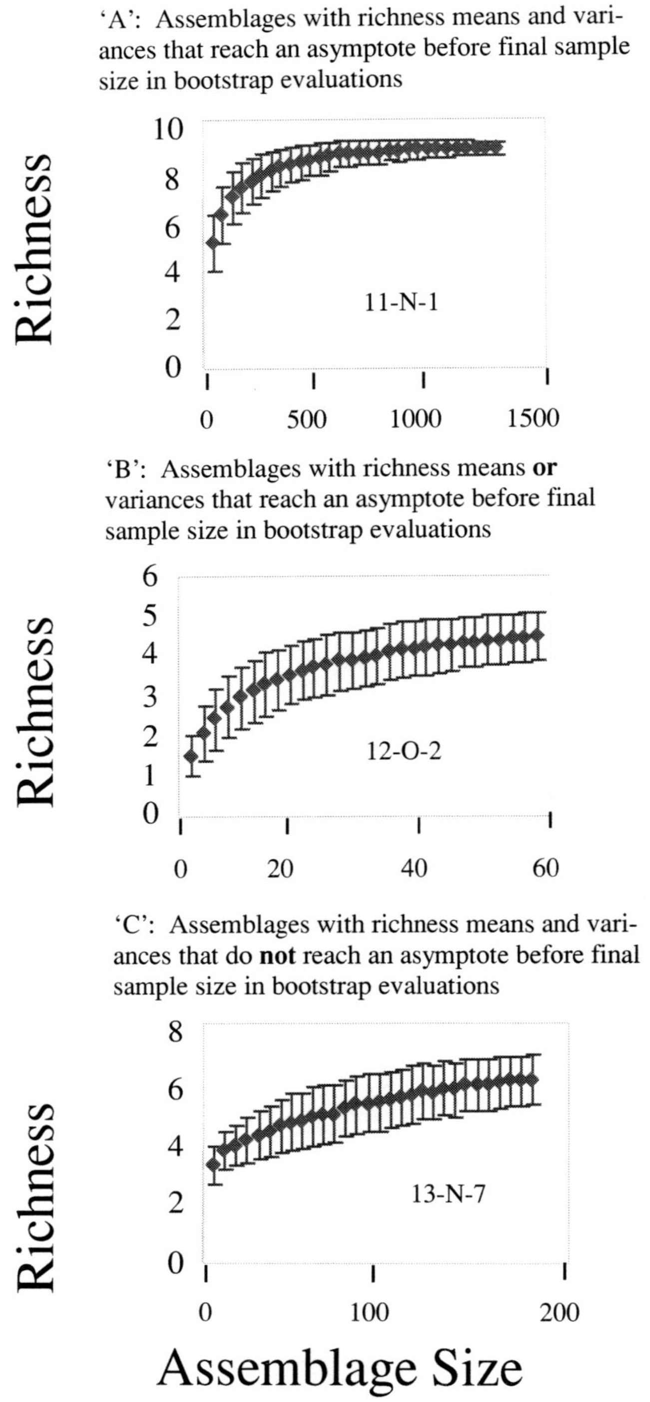

Figure 3.7. Bootstrap Richness and Sample Size Relationship Examples, St. Francis and Memphis Area Data.

As shown in the Figure, three types of assemblages were identified on the basis of the mean and standard deviation of bootstrapped richness values with increasing sample size. 'A' assemblages are those with richness means and variances that reach an asymptote before final sample size in bootstrap evaluations. 'B' assemblages have richness means or variances that reach an asymptote before final sample size in bootstrap evaluations. 'C' assemblages are those assemblages with richness means and variances that do not reach an asymptote before final sample size in bootstrap evaluations.

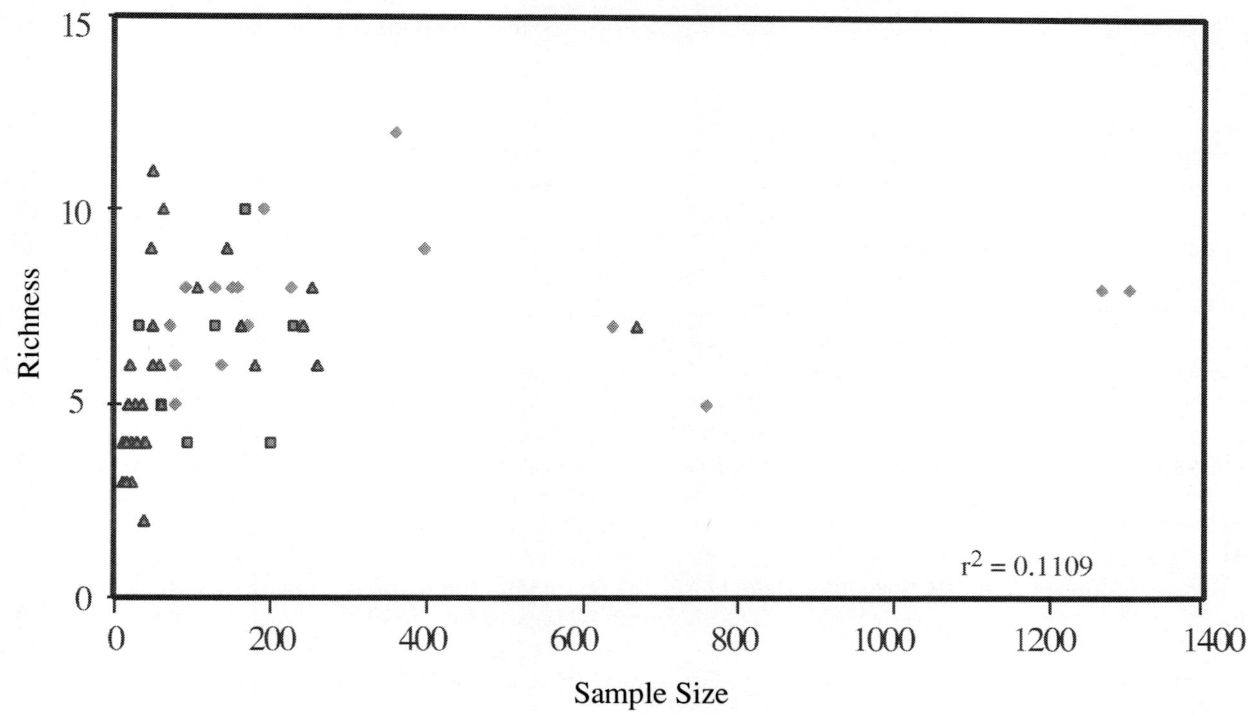

Figure 3.8. Relationship of Richness and Sample Size for All PFG Assemblages in the St. Francis and Memphis Area.

The diamond (◆), square (■), and triangle (▲) symbols refer to the bootstrap assessed 'A,' 'B,' and 'C' assemblages, respectively, as discussed in the text.

Microsoft Excel that allows one to order and reorder assemblages into various groups while displaying confidence intervals and testing hypotheses of fit to the model (Appendix B). All of the seriation results presented here were done using this tool.

ITERATIVE PAIRWISE SIGNIFICANCE TEST

Lacking in this analysis, however, is a method for determining the quality of the exact orderings that are created. Though the iterative approach yields a set of assemblages that are relatively robust in terms of samples sizes and classes, there must be a means by which the ordering of any two assemblages can be determined to be statistically defensible. Pairwise Student's *t*-tests are possible between individual data pairs within assemblages, but two problems emerged. First, masses of repeated *t*-tests quickly lose statistical power since the values are not independent in an array of frequencies. Second, there is no reason to suppose that the difference in frequencies within a seriation has a sampling distribution that is approximately normal, since it is not precisely the difference of means. In fact, seriation frequencies are neither sample means nor medians, but are moments of an unknown sampling distribution.

Thus, parametric statistical tests are inappropriate for testing the hypothesis that the rows in an assemblage have been correctly ordered. Statisticians increasingly use bootstrap techniques in situations such as this, to estimate the sampling distribution and moments of statistics that have unknown or poorly behaved distributions (Efron and Tibrishrani 1993). I have chosen an algorithm for measuring the ability of randomly chosen resamples to be ordered with the same directionality as the original assemblages. By resampling a large number of times, I can calculate the probability that assemblages generated as a sample of the original assemblages are ordered in the same direction. In this way, I can determine the confidence that can be attributed to orders provided by seriation solution.

In detail, the algorithm is as follows. First, I take the raw class frequency data and recode them as proportions of the total assemblage size. The directionality of the original samples is determined by subtracting the frequencies of each class, and saving the pattern of positive, negative, and equal directions for each (Figure 3.9). Then, using a simple roulette wheel algorithm, I resample "sherds" from the original samples and determine to which class each resample belongs and increment the count of such sherds in my generated assemblage. I generate random assemblages of the same size as the original samples. I then determine the directionality of each pair of random assemblages and compare this pattern to the original. If the pattern of positive, negative, and equal frequencies matches that of the original samples, a match is scored for that random assemblage (Figure 3.10). If no match, then no score is recorded (Figure 3.11). This process is repeated 1000 times to determine the total number of matches out of 1000 resamples for assemblages bootstrapped randomly from the initial samples. The null-hypothesis for this test states:

H_o = The directionality of the differences between the proportions of each type in two assemblages are as indicated by the original numbers.

H_a = The directionality of the differences between the proportions of each type in two assemblages are not as indicated by the original numbers.

The proportion of matches to bootstrapped attempts forms the p value for my hypothesis test. If, for example, the bootstrapped assemblages meet the original pattern 950 times out of a 1000, the test results in a p-value of 0.950. In this example, we would fail to reject the null hypothesis at a 95% confidence level (α =0.05).

To generate an error term for this p-value, I perform the 1000 resample test 10 times for each pair of assemblages, finally yielding a mean and variance for the p-value representing our hypothesis test about two specific assemblages. This algorithm is then repeated for each unique pairing of assemblages. The method is computationally intensive and requires several hours to run through the large number of unique combinations of assemblage pairs between sets of assemblages. Because of the unique nature of the algorithm, standard bootstrapping packages were not appropriate and I performed the analysis using a custom program (Appendix C).

SUMMARY

Biologists studying human interaction have discussed cultural transmission over the past several decades. While this work potentially provides theoretically robust avenues for the investigation of cultural phenomena by archaeologists, few researchers have implemented the methods in the study of the archaeological record. However, with our focus on the empirical products of transmission within and among populations through time, archaeologists are ideally situated to contribute to the development of methods and concepts needed in this effort. Archaeologists are not necessarily limited to playing the role of consumers of theory built by other disciplines.

Over the last fifty years, archaeology has accumulated a respectable set of methods and tools for studying cultural transmission. Culture-historical methods that used homologous similarity and that were materialist in perspective have provided a starting point for studying transmission (*e.g.*, Lyman, *et al.* 1997). In particular, seriation has been the backbone of archaeological efforts for studying continuous variation in stylistic similarity (Dunnell 1970, 1981, 1985b; Teltser 1995a). Ford's work in the Mississippi river valley (*e.g.*, Ford 1935a, 1936a, 1936b, 1938, 1949, 1962), for example, documents continuous evolution of stylistic traits without resorting to periodization schemes (*e.g.*, foci or phases).

	Type 1	Type 2	Type 3	Type 4	Type 5	Type 6	
Assemblage A	0.480	0.131	0.000	0.012	0.177	0.200	$N_A = 352$
	A > B	A < B	A < B	A > B	A > B	A > B	
Assemblage B	0.330	0.374	0.143	0.000	0.022	0.131	$N_B = 523$

Figure 3.9. Iterative Pairwise Significance Testing Procedure.

The first step is to convert raw counts into percentages and compare directions of each frequency. Here N is the size of the assemblage and the values in each box are the frequency of the types for each assemblage (Assemblages A and B).

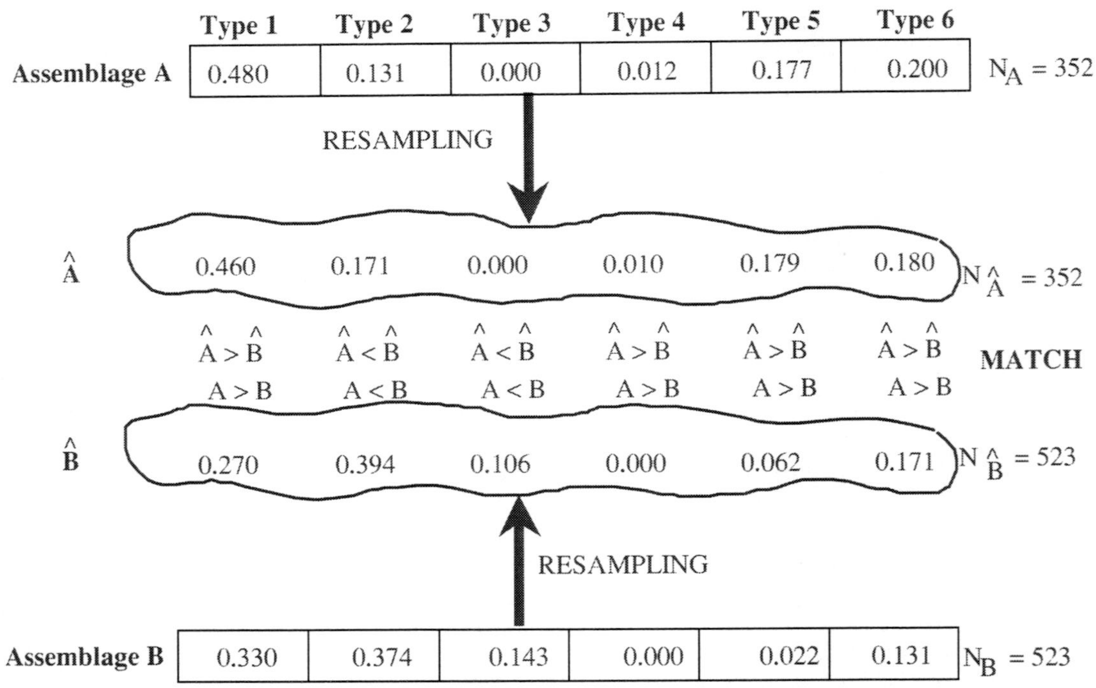

Figure 3.10. Iterative Comparisons of Assemblages.

After drawing new assemblages, the new bootstrapped values are compared. In this case the the directions of the resampled assemblages match the original assemblage directions, so a match is scored. In this figure, *N* is the size of the assemblage and the values in each box are the proportion of the types for each assemblage (Assemblages A and B). The size of the sample is indicated as *N* and resampled assemblages frequencies are labeled as $\hat{A}$ and $\hat{B}$.

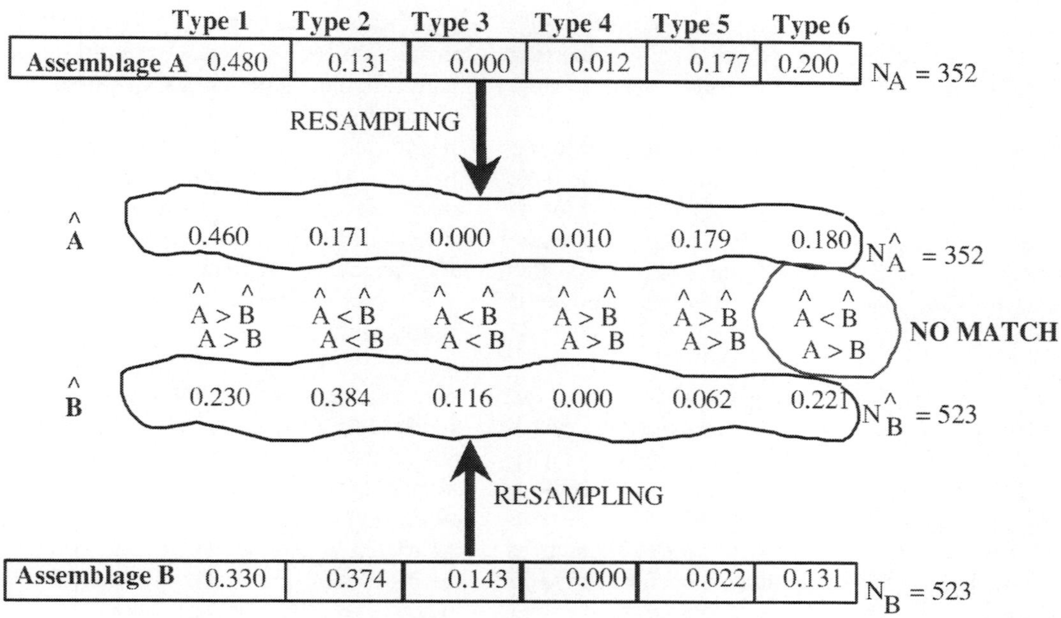

Figure 3.11. Iterative Comparisons of Bootstrapped Assemblages.

In this example, the directionality of resampled assemblages does not match original assemblages, so no match is scored. The total number of matches out of the number of resamples provides the *p*-value. The size of the sample is indicated as *N* and resampled assemblages frequencies are labeled as $\widehat{A}$ and $\widehat{B}$.

Until recently, however, little progress has been made among archaeologists in the construction of a *theoretical* framework that could account for the success of culture historical methods. The rationale given by PFG for their chronological claims based on the frequency of decorated types, for example, was primarily a series of empirical statements about the way descriptions of the archaeological record using particular kinds of measurement units can be ordered through time. PFG are clear, however, that these assumptions must be explained; that they are *not themselves explanations of the archaeological record.* There was a general recognition that the particular configuration of types as measurement units is responsible for allowing archaeological material to be reliably ordered through time. Although PFG recognized the need for theory development, no coherent rationale was available at the time that could account for the behavior of types. For the most part, they and other culture historians pointed to "style" as an explanation but could only offer common sense accounts for why "style" behaved the way it does.

Unfortunately, subsequent archaeologists who utilized the information collected during the PFG study did not always recognize that their explanations were incomplete. Phase units, for example, although based on frequencies of stylistic, decorated ceramic types collected by PFG, are simple descriptions of the empirical record with limited theoretical warrant. Few archaeologists made this connection and, as a result, phases became fixed in archaeological lore as meaningful units and were assumed equivalent to their common sense interpretations. The lack of a theoretical framework for phases has resulted in a general inability to analyze this assumption and to evaluate whether phase units are in any way related to the social and ethnographic meanings with which they have been equated.

In order to build an explanation for the behavior of culture-historical types and the results of the PFG study, in this chapter I presented the basics of a theoretical model for studying stylistic attributes. Based on the work of Dunnell (1978a) and evolutionary theory, the model treats stylistic variability as neutral with respect to selection. As a result, stylistic traits have distributions that are a function of transmission between individuals. Using a definition of neutrality derived from biology, I presented a pair of techniques for evaluating null hypotheses about the distribution of traits based on expectations of the neutral model. These tests allow one to statistically determine the degree to which one can claim that the distribution of traits in a population are considered to be derived from processes associated with transmission of non-selected (*i.e.,* selectively neutral or stylistic) traits.

Once the link between stylistic traits and the concept of neutrality in population genetics is made, culture-historical types can be understood as measurement tools for examining patterns of transmission. Seriation is easily expanded into a means of assessing interaction between populations in time and space. The results of my simulation showed that panmictic populations whose members are free to interact equally over the entire space produce perfect seriations, whereas any restrictions on the radius of interaction destroy our ability to include the entire test population in one seriation. In the case of populations in which individuals were distributed unevenly across space and interaction is local, frequencies of traits in one area of the population are out of "sync" with frequencies in other regions, preventing any single ordering from yielding a perfect solution. Areas within which samples form unimodal curves potentially represent fully interacting populations. Thus, the "failure" of a set of assemblages to seriate can be exploited to test hypotheses about the structure of interaction among populations in a region. With a theoretical basis for understanding seriation as a dating method, the "noise" arising from differences in space becomes a potential source of information on the transmission of homologous traits through space. Seriation, methods for evaluating expectations of the neutral model, and bootstrapping techniques to assess the role of sample size in the generation of differences in frequencies, become the primary tools with which archaeologists can evaluate hypotheses about past transmission.

We can also explain why Ford's "local area" criterion divided the Mississippi river valley into five regions allowed PFG's data to be successfully seriated. Each subset of assemblages, recognized as *phases*, potentially describe the interaction of populations. What one cannot tell is whether the spatial boundaries of PFG areas and Phillips' phases are the products of arbitrary divisions of continuous variability in space (as PFG's were in time), or whether they represent discontinuities compatible with the notions of "peoples," "groups," or "societies" often assumed in modern usage. Further empirical evaluation of the seriations and assemblage data are required. Consequently, in the next chapter, I return to the PFG data from their survey of the Mississippi river valley. With a robust theoretical framework, new tools, and statistical sophistication beyond the means of PFG, it is now possible to build a more detailed model of past populations in the region. This model takes us beyond simplistic units such as phases and provides hypotheses about continuous patterning of interaction in space and time for further empirical testing.

CHAPTER 4: ANALYSIS OF PHILLIPS, FORD AND GRIFFIN (1951) DATA

Using the methods presented in the previous chapter, we can now evaluate the PFG data and examine the assumptions underlying the interpretations of spatial patterns (*i.e.,* phases) based on them. If transmission among prehistoric populations were biased spatially as suggested by Ford's division of the assemblages into the St. Francis and Memphis areas and as interpreted by phase designations, we would expect distinct spatial patterning to the frequencies of stylistic types between groups of assemblages. After evaluating the data for sample size and neutrality criteria, this expectation dictates that not all the PFG data should form a single statistically defensible seriation solution. Instead, the largest seriations that can be made should consist of only a subset of the assemblages. If Ford's divisions are archaeologically meaningful rather than arbitrary, then the boundaries of these groups should correspond to the St. Francis and Memphis areas.

During the course of their survey, PFG (1951) collected 346,099 sherds from 383 locations in the Mississippi river valley in an area that spanned from just north of Louisiana to the Missouri state line. Figure 4.1 shows their survey area and the location of sites included in these seriations and in my reanalysis. Sites are numbered sequentially within the 15-minute quadrangle in which they are located. Each sherd was assigned to a culture-historical type as described in Chapter 2. Using these tabulated data, Ford divided the survey area into five "local areas" and constructed five separate seriations in order to minimize the effects of geography that were readily apparent in the latest time horizons (PFG 1951:224).

I was fortunate to be able to examine the original data for the Memphis and St. Francis seriations.[1] Coded on worksheets in Ford's own hand, the original data are invaluable for modern analyses, containing as they do the sherd counts and assemblage sizes. The data include frequencies for both shell- and clay-tempered ceramics, usually interpreted as Mississippian and Woodland, respectively. Woodland materials were not the focus of fieldwork by Phillips and colleagues and were usually recorded only when found underlying or mixed with the shell-tempered ceramics. Because the sample of Woodland materials is problematic, I restricted my analysis to the shell-tempered ceramics.

I excluded plain ceramics from the tabulations (*e.g.,* Neeley's Ferry, Bell Plain). The frequency of plain ceramics measures the *amount* of style in an assemblage. Since decorations must entail additional costs over undecorated pottery, amount of decoration is almost certainly under selection. The differences among decorations are much more likely to be negligible and, therefore, neutral (Dunnell 1978a; King and Jukes 1969). In addition, because decoration is not evenly distributed across individual vessels, the abundance of plain sherds will also be related to the amount of decoration on vessels, and thus breakage will have differentially affected the frequency of plain ceramics derived from different vessels.

I also excluded types that were present in only one assemblage since they do not inform on interaction frequency. In addition, I eliminated excavated assemblages from my analysis. These assemblages typically have idiosyncratic compositions depending upon the particular excavation location due to spatial autocorrelation effects arising from related fragments (Dunnell and Dancey 1983) and community structure (Dunnell *et al. 1971*). Since archaeological deposits are rarely, if ever, spatially homogenous even if described with stylistic (*i.e.,* neutral, *sensu* Dunnell 1978a) types (Dunnell 1981), frequencies calculated from spatially limited samples will differ, often greatly, from the frequencies in the assemblage as a whole. As a result, unless refitted (*e.g.,* Newell and Krieger 1949), excavated samples cannot be assumed representative of the entire deposit. Ceramics obtained from the surface, on the other hand, have a greater chance to exhibit representative frequencies due to post-depositional processes, such as plowing, that produce a locally randomized sample.

SERIATION AND SPATIAL STRUCTURE

I began my analyses of the PFG data by examining the solutions produced by Ford for the St. Francis and Memphis areas using only their surface assemblages and tabulating just the decorated, shell-tempered types. These seriations are shown in Figures 4.2 and 4.3. Clearly, there are numerous departures from strict unimodality, some probably caused by sampling error, others from the effects of space or assemblage duration (Dunnell 1970). In the remainder of my analysis, I explore the degree to which these potential effects produce deviations from the expected seriation model.

I broke the assemblages into three groups based on the character of the bootstrap analysis. In Figure 3.8, I illustrated each of these groups with a representative bootstrap sampling graph. Conclusions based on the assemblages that meet both the mean and variance criteria thus should be archaeologically meaningful (all other things being equal), while conclusions based uncritically on all of the assemblages (including the assemblages that met neither the mean nor the variance criteria) should be treated with extreme caution.

I began by combining all of the PFG assemblages ('A,' 'B,' and 'C' assemblages) from the St. Francis and Memphis areas together to form a single analytic set. In this way, I was able to examine the decisions made by PFG in their division of the study area into spatial units. I then seriated

[1] The late J.B. Griffin kindly supplied the data sheets to R.C. Dunnell.

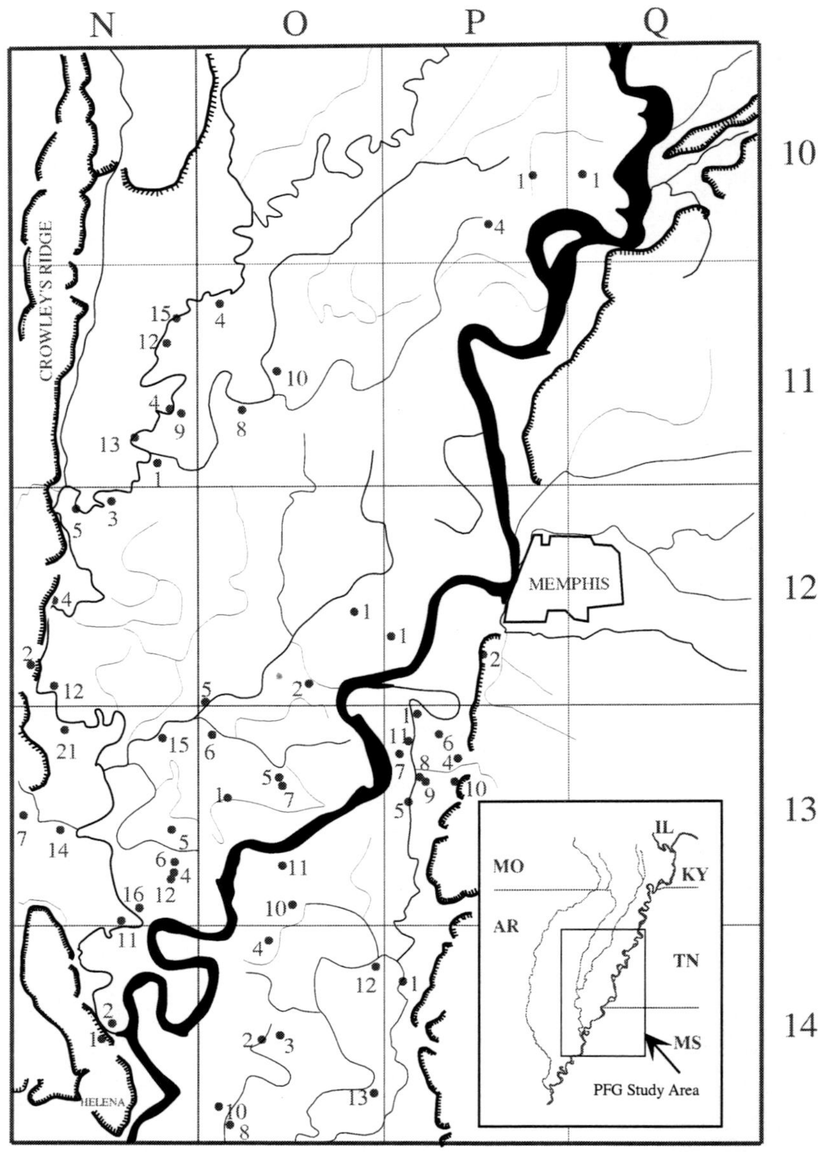

Figure 4.1. PFG St. Francis and Memphis Survey Areas

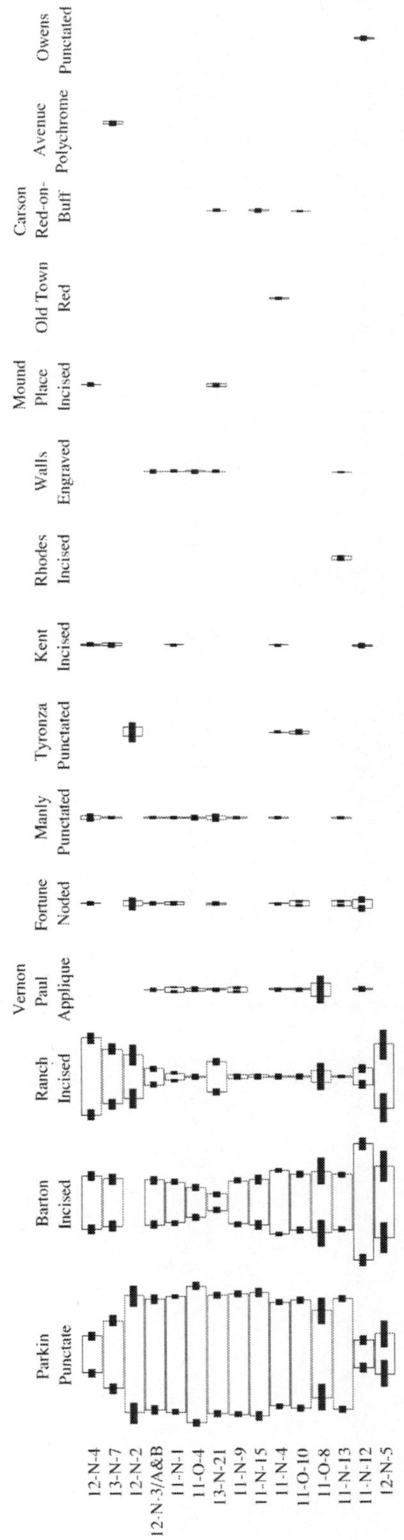

Figure 4.2. Ford's Seriation of the St. Francis Area Data.

Removing excavated samples and eliminating non-decorated types in the PFG data does not make the seriation conform more closely to the unimodal model.

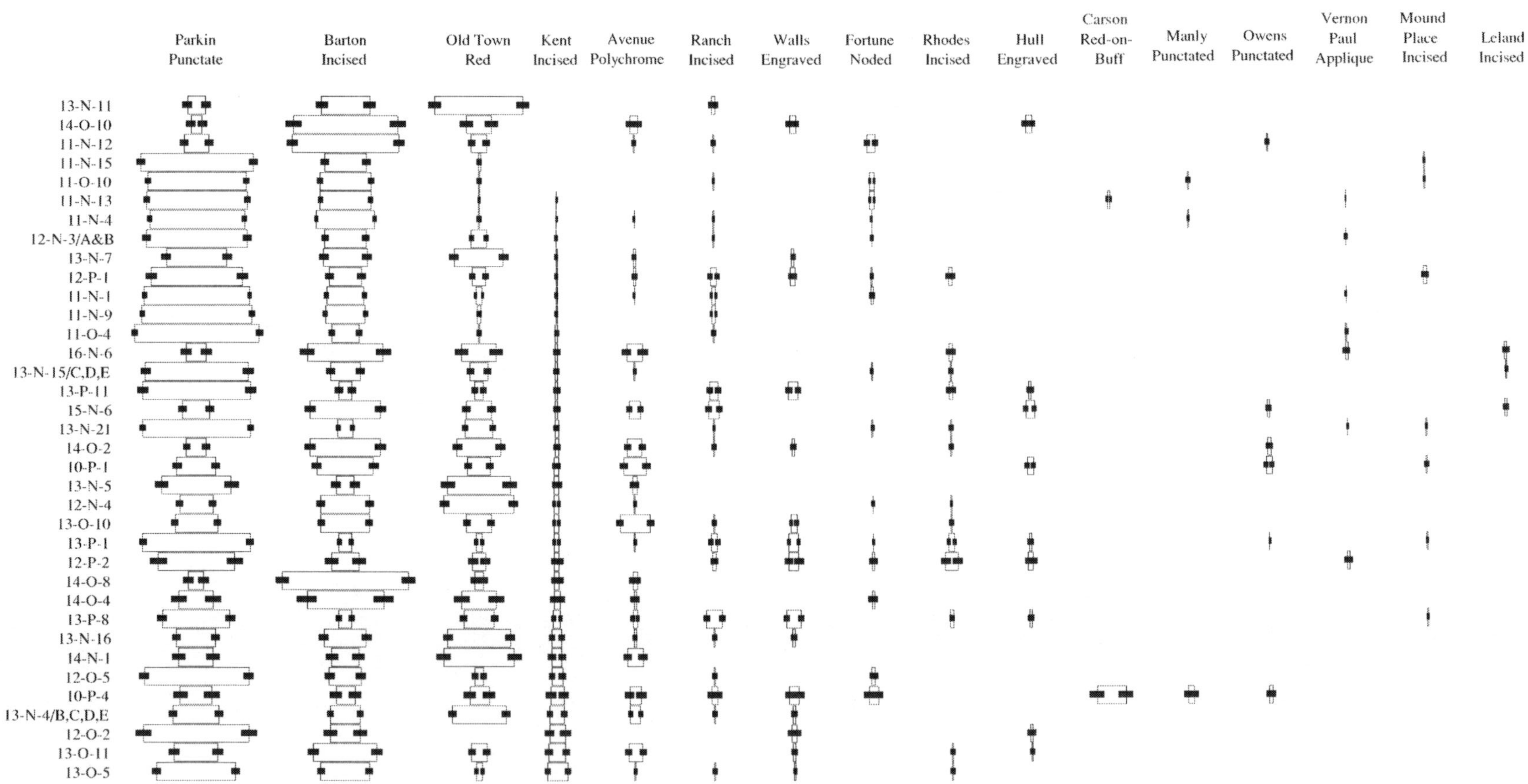

Figure 4.3. Ford's Seriation of the Memphis Area Data.

The order is Ford's and uses only surface assemblages and decorated, shell-tempered ceramic types. See note on Figure 4.2.

the assemblages into the largest groups possible without allowing violations of unimodality within calculated confidence intervals (α=0.005). If a seriation containing a group of assemblages displayed significant departures from unimodality, the group was broken into two groups in which the distribution of each was unimodal. Obtaining seriation solutions in this way is an iterative process, involving a series of hypothesis tests about particular orderings. Only the resulting seriation groups are shown in Figure 4.4. The groups shown in Figure 4.4 are the largest possible within the restriction that departures from unimodality do not occur. My analysis concludes that this is the minimum number of groups that can be formed since combining any group or portion of a group results in the rejection of the unimodal expectation of the seriation model.

When the assemblages comprising each group are mapped, a strong spatial pattern emerges (Figure 4.5). We can quickly determine that this spatial solution is very unlikely to have emerged by chance alone. This is done by calculating the probability (*p*) of randomly drawing a spatially cohesive set of assemblages by chance alone as the product of a series of combinatorial probabilities. In this calculation (Equation 4.1), *p* is calculated as:

Equation 4.1.

$$p = \frac{1}{\dfrac{N!}{n_1!(N-n_1)!} * \dfrac{(N-n_1)!}{n_2!(N-(n_1+n_2))!} * \dfrac{(N-(n_1+n_2))!}{n_1!(N-(n_1+n_2+n_3))!} \cdots * \dfrac{N-\sum_{i=1}^{n-1} n_i!}{n_1!(N-\sum_{i=1}^{n} n_i)!}}$$

N is the total number of assemblages and n_i is number of assemblages in the *ith* group. Using the PFG assemblages that meet the mean and variance criteria, we find that the probability of arriving at this spatially contiguous pattern by chance alone is extremely small ($p = 4.011 \times 10^{-12}$).

Thus, the groups within which assemblages seriate without violations of unimodality are spatially cohesive sets. This result is consistent with a hypothesis that the PFG types that the PFG types are stylistic and that the spatial patterns are the result of transmission (*i.e.*, lineages). The assemblages in the seriations are not randomly distributed across the region as one might expect if the PFG assemblages were a sample of a single large interacting population. Neither do many assemblages seem to belong to more than one group, with the exception Groups 2A and 2B (Figure 4.5). Assemblages that belong to more than one seriation set will occur whenever these assemblages are ancestors to other assemblages which have split into two or more distinct lineages or are the product of "hybridization" (*i.e.,* coalescence of formerly distinct lineages). They might also occur where interaction is hierarchically structured (Dunnell and Lipo 1999). The spatial groups of assemblages, for the most part, divide neatly into the spatial division recognized by Ford as the St. Francis and the Memphis areas. Group 3, however, spans across both areas and includes assemblages in the St. Francis area (12-N-4 and 12-N-2) and the Memphis area (13-N-21 and 13-N-7). The groups also spatially resemble, to a rough approximation, the general shapes and positions of culture historical phases, as one might expect. Parkin phase (P. Morse 1981, 1990) corresponds to Groups 2A and 2B, Nodena (D. Morse 1973, 1990) to Group 1, Walls phase (Smith 1990) to Group 5, Kent (House 1991; Smith 1990) to Group 7, and Commerce phase (Smith 1990) to Group 8.

That some correspondence to culture historical phases exists is not surprising, since culture historians were generally interested in delineating populations related by homologous similarity and I used PFG types to describe the assemblages. Based on transmission theory as outlined above, however, culture-historical phases bear a partial relation to lineages. Some of the spatial cohesion of a phase comes from real structure due to the interaction history of populations, with the remaining contribution coming from several sources: temporal differences between the deposit (duration differences or absolute differences in the period of deposition), sampling error (which must be controlled), and the effects of viewing continuous variation through the "filter" of a classification.

I must be able to discount all other sources of variability before I can conclude securely that the groups seen in Figure 4.5 represent the structure of population interaction in the past. If transmission through the region is continuous but variable across space, several factors could account for the clumped appearance of assemblages that seriate well together. The first and foremost of these factors is sampling error. As seen in the Chapter 3, differences in evenness and richness between sets of assemblages can be a function of transmission or simply due to sampling error. Assemblages with small samples are likely to have fewer types and thus be more similar to other assemblages than might be otherwise true given larger samples. Thus, to assess the possibility that sampling error and changes in assemblage composition due to sample size are affecting the size, position, and inclusiveness of seriated groups, I performed iterative seriations with only those assemblages marked as 'A' in bootstrap analysis. These assemblages each have richness values with asymptotic means and decreasing standard deviation when the samples were resampled with increasingly large sizes. This result indicates that the samples are reliable in their frequency characteristics. The seriation results are presented in Figure 4.6, and the geographic distribution of the seriation groups in Figure 4.7.

The clusters depicted in Figure 4.7 are remarkably similar to those presented in the map using all assemblages (Figure 4.5). Many assemblages were dropped, however, due to sample size problems, making exact comparisons of relations difficult. Group 5, Walls Phase assemblages, for example, shows up as a cohesive group that seriates well,

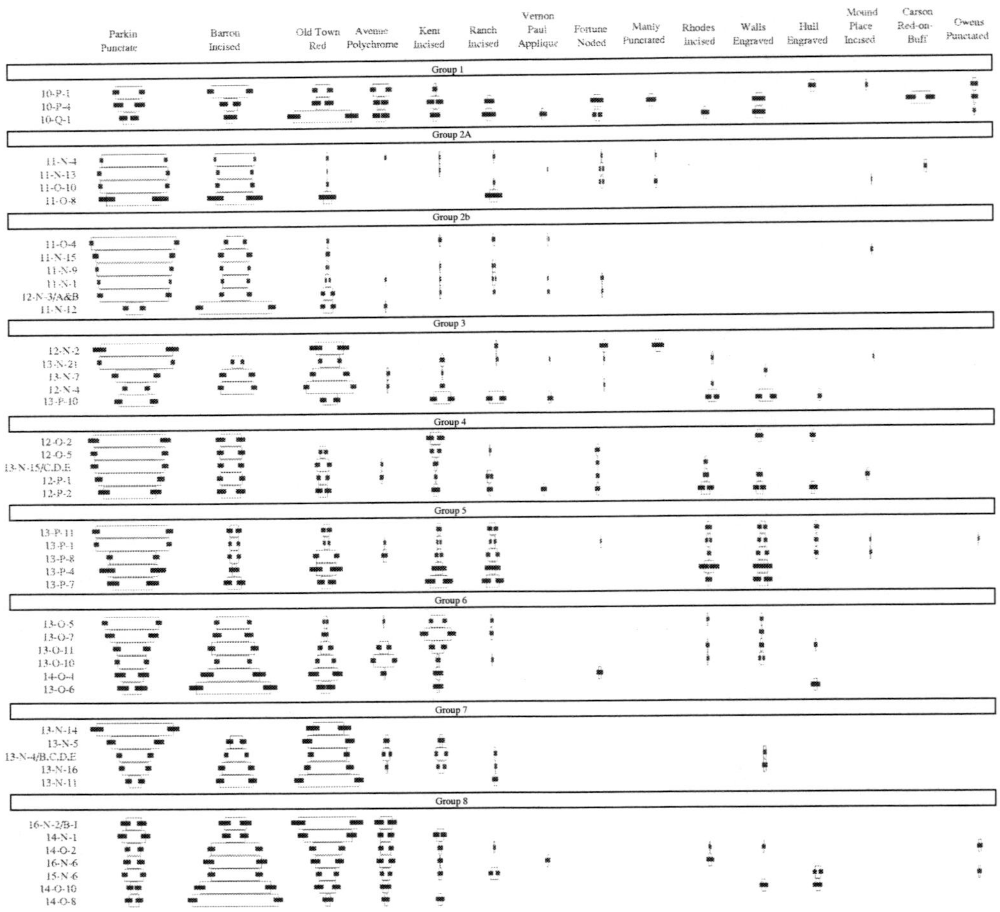

Figure 4.4. Iterative Deterministic Seriation Solution Groups for the Combined Memphis and St. Francis Assemblages ('A,' 'B,' and 'C').

The eight solutions (1-8) are the largest seriation groups that can be formed with the 'A,' 'B,' and 'C' assemblages from the combined St. Francis data set and Memphis data set and still statistically meet the expectations of the model. Analysis concludes that these are the minimum number of groups that can be formed since combining any group or portion of group together results in the rejection of the unimodal expectation of the seriation model.

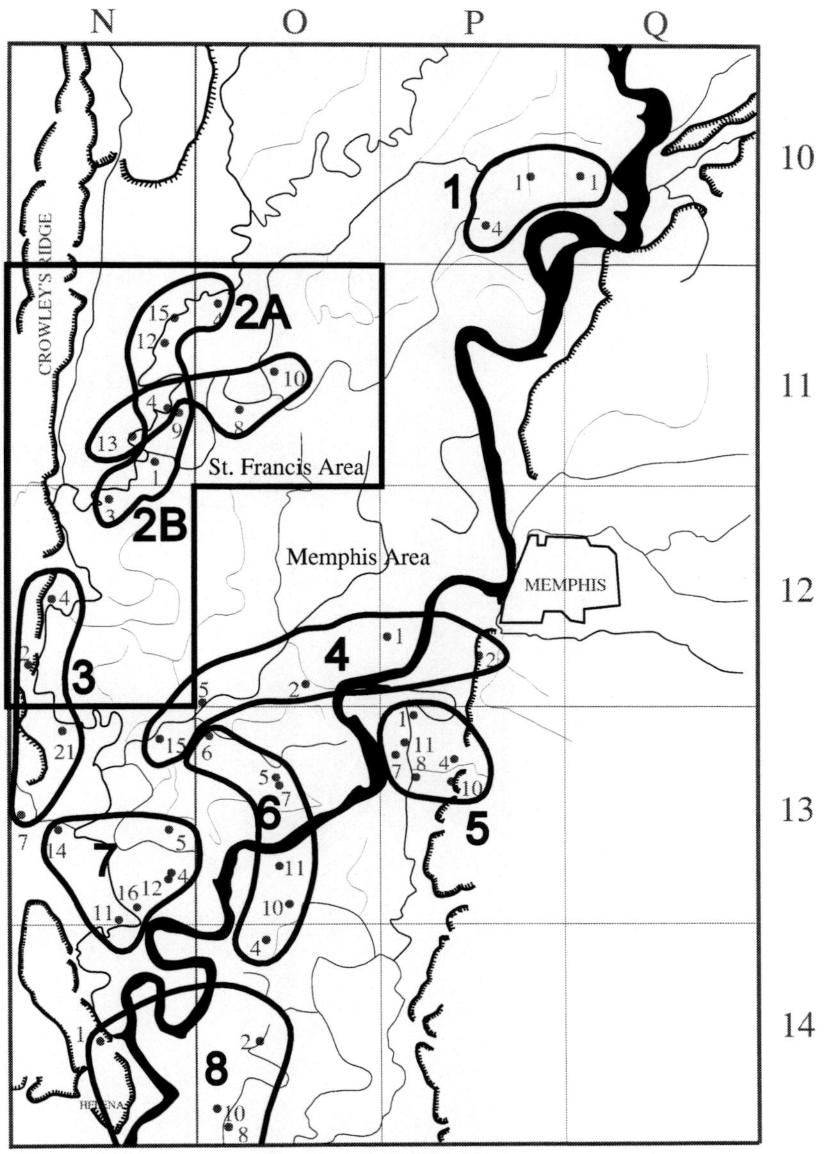

Figure 4.5. Geographic Distribution of Solution Clusters of Combined St Francis and Memphis PFG Assemblages ('A,' 'B,' and 'C').

Numbers (1-8) refer to the Figure 4.4 seriation solution. The location of the St. Francis area is shown in bold lines. The remainder of the area was considered to be the Memphis area by PFG.

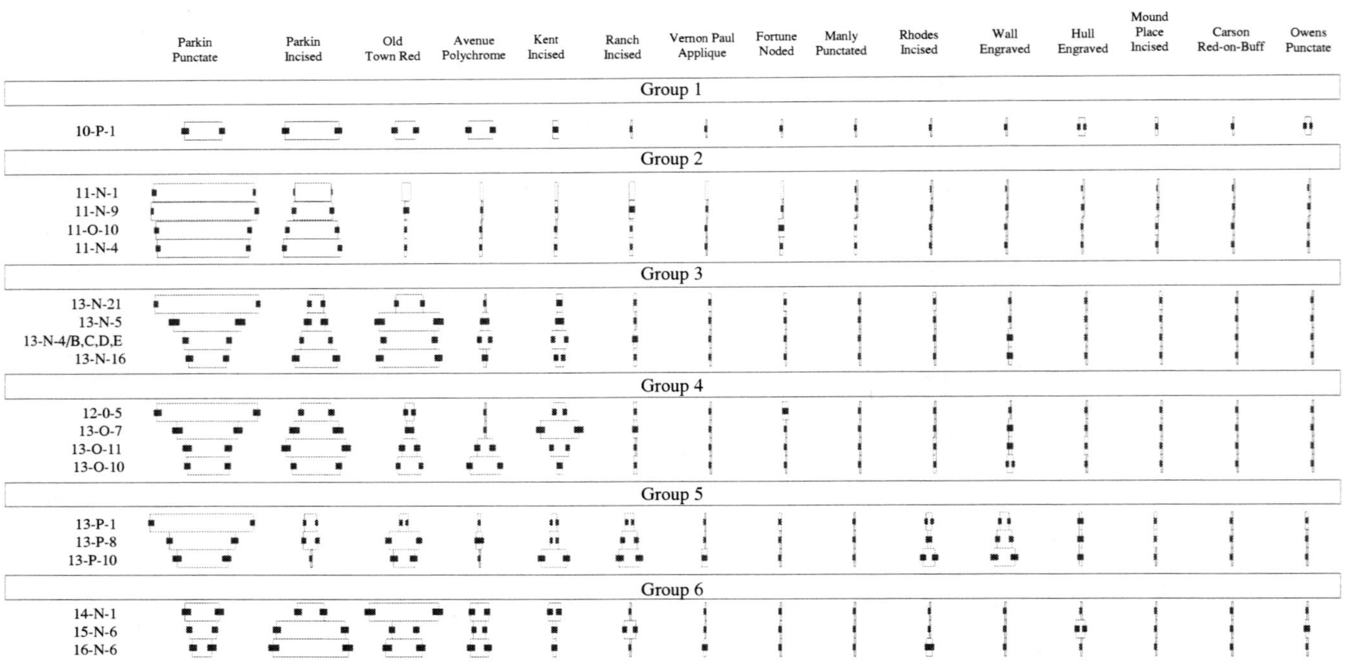

Figure 4.6. Seration Solutions for 'A' Assemblages.

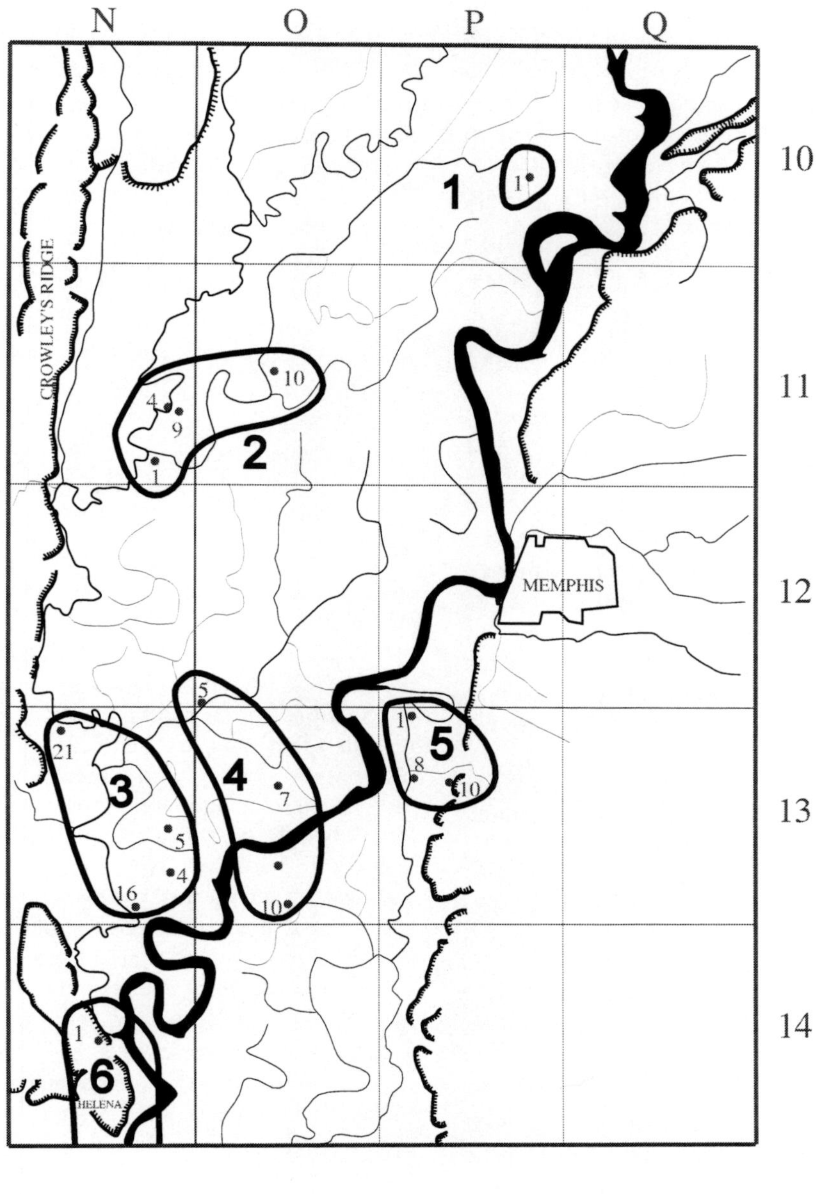

Figure 4.7. Geographic Distribution of Seriation Clusters of the St. Francis and Memphis Using 'A' Assemblages.

This geographic distribution of seriation clusters uses only assemblages that meet the asymptotic means and variance criteria ('A') as measured by bootstrap assessment. All groups are spatially contiguous and non-overlapping. It is likely that the differences between these solutions and the solution in Figure 4.5 are due to differences in data quality.

although the relations between seriation groups and phases become less clear in the area traditionally considered to be the Kent (Group 3) and southern Parkin Phase (Group 2) areas. In these areas, the small number of available assemblages may have artificially linked assemblages from these separate seriation sets together. On the other hand, the phases themselves might be simply arbitrary spatial and temporal slices of a single interacting population that is much larger in size (*e.g.,* covering all of 11-N, 12-N, and 13-N, along with some of 13-O). Distinguishing these possibilities, and understanding the relation of the groups mapped in Figure 4.7 to the underlying interaction history of the region, requires that I examine how these groups and seriations are artifacts of the classificatory scheme used in PFG.

NEUTRALITY EVALUATIONS

Are the groups identified in the seriation analyses the result of strong selective differences between the areas? Traditionally, many researchers treat the culture-historical types created by PFG for the Mississippi river valley collections as examples of "neutral" archaeological descriptions. Other than qualitative arguments, however, neutrality has not been tested independently of seriation. Fortunately, using the methods discussed in Chapter 3, we can now examine the neutrality hypothesis by determining whether the frequencies of culture-historical types used in the PFG study match the expectations of the neutral model. Ewen's test, for example, provides one means of testing the statistical properties of the PFG types by comparing these frequencies with the distribution expected in a neutral, infinite alleles model (Hurt 1999; Lynch and Baker 1993; Neiman 1995). In this method, a meme identity value is calculated using the frequency of each type in a sample. The expected meme identity is then derived using Equation 3.3. In order to generate error terms to evaluate whether the expected and the observed identities are statistically the same, confidence intervals of diversity estimates were calculated using a resampling procedure that generated one thousand replicates (Wu 1986). From this distribution, the 2.5^{th} and the 97.5^{th} percentiles were used as the 95% confidence limits. I made the calculations for this statistic using a Turbo Pascal program written by Alejandro Lynch (Lang 1995).

In Table 4.1, I show the relation between the observed numbers of types with the calculated expected number of types, a relation referred to as meme identity. In PFG assemblages, the observed distribution of classes consistently fell within the expected interval derived from a stochastic model of trait frequencies, consistent with the hypothesis that PFG ceramic classes measure neutral variation. The error terms, however, are relatively large and suggest that within-sample variability may be due to the result of immigration, mutation, and random extinction of types. Furthermore, there may be other factors driving variability in the frequency of historical types than is predicted explicitly by the neutral model including selection, sample size, and the relatively *ad hoc* construction of the PFG types.

The random walk test described in Chapter 3 provides a means for examining if the drift is driving changes in frequencies, which could be the case if populations were relatively small and types neutral. After ordering the assemblages from north to south, I determined if the sum of the differences between the assemblages was statistically distinguishable from zero at the α=0.05 level of significance. Since environmental difference is critical to selection, hypotheses tested here are:

H_0 = Classes describe the assemblages neutrally with respect to location. In this case, the summed mean differences between the frequencies of types in pairs of assemblages are not distinguishable from zero.

H_1 = Classes do not describe the assemblages neutrally with respect to location. The summed mean differences between the frequencies of types in pairs of assemblages are distinguishable from zero.

The analysis was conducted using the neutrality analysis software (see Appendix D). The results of these analyses are shown in Table 4.2. Of the culture-historical types used in the PFG study, only Owen's Punctated does not have a distribution across the assemblages that meet the expectations of the neutral model. There are two potential reasons that Owen's Punctated does not meet the distribution. First, Owen's Punctated appears only in a few very late assemblages and may not have a distribution that is large enough to evaluate neutrality. Second, this type was labeled a 'tentative' type by PFG because it was only assigned to sherds that were small and fragmentary. PFG admit, "it is to be doubted whether we are actually dealing with a 'type' or merely a collection of sherds which share similar *observable* characteristics" (PFG 1951:136, emphasis in PFG). Consequently, the distribution of Owen's Punctated may be randomized by combining sherds from two or more stylistic types, or conversely, it may signal a variety of functional criteria that produce sherds with the size and characteristics as those described as Owen's Punctated. The former seems likely in view of PFG's discussion. Except for Owen's Punctated then, the general conclusion of these analyses is that the PFG types behave as if they are neutral in the assemblages they used.

SIGNIFICANCE EVALUATIONS

Lacking in this analysis, however, is a determination of the quality of the exact orderings created. Though I have confidence that the iterative approach has yielded a set of assemblages that are relatively robust in terms of samples sizes and classes, it is necessary to assess whether the ordering of any two assemblages is statistically defensible. Using a program written to perform this test (Appendix C), I

Table 4.1. Calculated Values of Observed and Expected Meme Identity and 95% Confidence Intervals for Each Assemblage in St. Francis and Memphis Area Assemblages.

Site Name	Observed Meme Identity	Expected Meme Identity	95% CI
Carson Lake	0.288	0.4148	0.177, 0.698
Parkin	0.505	0.5345	0.253, 0.842
Turnbow	0.494	0.4649	0.228, 0.782
Williamson	0.503	0.5236	0.190, 0.860
Fortune	0.580	0.6365	0.285, 0.876
Neeley's Ferry	0.499	0.5316	0.263, 0.867
Vernon Paul	0.549	0.6207	0.271, 0.956
Barton Ranch	0.489	0.5236	0.261, 0.862
Cummins	0.652	0.5099	0.266, 0.889
Rose Mound	0.477	0.4523	0.224, 0.773
Big Eddy	0.334	0.5236	0.219, 0.898
Pouncey	0.481	0.4760	0.240, 0.781
Cramor Place	0.477	0.4715	0.239, 0.784
Mound Place	0.388	0.3559	0.166, 0.602
Shelby Place	0.274	0.2427	0.125, 0.449
Grant Place	0.411	0.6211	0.301, 0.896
Nickel	0.408	0.4975	0.242, 0.823
Starkley	0.293	0.4829	0.245, 0.831
Castile Landing	0.502	0.3982	0.190, 0.617
Kent Place	0.273	0.4998	0.236, 0.864
Davis	0.375	0.5811	0.275, 0.874
Clay Hill	0.326	0.5522	0.259, 0.892
Hollywood	0.289	0.4346	0.230, 0.813
Commerce	0.304	0.4149	0.184, 0.789
Belle Meade	0.366	0.4543	0.223, 0.787
Beck	0.305	0.4038	0.217, 0.684
Walls	0.480	0.3320	0.160, 0.647
Irby	0.186	0.3423	0.173, 0.572
Woodlyn	0.479	0.3545	0.169, 0.589
L. Cormorant	0.262	0.3545	0.163, 0.608
Moore	0.339	0.5744	0.262, 0.836
West	0.496	0.4671	0.239, 0.823
Owens	0.340	0.3689	0.158, 0.695
Dundee	0.670	0.5848	0.271, 0.900
Montgomery	0.297	0.4329	0.197, 0.774
Oliver	0.373	0.4131	0.189, 0.717

Table 4.2. Analysis of Neutrality on Culture-Historical Types from Memphis and St. Francis Assemblages.

Type	Mean Difference between Assemblages	Std. Dev.	Sample Size	# of Assems.	C.I. Lower 95%	C.I. Upper 95%	C.I. Includes Zero?
Parkin Punctated	0.00727	0.2785	5210	36	-0.086	0.100	YES
Barton Incised	0.01066	0.2102	2623	35	-0.060	0.082	YES
Ranch Incised	0.00148	0.0458	159	23	-0.018	0.021	YES
Vernon Paul Appliqué	-0.00043	0.0182	13	8	-0.013	0.012	YES
Fortune Noded	-0.00007	0.0215	74	13	-0.012	0.012	YES
Manly Punctated	-0.00069	0.0119	13	2	-0.018	0.016	YES
Kent Incised	0.00097	0.0711	225	31	-0.025	0.027	YES
Rhodes Incised	0.00024	0.0347	52	14	-0.018	0.019	YES
Walls Engraved	0.00140	0.0479	93	16	-0.022	0.025	YES
Hull Engraved	-0.00035	0.0371	32	10	-0.024	0.023	YES
Mound Place Incised	0.00039	0.0117	9	7	-0.008	0.009	YES
Old Town Red	0.00826	0.2058	901	35	-0.061	0.078	YES
Carson Red-on-Buff	0.00264	0.0183	12	1	-0.034	0.039	YES
Owens Punctated	0.02246	0.0152	13	5	0.0089	0.036	NO

generated p-values and error terms for each pair of assemblages with the algorithm described in Chapter 3. The p-values are listed in Table 4.3 and the resulting distribution of pairwise significance values is shown in Figure 4.8. It is apparent from the p-value distribution that many possible orderings are not warranted. Superimposed on an essentially normal distribution of significance values, however, is a cluster of comparisons which are statistically significant and whose p-values are greater than 0.80. Two distributions of p-values are expected for assemblage pairs within clusters (<0.80) and assemblage pairs between clusters (>0.80). Consequently, these significant comparisons will allow me to examine the sufficiency of the iterative solution.

In examining the pairwise comparisons within a perfect seriation group, I noted that the significance values increase with increasing distance between assemblages in the ordering. This is, in fact, simply a way of restating the similarity principle upon which probabilistic seriation methods are founded. In my case, however, the "similarity" values are full significance values with error terms derived from testing our ability to properly distinguish the frequencies of cultural-historical types in the data set using resampling. Thus, we should accept seriation orderings in which the pairwise significance values increase monotonically as one begins at one end and starts moving to the other end of the seriation. I evaluated all of the seriation groups with this criterion in mind, and in addition, attempted to reform larger groups that retain monotonicity of significance values (Figure 4.9).

The results of this reanalysis show that while some details of the preliminary orderings changed, I was unable to form larger groups that retained monotonicity (Figure 4.10). In particular, I found that two assemblages, 13-N-21 (Castile Landing), and 12-O-5 (Cramor Place), do not seriate with other assemblages when I examined the results of the pairwise tests. These assemblages can no longer be combined with other groups and their relation will remain enigmatic until larger samples can be obtained. Group 2, surrounding the Parkin site and comprising several Parkin-phase assemblages, retains its integrity in this analysis. Furthermore, the ordering has no clear internal spatial patterning, thus I cannot reject the hypothesis that the Parkin seriation represents a temporal ordering of assemblages.

Two groups, labeled Groups 3 and 5 in Figure 4.10, display orderings that are difficult to ascribe to either time or space. The pairwise significance values are low, thus providing no particular confidence that the ordering represents anything other than poor sampling. Figure 4.11 displays the assemblages in the order that best fits the unimodal model, but we should keep in mind that no statistical warrant exists for an ordering. It is possible that the samples are contemporaneous but additional data from the deposits themselves would be required to test this hypothesis.

Group 4 has a similar problem, in that it is difficult to ascribe the ordering to any particular cause due to poor significance values. However, the ends of the seriation can be distinguished from each other, indicating at least some confidence in placing these two assemblages in their displayed position. Given the distribution of assemblages in a north-south line, the ordering within the group is potentially temporal and reflects transmission between sites in this direction. Of Groups 1 and 6, little can be said. Group 1 is composed of the Nodena site, the only site from the area left after the sample-size tests eliminated adjacent assemblages as unstable. Group 6 is mostly outside of the study area, and there are no data to substantiate either a spatial or temporal ordering.

Given my confidence in the groups determined by these new seriation analyses, it is instructive to examine between-group significance values as a measure of the relation between the seriation solutions I am positing. Group 1, the Nodena site, is markedly different from most of the other groups,

Table 4.3. *p*-values for Iterative Pairwise Significance Test for Pairs of PFG 'A' Assemblages.

Assemblage 1	Assemblage 2	*p*-value	Std. Dev.
10-P-1	11-N-9	0.928	0.019
10-P-1	11-N-1	0.912	0.028
10-P-1	11-O-10	0.842	0.023
10-P-1	11-N-4	0.642	0.057
10-P-1	13-N-4	0.682	0.036
10-P-1	13-N-16	0.352	0.049
10-P-1	13-N-5	0.574	0.069
10-P-1	13-O-7	0.390	0.063
10-P-1	13-O-11	0.228	0.028
10-P-1	13-O-10	0.082	0.018
10-P-1	13-P-1	0.626	0.023
10-P-1	13-P-8	0.686	0.060
10-P-1	13-P-10	0.540	0.055
10-P-1	14-N-1	0.676	0.038
10-P-1	16-N-6	0.546	0.054
10-P-1	15-N-6	0.350	0.081
10-P-1	12-O-5	0.588	0.024
10-P-1	13-N-21	0.496	0.073
11-N-9	11-N-1	0.198	0.033
11-N-9	11-O-10	0.838	0.086
11-N-9	11-N-4	0.774	0.027
11-N-9	13-N-4	0.874	0.050
11-N-9	13-N-16	0.558	0.064
11-N-9	13-N-5	0.716	0.054
11-N-9	13-O-7	0.210	0.024
11-N-9	13-O-11	0.842	0.032
11-N-9	13-O-10	0.868	0.031
11-N-9	13-P-1	0.696	0.087
11-N-9	13-P-8	0.994	0.009
11-N-9	13-P-10	1.000	0.000
11-N-9	14-N-1	0.952	0.026
11-N-9	16-N-6	0.654	0.047
11-N-9	15-N-6	0.806	0.015
11-N-9	12-O-5	0.544	0.048
11-N-9	13-N-21	0.498	0.052
11-N-1	11-O-10	0.894	0.029
11-N-1	11-N-4	0.700	0.037
11-N-1	13-N-4	0.904	0.015
11-N-1	13-N-16	0.596	0.042
11-N-1	13-N-5	0.772	0.052
11-N-1	13-O-7	0.376	0.030
11-N-1	13-O-11	0.904	0.038
11-N-1	13-O-10	0.908	0.018
11-N-1	13-P-1	0.406	0.038
11-N-1	13-P-8	0.984	0.005
11-N-1	13-P-10	0.996	0.009
11-N-1	14-N-1	0.936	0.022
11-N-1	16-N-6	0.630	0.039
11-N-1	15-N-6	0.804	0.032
11-N-1	12-O-5	0.332	0.056
11-N-1	13-N-21	0.702	0.037

Assemblage 1	Assemblage 2	*p*-value	Std. Dev.
13-N-4	13-P-8	0.908	0.045
13-N-4	13-P-10	0.866	0.017
13-N-4	14-N-1	0.370	0.038
13-N-4	16-N-6	0.774	0.043
13-N-4	15-N-6	0.638	0.049
13-N-4	12-O-5	0.386	0.025
13-N-4	13-N-21	0.674	0.055
13-N-16	13-N-5	0.234	0.028
13-N-16	13-O-7	0.198	0.036
13-N-16	13-O-11	0.238	0.038
13-N-16	13-O-10	0.256	0.042
13-N-16	13-P-1	0.516	0.058
13-N-16	13-P-8	0.586	0.066
13-N-16	13-P-10	0.832	0.028
13-N-16	14-N-1	0.174	0.044
13-N-16	16-N-6	0.576	0.044
13-N-16	15-N-6	0.840	0.044
13-N-16	12-O-5	0.262	0.028
13-N-16	13-N-21	0.504	0.063
13-N-5	13-O-7	0.210	0.042
13-N-5	13-O-11	0.846	0.044
13-N-5	13-O-10	0.494	0.033
13-N-5	13-P-1	0.406	0.030
13-N-5	13-P-8	0.224	0.046
13-N-5	13-P-10	0.710	0.029
13-N-5	14-N-1	0.574	0.029
13-N-5	16-N-6	0.572	0.040
13-N-5	15-N-6	0.520	0.050
13-N-5	12-O-5	0.482	0.026
13-N-5	13-N-21	0.290	0.041
13-O-7	13-O-11	0.368	0.048
13-O-7	13-O-10	0.272	0.047
13-O-7	13-P-1	0.744	0.046
13-O-7	13-P-8	0.720	0.043
13-O-7	13-P-10	0.500	0.044
13-O-7	14-N-1	0.416	0.034
13-O-7	16-N-6	0.398	0.061
13-O-7	15-N-6	0.634	0.057
13-O-7	12-O-5	0.250	0.062
13-O-7	13-N-21	0.308	0.059
13-O-11	13-O-10	0.388	0.044
13-O-11	13-P-1	0.968	0.026
13-O-11	13-P-8	0.974	0.011
13-O-11	13-P-10	0.750	0.080
13-O-11	14-N-1	0.378	0.013
13-O-11	16-N-6	0.492	0.047
13-O-11	15-N-6	0.498	0.044
13-O-11	12-O-5	0.452	0.041
13-O-11	13-N-21	0.826	0.019
13-O-10	13-P-1	0.594	0.047
13-O-10	13-P-8	0.738	0.036

Table 4.3 (continued).

Assemblage 1	Assemblage 2	*p*-value	Std. Dev.	Assemblage 1	Assemblage 2	*p*-value	Std. Dev.
11-O-10	11-N-4	0.276	0.037	13-O-10	13-P-10	0.522	0.044
11-O-10	13-N-4	0.784	0.086	13-O-10	14-N-1	0.674	0.084
11-O-10	13-N-16	0.422	0.036	13-O-10	16-N-6	0.616	0.037
11-O-10	13-N-5	0.734	0.055	13-O-10	15-N-6	0.462	0.026
11-O-10	13-O-7	0.246	0.060	13-O-10	12-O-5	0.544	0.009
11-O-10	13-O-11	0.798	0.044	13-O-10	13-N-21	0.278	0.079
11-O-10	13-O-10	0.504	0.052	13-P-1	13-P-8	0.330	0.035
11-O-10	13-P-1	0.954	0.017	13-P-1	13-P-10	0.928	0.031
11-O-10	13-P-8	0.996	0.009	13-P-1	14-N-1	0.850	0.023
11-O-10	13-P-10	1.000	0.000	13-P-1	16-N-6	0.860	0.046
11-O-10	14-N-1	0.966	0.011	13-P-1	15-N-6	0.682	0.066
11-O-10	16-N-6	0.602	0.044	13-P-1	12-O-5	0.506	0.061
11-O-10	15-N-6	0.942	0.026	13-P-1	13-N-21	0.526	0.038
11-O-10	12-O-5	0.382	0.044	13-P-8	13-P-10	0.594	0.075
11-O-10	13-N-21	0.532	0.033	13-P-8	14-N-1	0.684	0.032
11-N-4	13-N-4	0.750	0.032	13-P-8	16-N-6	0.650	0.039
11-N-4	13-N-16	0.524	0.057	13-P-8	15-N-6	0.482	0.040
11-N-4	13-N-5	0.750	0.054	13-P-8	12-O-5	0.810	0.032
11-N-4	13-O-7	0.284	0.029	13-P-8	13-N-21	0.372	0.052
11-N-4	13-O-11	0.618	0.030	13-P-10	14-N-1	0.974	0.011
11-N-4	13-O-10	0.542	0.037	13-P-10	16-N-6	0.906	0.032
11-N-4	13-P-1	0.898	0.022	13-P-10	15-N-6	0.672	0.062
11-N-4	13-P-8	0.990	0.000	13-P-10	12-O-5	0.986	0.013
11-N-4	13-P-10	0.988	0.016	13-P-10	13-N-21	0.778	0.058
11-N-4	14-N-1	0.994	0.005	14-N-1	16-N-6	0.554	0.038
11-N-4	16-N-6	0.646	0.041	14-N-1	15-N-6	0.466	0.080
11-N-4	15-N-6	0.932	0.019	14-N-1	12-O-5	0.286	0.055
11-N-4	12-O-5	0.432	0.049	14-N-1	13-N-21	0.848	0.038
11-N-4	13-N-21	0.520	0.012	16-N-6	15-N-6	0.250	0.042
13-N-4	13-N-16	0.276	0.034	16-N-6	12-O-5	0.646	0.061
13-N-4	13-N-5	0.734	0.025	16-N-6	13-N-21	0.550	0.042
13-N-4	13-O-7	0.254	0.046	15-N-6	12-O-5	0.978	0.011
13-N-4	13-O-11	0.216	0.082	15-N-6	13-N-21	0.310	0.028
13-N-4	13-O-10	0.334	0.066	12-O-5	13-N-21	0.352	0.041
13-N-4	13-P-1	0.988	0.011				

including the Parkin cluster (Group 2 and from 12-O-5). The only relation it bears to the other clusters, albeit ambiguous, is to Group 5, a group that occupies the area traditionally assigned to the Walls Phase. This makes some degree of sense, given the fact that both were originally lumped into one phase before being broken out into separate phases by Phillips (1970). Group 2, in the "Parkin" region, is generally distinct from most of the clusters to the south, with the exception of assemblage 12-O-5 (Cramor Place). As I mentioned previously, the relations of 12-O-5 to surrounding seriation groups suggest the need for additional research, requiring larger samples and additional intervening assemblages.

SERIATIONS AT MULTIPLE SCALES: CLASSIFICATION EFFECTS

Using the discussion developed in Chapter 3, we can also examine the effects that the number of stylistic classification dimensions has on the size, position, and inclusiveness of seriated groups. In these analyses, it is useful to perform seriation analyses using a hierarchical series of classifications (Dunnell 1970). Ideally, the classifications used will be hybrids and rigorously dimensional in nature (Dunnell 1971:70-76), in order that systematic differences in level can be achieved by collapsing dimensions of variability in an orthogonal manner.

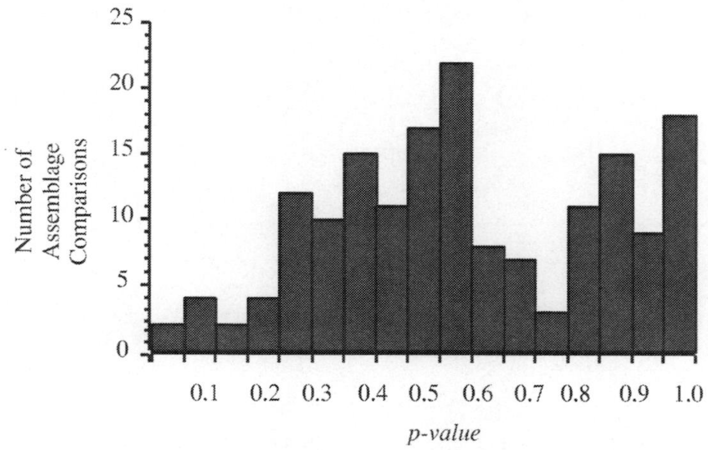

Figure 4.8. Distribution of *p*-values for the Iterative Pairwise Significance Test.

The distribution of *p*-values is not normal but bi-modal in distribution. This result suggests, as is expected from the seriation solutions, that *p*-values differ for assemblage pairs *within* clusters and assemblage pairs *between* clusters.

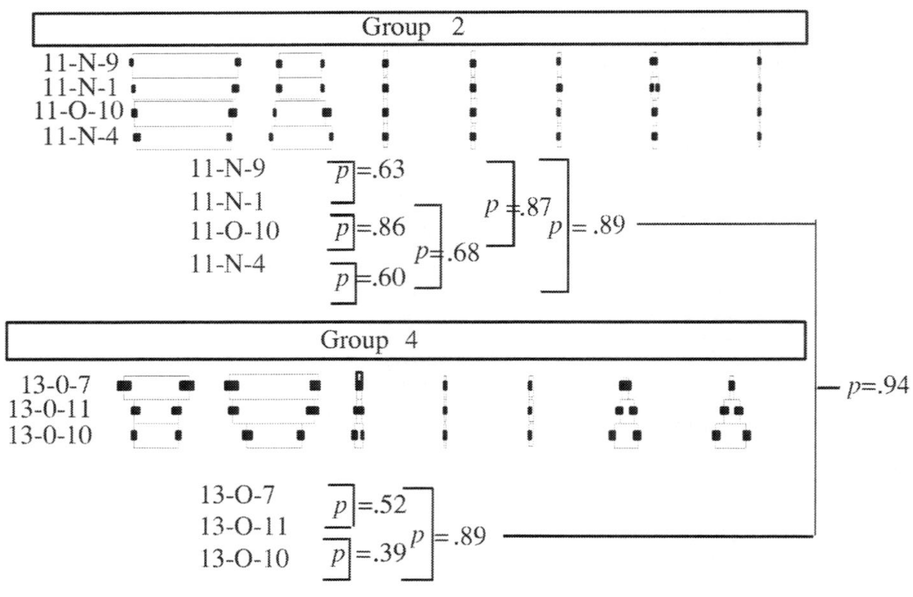

Figure 4.9. Iterative Pairwise Significance Testing.

The *p*-values are calculated with the iterative pairwise significance-testing program and are shown for pairs of assemblages with brackets. The more easily a pair of assemblages can be distinguished given the effects of sample size, the greater the *p*-value. The *p*-values should increase, therefore, the further assemblages are apart from one another. As an example, the paired results for two seriation groups (Group 2 and 4) are shown here. Horizontal bars show the proportions for the cultural historical types Parkin Punctated, Barton Incised, Old Town Red, Ranch Incised, Rhodes Incised, Kent Incised, and Walls Engraved from left to right.

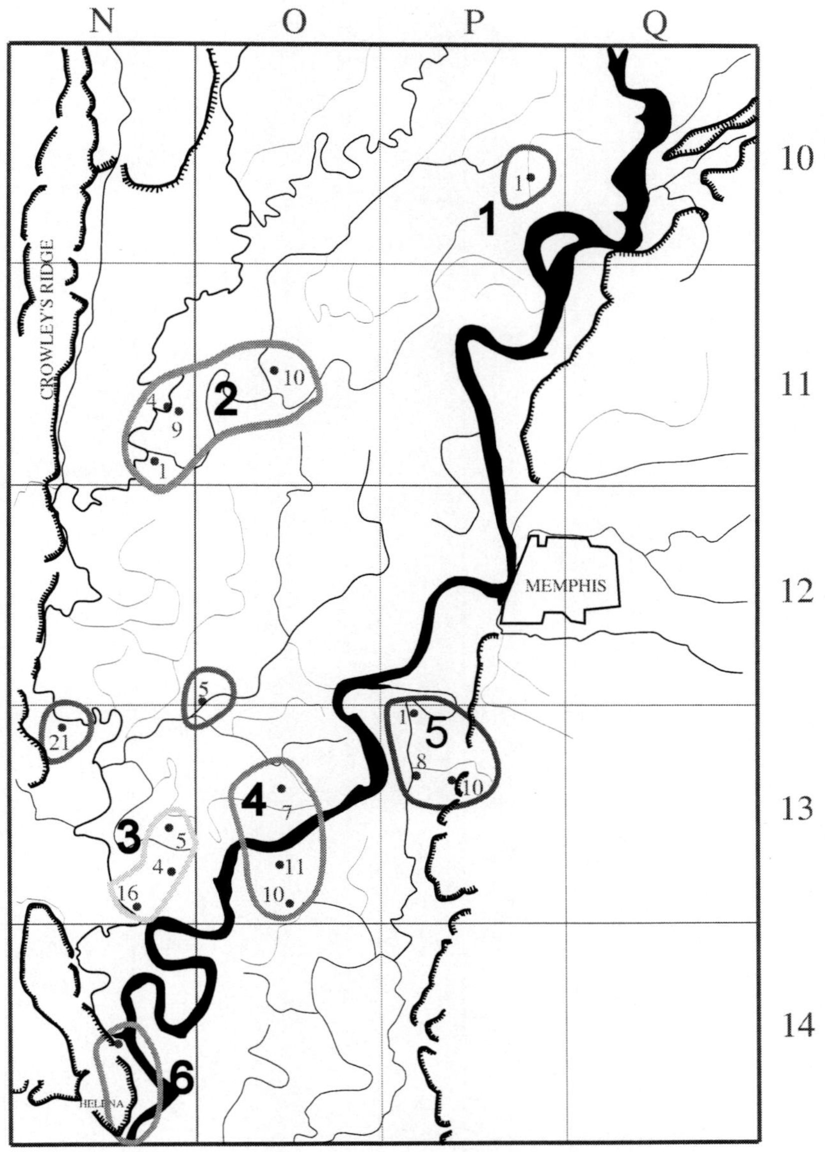

Figure 4.10. Spatial Solution with Iteratively Tested Seriation Groups.

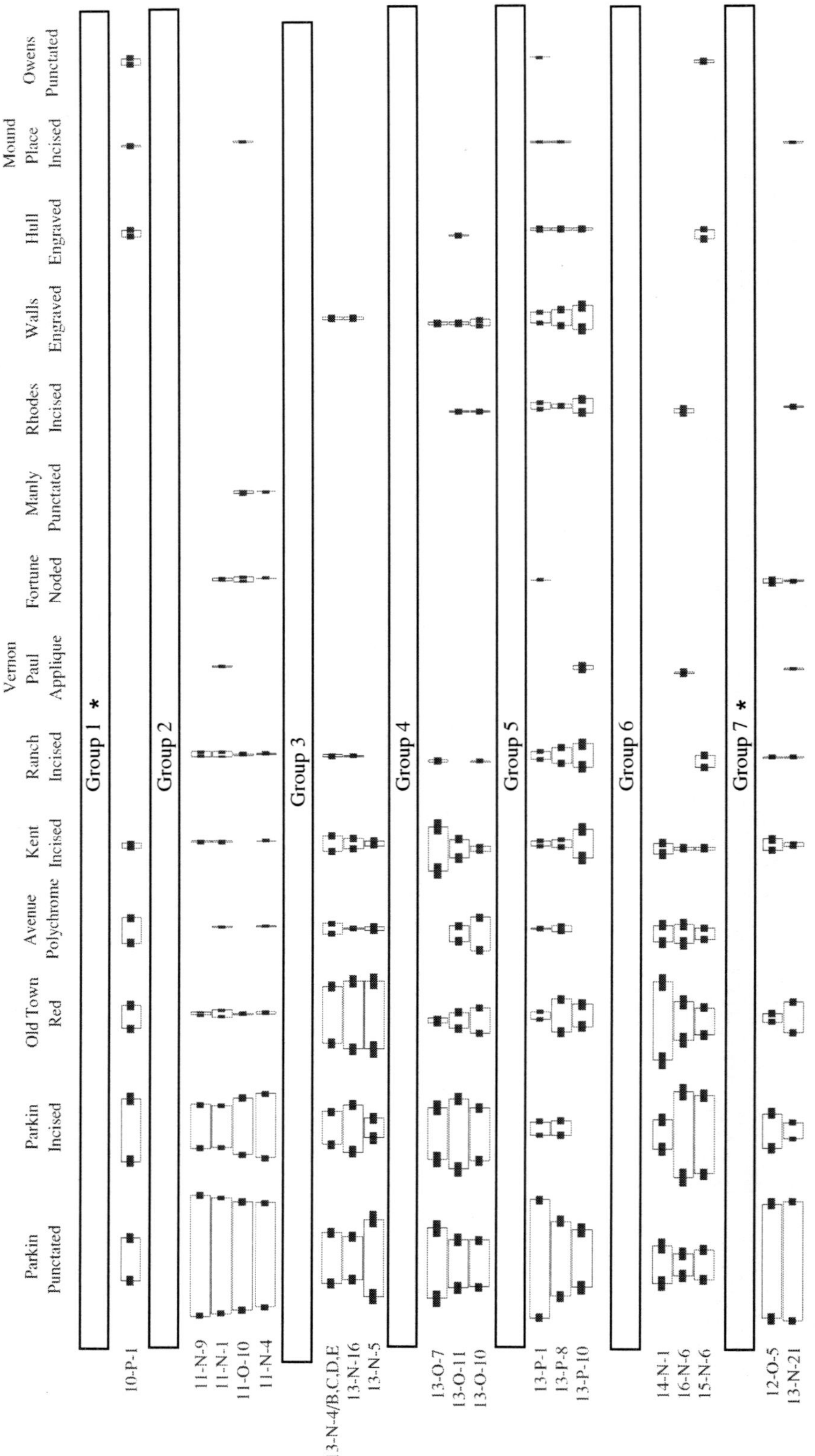

Figure 4.11. Seriation Groups after Iterative Pairwise Significance Testing.

* No order implied with 1, 2 assemblages.

(1) Absences where confidence intervals of adjacent frequencies include zero and do not violate assumption of continuity.

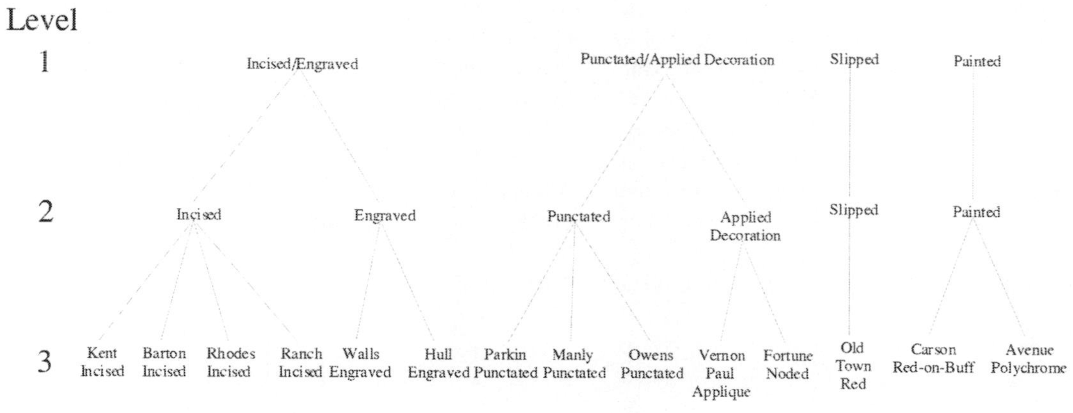

Figure 4.12. Three Hierarchical Levels of Classification Used to Assess Effects of Classification Level on Group Membership.

Here, I aggregated the original PFG types (the bottom level) to create two coarser classification levels (middle and top levels). The bottom level contains the original PFG types. Aggregating types with the same decoration technique created the middle level. The upper level represents types of the same form; linear designs for incised/engraved, circular designs for punctated/applied. Here, slipped (paint all over the surface) and painted (paint in patterns) designs remain separate.

Figure 4.12 depicts the relation between the original PFG types, along the top, and two successive levels of collapsing based on ever-more inclusive categories of surface decoration attribute classes. For each set of newly collapsed classes, I recalculated frequencies for each assemblage and built seriations with the two collapsed sets of classes. The spatial distribution of seriations is depicted in Figure 4.13.

The effects of changing the level of classification for the PFG assemblages in the direction of coarser classes (fewer dimensions) is simply to create larger groups of assemblages, and in some cases allowing coalescence of formerly separate assemblages into more inclusive groups. These "sister" groups may well represent the gradual differentiation of a single lineage into two. The groups formed, however, still retain their spatial contiguity, suggesting that the seriations performed here may reflect structure in the interaction history of the region with obvious sample distribution problems, and thus may map the approximate location of lineages. Should the data support this conclusion, one must then evaluate whether the relation between "filters" of successive sizes indicates that interaction is widespread and continuous, and the appearance of discrete communities comes from slicing a continuous range of interaction frequencies using an arbitrary measuring unit for interaction (Figure 4.14). If, on the other hand, social boundaries existed in the patterns of interaction, these will be visible as discontinuities in the relation between the numbers of dimensions and number of groups as the level of classification is changed. Differentiating these alternatives requires new, more complex classifications, and much larger sample sizes (Lipo and Madsen 1996) to support the use of such a classification.

CONCLUSIONS

In Chapter 3, I presented a simple theory of cultural transmission that utilized deterministic frequency seriation as a means for examining interaction history in the archaeological record. This theory and derived methods permitted me to evaluate assumptions of spatial patterning inherent in the spatial divisions of the PFG survey data and phase interpretation based on these data. The nature of neutral traits and cultural transmission within populations leads us to expect that cultural lineages will not be discrete bounded entities but temporal and spatial areas of greater similarity due to increased probabilities of interaction. Examining the large PFG ceramic data set from the St. Francis and Memphis areas, I found that the largest seriations that could be produced were statistically defensible clusters of spatially coherent assemblages. An analysis of the composition of culture-historical types within and between assemblages supports the notion that these classes are measuring homologous similarity as they meet the expectations of the neutral model. Overall, these results support the notion that there is a spatial component to the interaction and the assumptions made by Ford in his division of the survey into spatial units or that both analyses have been constrained by the samples (*i.e.*, distribution, size) which they share. Although there are differences in the details of the groups, outlines of the seriation groups generally match the outlines of "phase" boundaries. Significantly, the theoretical warrant for these seriation groups differs substantially from traditional phase interpretations since cultural transmission theory emphasizes that stylistic frequencies are a function of transmission through both time *and* space.

Explaining these seriation results further requires a detailed understanding of the factors involved in producing the frequencies of culture-historical types used to create the seriations. Stylistic similarity is calculated using class frequencies. Therefore, the process of data collection, measurement, and analysis must be controlled in such a way that the methods used to gather data, the units used to make the measurements, the size of the samples, comparability between classes, and other abundance estimation procedures do not produce unestimated variability in the class frequencies. Each step of the measurement process has the potential to increase the class frequency error terms. If these terms become too large, or worse, are unknown, use of these frequencies to describe transmission processes is unwarranted.

In the case of the PFG study, the authors were concerned with minimizing the contribution of unwanted (*i.e.*, non-temporal) variability to abundance estimates. However, they lacked the analytic tools to insure that their efforts produced comparable and representative samples. For example, PFG were well aware that sample size affected class frequencies but were only able to claim that a sample of fifty sherds was adequate (PFG 1951:223). From the discussion presented here and in Chapter 3, it is clear that this value is characteristically too low, though the number depends primarily upon the evenness of the population sampled. Thus, many samples in the PFG data set are too small to be useful. PFG were also careful to separate surface collections into areas that appeared compositionally distinct. They did not provide any warrant, however, for making these divisions nor do we have any way of testing their intuition. In addition, PFG were aware that collection methods could produce unwanted variability in sherd frequencies. Consequently, they instructed collectors to gather every sherd they encountered and routinely paid these workers by sack to encourage them to be systematic in their collection (PFG 1951:43). Unfortunately, PFG had no means of evaluating the degree to which these careful steps produced comparable results. Finally, the process of counting, type assignment and tabulation is unrecorded in PFG and accompanying documentation; no error terms were generated for type assignments of the samples.

Since the PFG data were collected without controlling many potential sources of variation, and since the original collections have lost their integrity (Phillips 1970) — precluding evaluation — it is not possible to rule out factors extraneous to community interaction that may be responsible for the variation. Without additional data, it is not possible to distinguish interaction patterns from variability caused by how the sample was collected, the types defined, and type frequencies calculated. If, for example, assemblages differ in

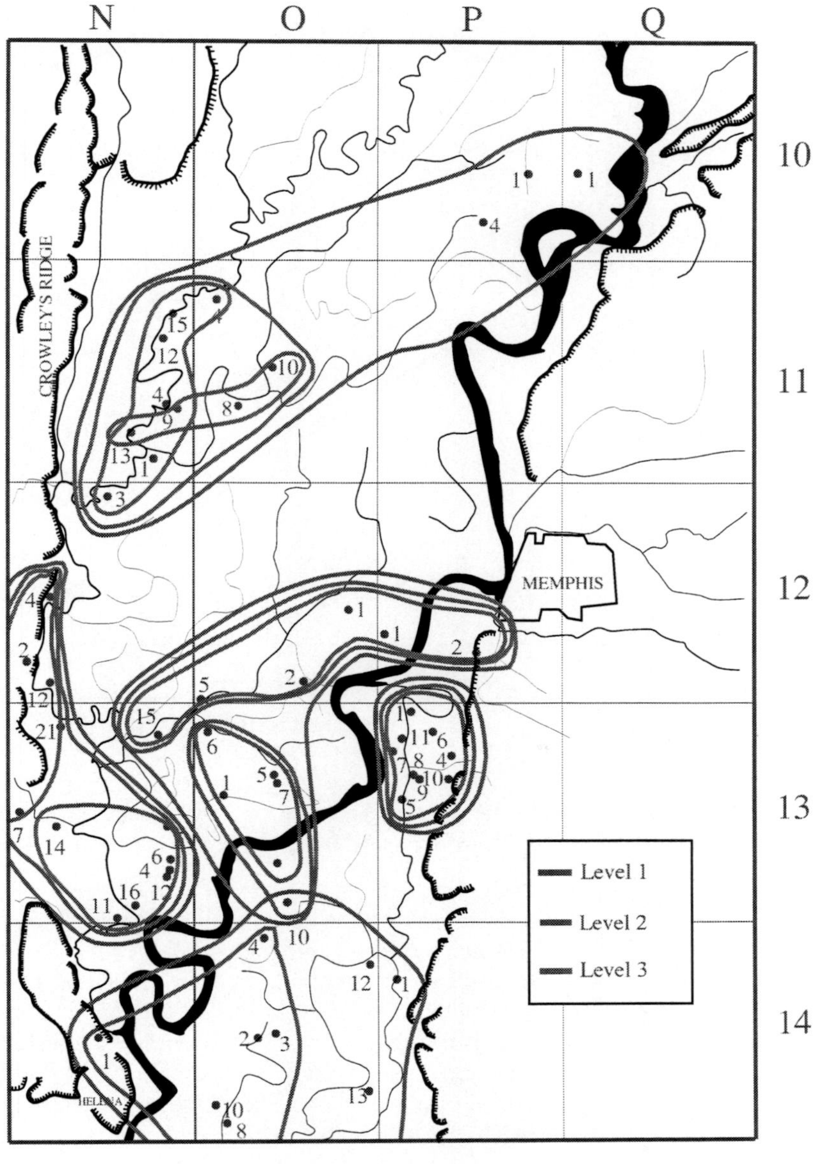

Figure 4.13. Geographic Distribution of Maximal Seriation Solution Clusters of PFG Assemblages Using Three Classification Levels.

Levels are specified in Figure 4.12. The effects of increasing the level of classification for the PFG assemblages in the direction of coarser classes (fewer dimensions) is to simply create larger groups of assemblages, and in some cases allowing coalescence of formerly separate assemblages into more inclusive groups. The groups formed retain a great deal of spatial contiguity. This result suggests that the seriations performed here may reflect structure in the interaction history of the region and thus may map the approximate location of lineages. More detailed classifications and larger sample sizes, however, are needed to test this conclusion.

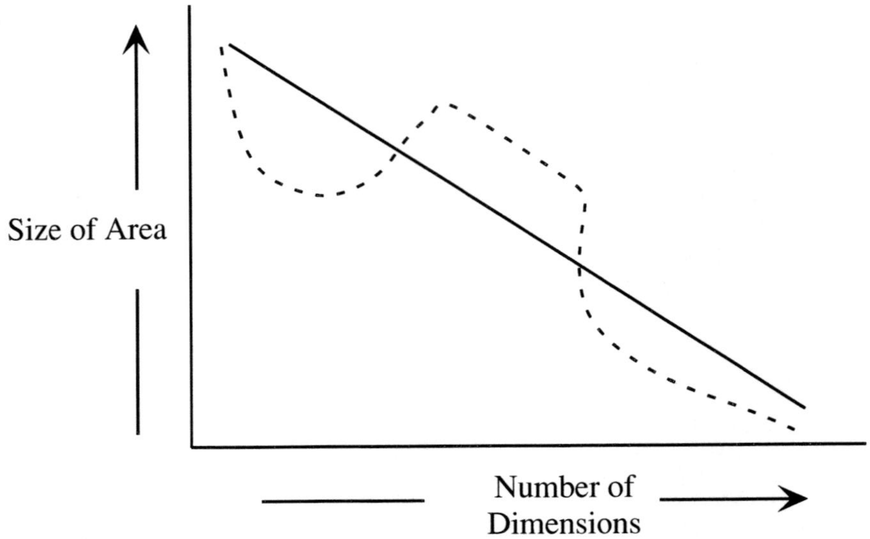

Figure 4.14. The Relationship between the Size of Area Covered by Groups of Assemblages and the Number of Dimensions Used to Measure Similarity between Assemblages in Cases of Continuous and Discontinuous Interaction.

The x-axis consists of the number of dimensions used to construct measurement classes. On the y-axis, area is measured by the total distance between similar assemblages measured with classes of a particular level. If transmission is continuous (solid line), there should be an inverse linear relationship between the number of dimensions used in the classification. The fewer the dimensions, the larger groups that will be formed in the seriation analyses. In discontinuous transmission (dashed line), the size of the groups should not be linear. Rather, at particular classification levels, the size of the groups will change in a non-linear fashion. There will be, therefore, inflections in the group sizes created with different classification levels.

the overall frequency of decorated ceramics (a factor almost certainly under the control of selection), or if differences in settlement patterns or geomorphology create differences in the kinds of sites available in the surface record, some boundaries would be due to these factors rather than transmission. Any factor that creates differential representation of ceramic types across several assemblages will potentially structure the distribution of assemblages across a seriation independently of space or time and thus yield either systematic or random errors in mapped representations of communities. Resolving these issues requires additional fieldwork in the study area.

To conclude that the groups of assemblages formed in this analysis reflect interaction history and thus community structure, it is also necessary to consider less easily controlled sources of variation. For example, Phillips and colleagues (1951) attempted to gain "uniform" coverage of the study area (PFG 1951:43). However, there is not enough information to determine if spatial gaps reflect the record, if particular areas of the survey area were collected more intensely than other areas, or if there were differences in the way in which each location was collected. From the PFG data alone, it cannot be determined whether the seriation groups I detected are a reflection of past interaction history or the consequence of simply "missing" assemblages intermediate in space (and in class frequencies) between those analyzed here. An analysis of the geomorphology of the region could provide valuable information about factors that may have affected the distribution of settlements.

Significantly, even if there are serious problems with the comparability of PFG assemblages and frequencies (and there almost certainly are), the fact that the groups outlined here closely match culture historical phases suggests that both culture historical methods and this approach to seriation are measuring the same variability at least indirectly. It is certainly not the case that the PFG data set is unusable, but we need to determine in what ways and to what degree it can be used.

The need to evaluate the PFG data motivated the initiation of a field project to recollect deposits previously visited by PFG in their study. By recollecting these deposits using a procedure explicitly designed to control variation in abundances of types and measuring other variables that can contribute to differences in type frequencies, it is possible to evaluate the PFG results for systematic biases and random errors. These new data permit the evaluation of the hypothesis that the seriation groups represent lineages measured indirectly via the frequency of interaction. The project design and results of this work, therefore, are reported in the next chapters.

CHAPTER 5: NEW ASSEMBLAGES

Figure 5.1 schematically depicts sources of variability that must be controlled to explain stylistic type frequencies with models of cultural transmission. At the top of the figure is problem under investigation, defined within a theoretical framework of cultural transmission. The construction of the problem must take into consideration dynamic and empirical sufficiency as well as tolerance limits for specific hypotheses to be evaluated (Lewontin 1974). Based on the problem definition, the study universe can then be chosen and a research design formulated for generating data to falsify the questions posed by the problem definition. In the case of evolutionary theory, this means specifying the set of attributes needed for measuring variability between individuals, the degree to which this variation confers differing probabilities of replication and the ability of the variability to be transmitted with better than chance fidelity through successive replications. In addition, as discussed in Chapter 3 populations of individuals must be defined since variants of attributes are measured within groups rather than on observations of single individuals. Finally, a strategy for generating reliable and statistically significant empirical measurements of individuals must be determined. This strategy includes artifact definitions, minimum size requirements, and the collection techniques involved in generating archaeological assemblages.

The next step is to establish the classification, sorting, and counting criteria for describing assemblages with stylistic attributes and building frequencies of stylistic types. Measurement error, inherent in all observation, must be estimated to avoid attributing explanatory significance to variability that arose solely in consequence of the observation process. At the bottom of the figure are the target variables, those involving cultural transmission. Deposits vary in composition due to the rate of deposition (as indicated by the variable *dt*) and the length of the deposition process (*r*).

In addition, archaeological collection techniques, sampling methods, classification, and observation practices used to assign instances of classes drastically affect the frequencies that are identified. In particular, the quality of sampling and the overall sample size determine one's ability to make distinctions between possible orderings.

Finally, once the effects of taphonomic factors and archaeological collection strategies are controlled, it is possible to test hypotheses concerning the contribution that transmission through space and time makes to the composition of a particular set of assemblages. PFG's collections are problematic in this regard, as the data necessary to evaluate these "extraneous" sources of variation are lacking or incomplete. New collections are required, therefore, to take into account the full range of factors that can influence measures of stylistic frequencies. These new data permit me to evaluate the potential biases in PFG data, and determine the degree to which the original PFG data may be a reflection of sampling practices, or a statistically defensible portrayal of the stylistic structure of the archaeological record.

DATA SELECTION

While the approach taken here makes no particular demands in terms of data distribution, the context of these procedures imposes some restrictions. Since the interests in this research project are spatial rather than chronological, and since the methods employed are well established as chronological tools, I decided to seek a set of assemblages from a temporally restricted period that spans at least two contemporary "phases." These data support an examination of the key issues with a minimum of new data. I identified four areas large enough to contain the record of several phases and that have intact surfaces old enough to contain the full archaeological record for the period of interest. These areas are located in the St. Francis Basin area (St. Francis and Crittenden Counties, Arkansas), the southern end of Crowley's Ridge (Lee and St. Francis Counties, Arkansas), Nodena (Mississippi County, Arkansas) and the Sunflower River Basin (Bolivar and Washington Counties, Mississippi).

After extensive records research and several field visits during the summer of 1995 and the spring and summer of 1996, I chose the St. Francis basin area west of Memphis as the focus of my research (Figure 5.2). This area consist of a tangled maze of watercourses stretching from the Mississippi river west to the St. Francis river and from the Tyronza River to the north to the Mississippi to the south. This study area is centered squarely between Ford's St. Francis and Memphis local areas. At least fifteen, large, late prehistoric deposits that were part of the PFG study are located in this region (Figure 5.3) as well as Holden Lake (3CT240), a deposit that had not been previously investigated. These prehistoric settlements are assigned to four Late Mississippian "phase" areas – the Cranor (PFG's "Cramor Place"), Parkin, Walls, and Horseshoe Lake phases as defined by Phillips (1970) and Smith (1990) – and are included in several of the seriation groups defined in the analyses presented in Chapter 4. Descriptions of these deposits are listed in Table 5.1 and their locations are shown in Figure 5.3.

THE ST. FRANCIS BASIN

The St. Francis Basin area is a region of overbank deposition not affected subsequently by the erosive action of the Mississippi river. It is located in the Eastern Arkansas lowlands, an area of the Mississippi Embayment of the Coastal Plain physiographic province (Fenneman 1938:83). The Mississippi Embayment is a large, low, south-sloping trough bounded by the Appalachian uplift on the east and the Ozark and Ouachita highlands on the west. Crowley's

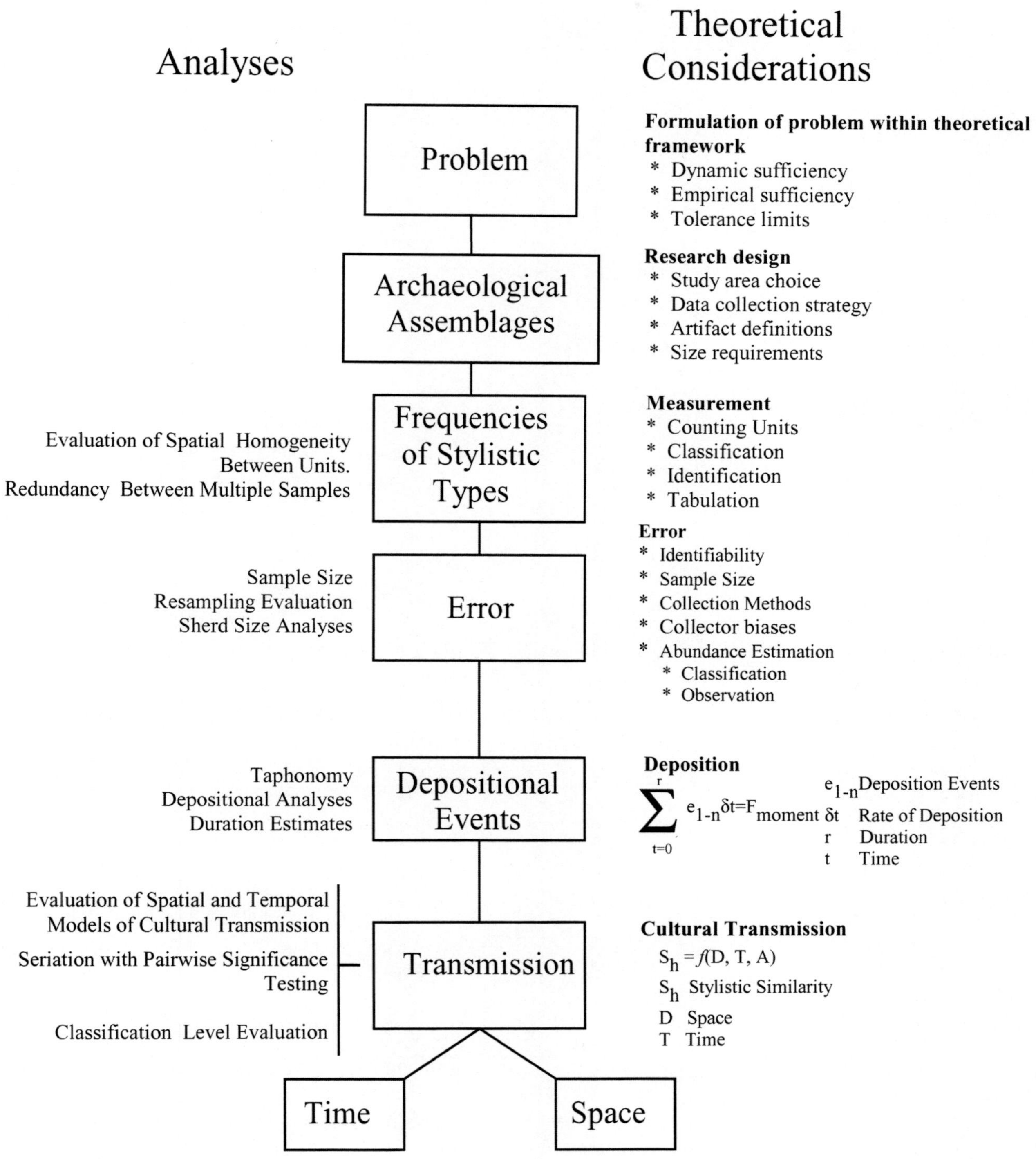

Figure 5.1. Sources of Variability to be Considered When Linking The Frequencies of Stylistic Types Models of Cultural Transmission.

Table 5.1. PFG Late-Mississippian Sites Known in the St. Francis Basin Area.[a]

Name	Phase Phillips (1970)	Phase Smith (1990)	Seriation Group	Description
Barrett (13-O-1)	Kent	Cramor	4	Large village site with large rectangular platform mound and small mounds
Beck (13-O-7)	Kent	Horseshoe Lake	4	St. Francis-type site with mounds in plaza arrangement
Belle Meade (13-O-5)	Kent	Horseshoe Lake	--	Large St. Francis-type site with large platform mounds and small mounds in plaza arrangement. Deep midden with burials.
Big Eddy (12-N-4)	Parkin	Kent	4	St. Francis-type site. Deep midden with burials.
Brackenseik [B] (12-O-6)	Walls	Horseshoe Lake	--	Large village site with large rectangular mound
Castile Landing (13-N-21)	Parkin	Kent	--	St. Francis-type-site. Deep midden with burials.
Cranor [C] (12-O-5)	Kent	Cramor	--	St. Francis-type site. Moderately abundant material.
Lakeside (13-N-18)	Kent		--	Village site with large and small mounds.
Manly (12-N-2)	Parkin		--	St. Francis type-site.
Mound Place (12-P-1)	Walls	Horseshoe Lake	--	Village site with large platform mounds
New Bethel (13-O-6)	Walls	Cramor	--	Village site with large platform mounds
Nickel (13-N-15)	Kent	Cramor	--	Large village site (12 acres) with rectangular platform mound and small mounds.
Nickey (13-O-4)	Kent	Cramor	--	Village site with large mound.
Pouncey (12-O-2)	Walls	Horseshoe Lake	--	Large village site with large platform mounds and small mounds in plaza arrangement
Rose Mound (12-N-3)	Parkin		2	St. Francis-type site. Abundant material and daub.
Turner Place (12-O-1)	Walls		--	Small mound and village site

[a] Table includes phase area designations and descriptions as given by Phillips (1970) and Smith (1990). Locations are shown in Figure 5.3.

[b] PFG report this location as "Brackenseed Place". However, local residents report that the location was owned by the Brackenseik family. I describe the location hereafter as Brackenseik.

[c] PFG report this location as "Cramor Place." Local residents report that the location was, in fact, inhabited by the Cranor family at the time of the PFG survey. Consequently, I describe the location as "Cranor" in this research.

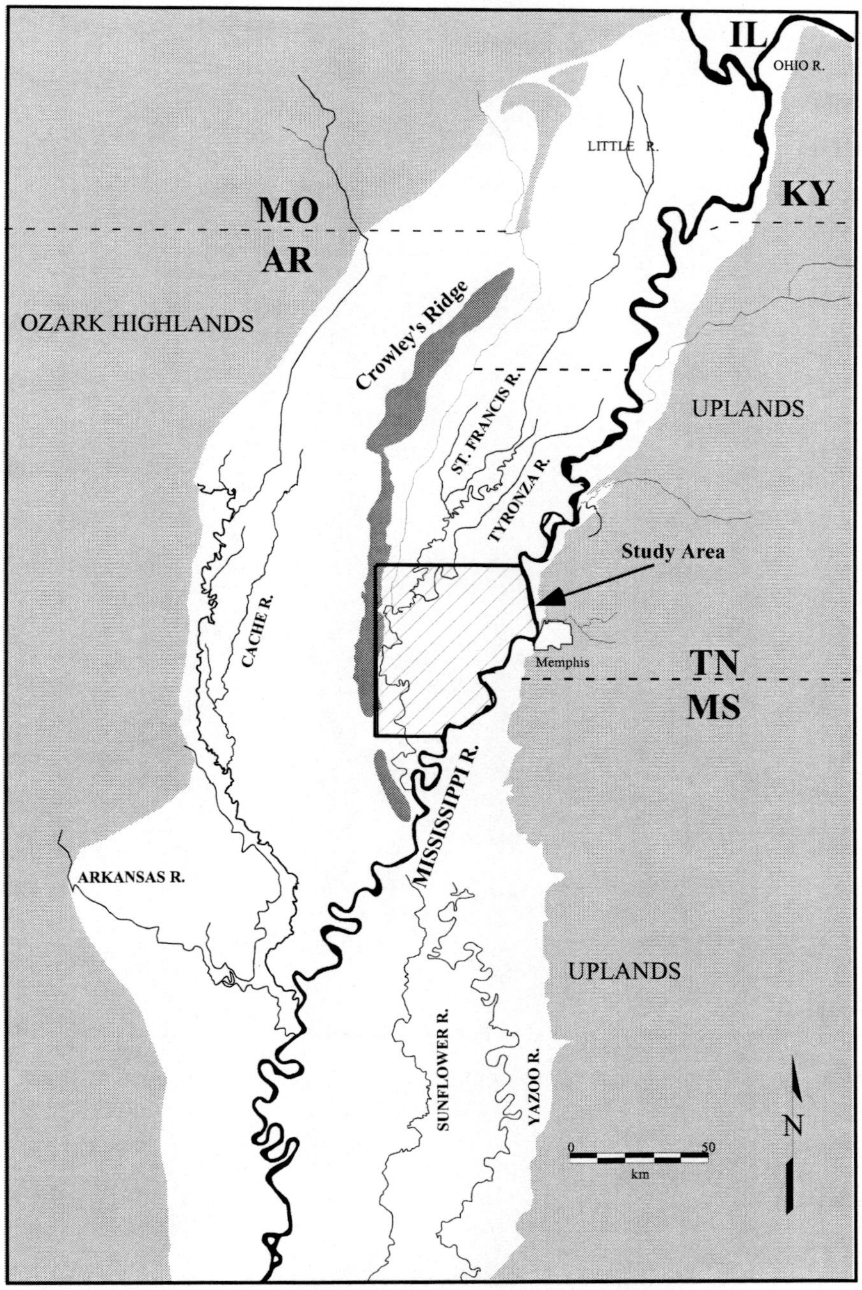

Figure 5.2. The Geographic Location of the Study Area.

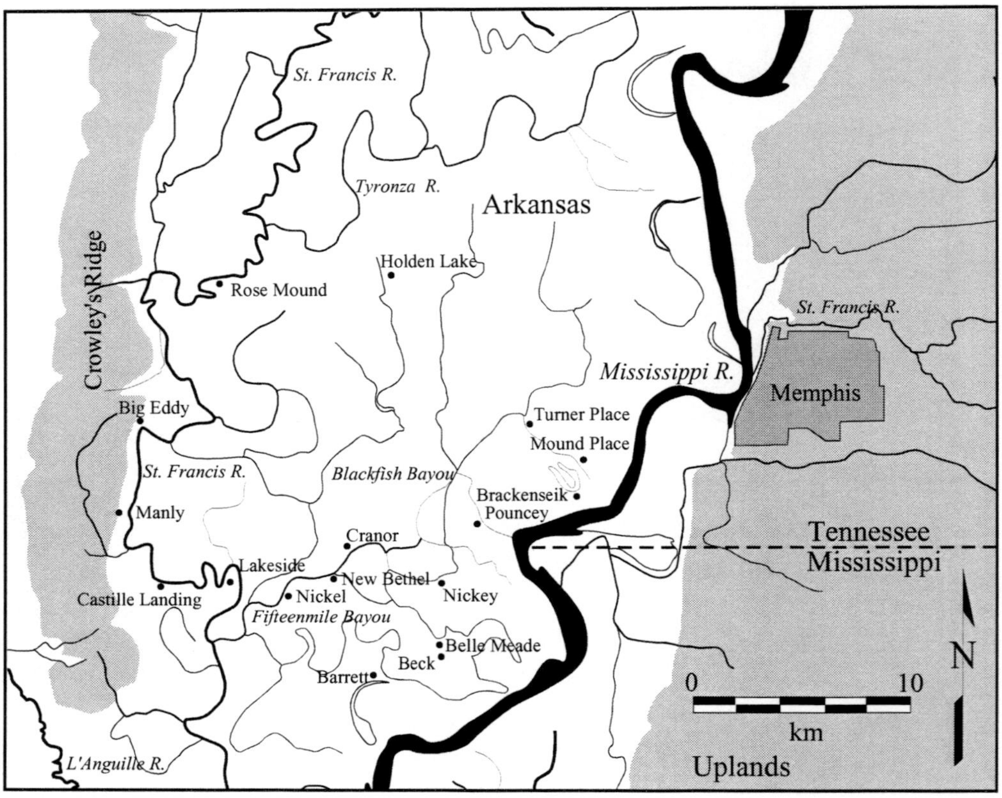

Figure 5.3. The Lower St. Francis Basin Study Area.

Ridge, topographically capped by terminal Pleistocene deposits, is a dissected remnant of the original plain through which the Mississippi river previously flowed (Autin *et al.* 1991; Fisk 1944; Saucier 1974). At the end of the Pleistocene, the Mississippi and Ohio rivers flowed down the west and east sides of Crowley's ridge, respectively. By 9,900 B.P., the Mississippi river shifted its course through the Thebes Gap and captured the Ohio River at Cairo, Illinois (Royall 1988:13-14). Once its course shifted, the Mississippi entered its current meander belt, and in the remaining part of the Holocene has been depositing alluvium in backswamp areas along its course (Saucier 1974). Consequently, aggradation is the most important factor in the present topography of the Mississippi Embayment.

A number of flood basins are found in the Mississippi Embayment, including the Yazoo basin to the east and the St. Francis and Tensas basins to the west and south (Fenneman 1938:91-93). Crowley's Ridge divides the St. Francis basin into two sections known as the Eastern and Western Lowlands (Morse and Morse 1983). These flood basins are formed because aggrading watercourses like the Mississippi have levees that slope away from the stream. Subordinate streams "flow down the slope in courses almost parallel to the Mississippi, but [are] tortuous and often interlacing, the gradient being too small to support a system of dendritic drainage" (Fenneman 1938:91). These basins, therefore, are composed of a vast network of slow-moving streams, watercourses, bayous, and swamps. Even the most significant watercourse of the basin, the St. Francis river, only enters the aggrading banks of the Mississippi river where the Mississippi river and Crowley's Ridge come together at Helena and the levee is interrupted. Prior to the construction of drainage and levee systems, the St. Francis basin was subject to seasonal overflow by the Mississippi river creating a very wet, yet rich, biotic environment. Historic floods in this area, such as one in 1927, inundated nearly the entire area with as much as forty feet of water (Daniel 1977).

The study locations were prehistorically situated in the lower portion of the St. Francis Basin on the southern edge of a vast backswamp (Fisk 1944). The area was dotted by flood-plain lakes and drained by the St. Francis River and large bayous such as Fifteen Mile Bayou, Blackfish Bayou and Carrouthers Bayou as well as many smaller watercourses. The network of watercourses in the Basin is comprised of underfit streams that were formed by overlapping sets of old channels of the Mississippi, St. Francis, and other local tributaries once they were abandoned. Prior to drainage, these bayous carried Mississippi river floodwaters in and out of the area.

Although created by the meandering Mississippi, the river has not been active in this area recently enough to impact features of interests (Autin *et al.*1991; Saucier 1974). Thus, aggradation is the dominant geologic force shaping the depositional environment and is restricted to low-lying areas within the region. It is likely that entire prehistoric communities are represented on the surface. Assemblages location will not inadvertently map surface age. During my preliminary visits, I was also able to ascertain that many of the sites in the study area are readily accessible.

ASSEMBLAGE SELECTION

The first concern for building a research plan is the choice of the assemblages from the original PFG data set. The recollection of all the PFG locations would obviously provide the greatest amount of information regarding the relationship between the PFG samples and the character of the archaeological record. A carefully selected subset of the PFG assemblages, however, will provide information about how the results of the seriation analyses are impacted by the process of collection if these assemblages are distributed across each of the spatial groupings as indicated by the seriation analyses. If significantly new groups result from new collections, it will suggest that the groupings obtained from the PFG seriation analyses are an accident of the distribution of PFG's samples. Thus, the study requires that assemblages be distributed across several phases.

Ceramic assemblages must meet a set of criteria to ensure that frequencies of stylistic types are not caused by factors unrelated to transmission probabilities. First, assemblages must be derived from functionally equivalent deposits. Choosing sites of similar function will reduce functional differences in assemblage composition. This criterion is met in the present case, in that only one kind of late Mississippian settlement is known, the so-called "St. Francis-type" settlement (Morse 1981; PFG 1951:329-330). This precludes any confounding effects of spatial partitioning by different activities.

ST. FRANCIS-TYPE SITES

The site plans for seven of the sixteen sites in the study area were identified by PFG as "St. Francis-type": Beck, Belle Meade, Big Eddy, Castile Landing, Cramor Place, Manly, and Rose Mound. St. Francis-type sites are defined by PFG as large, rectangular deposits of debris, often several meters deep and 100s of meters in horizontal dimensions. They appear to have been accumulated from a single, fortified village. Figure 5.4 shows a St. Francis-type mound (Rose Mound, 12-N-3). These accumulations are so massive that it is not uncommon for present-day farmers to plow their tops as separate field areas and for this reason plowing seems to have little impact on the integrity of the deposit. Often a wide ditch surrounds these massive mounds. The presence of a ditch and the concentration of the deposit in a rectangular shape have led most researchers to conclude that a stockade and ditch fortified these towns (PFG 1951:329). The deposits are middens: accumulations of sediment, faunal remains, ash, charcoal, burnt-clay daub, and "potsherds by the hundreds of thousands" (PFG 1951:329). It is thought that these deposits grew partly through gradual accumulation over time but also through managed redeposition. Debris was added to the mound not only through the construction and through the demise of structures, but also through the

Figure 5.4. A Classic Example of a St. Francis-Type Deposit, Rose Mound (12-N-3).

The photos were taken by PFG in the 1940s and show the Rose Mound as an enormous, low platform that is a distinct feature of the landscape. Rather than simply an earthen mound, however, St. Francis-type deposits are composed of household and village debris. From Philip Phillips, James A. Ford, and James B. Griffin, *Archaeological Survey in the Lower Mississippi Alluvial Valley, 1940-1947*, Papers of the Peabody Museum of American Archaeology and Ethnology, vol. 25, 1951, Figure 74c. Reprinted courtesy of the Peabody Museum of Archaeology and Ethnology, Harvard University.

addition of soil from the surrounding ditches and the local area. Since all of the St. Francis-type sites are located within the backswamps of the Mississippi river valley, the function of this form is undoubtedly related to keeping the population out of often massive floods. In general, St. Francis-type sites are slightly more elevated along the edges and lower in the center where it is thought plazas were situated. Some St. Francis-type sites have mounds situated on top of the platform, although this is not always the case.

The massive scale and richness of artifacts of the St. Francis-type sites have attracted attention from professional investigators and pothunters since the earliest days of archaeology in North America (*e.g.,* Brackenridge 1818; Holmes 1886a, 1886b; Moore 1910, 1911; Thomas 1887, 1894). Literally, hundreds of thousands of pottery vessels have been removed from the mounds by archaeologists and collectors. The richness of St. Francis-type sites made them of particular interest to this research project. First, the dense midden deposits make it likely that modest surface collection effort would produce large numbers of ceramics. Second, the nature of the deposits as constantly shuffled, redeposited, and managed by their inhabitants suggests that their composition will be relatively homogenous. Collections, therefore, will be more likely to generate representative samples.

FURTHER REQUIREMENTS

In addition to being St. Francis-type sites, the locations of assemblages must be distributed across the study region so that a high degree of similarity due to tight spatial clustering does not overwhelm all other factors. Furthermore, the assemblages must be relatively contemporaneous to minimize the effects of time on stylistic similarity. Unfortunately, few absolute dates are available for any Mississippian deposits in the central Mississippi river valley. It is, however, relatively easy to identify late prehistoric ceramics, so I began my analysis with information gathered through traditional means and chose sites that have stylistic markers of the latest prehistoric period. I also chose assemblages that have a similar ratio of decorated to undecorated ceramics, since large differences in the frequency of decorated ceramics, not accounted for by differences in sample size, might indicate differences in contemporaneity or duration. This strategy gave me a reasonable starting point.

In general, late Mississippian burials are not found in cemeteries but rather throughout the midden, particularly along the edges where the greatest concentration of house debris is found. This is especially important since investigators from Holmes onward (1886a, 1886b) have suggested that "burial pottery" may be different from other pottery. Segregation of burials into spatially distinct cemeteries, therefore, could produce assemblages with distinctive pottery compositions. Ceramics included in burials display use-wear but are biased towards smaller sizes and thus are integral parts of the overall assemblages from which they are drawn (Childress 1990; O'Brien 1994). Since the size of a ceramic vessel determines the range of decorations that are applied, frequencies of decorative types in assemblages will be driven, in part, by the presence or absence of burial deposits with accompanying ceramics. Problems in assemblage comparability can arise, therefore, if burial assemblages are not uniformly included because burials are absent or the graves are too deep to have a surface signature. Since St. Francis-type deposits characteristically contain burials and burial pottery, the choice of this kind of deposit contributes to meeting the criterion that all of the locations be functionally equivalent. In addition, comparability between deposits was ascertained by examining each site for the presence of human skeletal remains.

SITE SELECTION

It would have been ideal to study all of the St. Francis-type sites from the St. Francis Basin area that were described by PFG. A number of practical factors, however, prohibit choosing sites in this fashion. First, since the time of the PFG survey many of the sites have been radically altered by construction, agriculture, and looting. In several cases, (*e.g.,* Brackenseik, 12-O-6, Castile Landing, 13-N-21) sites have been nearly obliterated by levee construction. In others, (*e.g.,* Big Eddy, 12-N-4), commercial pot hunting has destroyed most of the deposits (Morse 1981:50). Second, for the sites to be considered for examination, they had to be readily accessible from the surface to allow the generation of large representative collections. Third, with the exception of the Parkin site (11-N-1), all of the deposits collected by PFG are located on private land. Collections require the permission and cooperation of landowners. Finally, the surfaces of the deposits to be collected require field conditions that allowed large ceramic assemblages to be made relatively efficiently. Thus, the ideal deposits were those currently undergoing tillage but had been little modified otherwise (Dunnell and Simek 1995). Additionally, the condition of the surface will also heavily impact the ability to make large collections. Good field conditions are found when the crops are very low and several inches of rainfall have fallen. These kinds of field conditions generally occur in the spring prior to the substantial summer crop growth.

I made two preliminary visits to the central Mississippi river valley in the summer of 1995 and spring of 1996 in order to gather information on potential locations for doing recollection. During these trips, I traveled across the St. Francis Basin area relocating PFG sites, meeting landowners, local residents, and local archaeologists. Whenever possible, I made surface collections in order to evaluate representative ceramics from each place I visited. After these trips, I settled on making extensive collections at six PFG locations in the St. Francis Basin study area: Belle Meade (13-O-5), Beck (13-O-7), Castile Landing (13-N-21), Cranor (12-O-5), Nickel (13-N-15), and Rose Mound (12-N-3). Existing records and my field studies suggested that all of these sites are roughly contemporaneous, equidistant, large, and functionally similar. In addition, each site has an abundance of decorated ceramics available for collection on the surface. Human bones from disturbed burials were also present.

Subsequently, I received landowner permission for survey work at each site. Based on information given to me by local collectors, I also located a large Mississippian deposit, at the Holden Lake site (3CT240) not previously described by PFG. The location of the Holden Lake site in the northeastern corner of the study area does not fall geographically into one or the other defined spatial groupings or a traditional phase area. Thus, Holden Lake collections will provide new information about the archaeological record in an area in which the PFG survey has no data.

COLLECTION CONSIDERATIONS

The regional-scale hypothesis being tested is that the phases represent arbitrary divisions of roughly continuous spatial variation. Alternatively, it is possible that these groups mark joints in the spatial structure of interaction. On one hand, interaction may have been continuous across this area and the differences in similarity detected by archaeologists using PFG data are simply the product of sampling and collection procedures. On the other hand, it is possible that transmission was structured in late prehistoric populations of St. Francis Basin. Here, one would expect that assemblage similarity would not simply correlate with distance once differences in time are considered. Consequently, my data were carefully gathered in order to control as many of the measurement procedures as possible. Collection procedures were designed to minimize the effects of variability caused by the data collection process and to provide a suitable standard with which to evaluate the quality of the PFG data.

SAMPLE SIZE

My basic fieldwork consisted of making large surface collections of ceramics that can be used to measure homologous similarity between assemblages. Similarity is a function of the richness, or the number of different categories, the evenness, or proportions of the categories and, very importantly, the number of things present (Jones *et al.* 1983; Kintigh 1984; Leonard and Jones 1989; Plog and Hegmon 1993; Rhode 1988; Shott 1989). As seen in Chapter 3 (*e.g.,* Figure 3.7), the minimum sample size needed to represent evenness and richness of classes is not the same for all samples. Despite Ford's recognition of the effects of sample size (*e.g.,* PFG 1951:223), 50 sherds may or may not be enough in any particular setting. Appropriate target assemblage size depends upon two factors: (1) the frequency of decorated ceramics in an assemblage and (2) the statistical resolution needed in distinguishing frequencies. My analysis of the PFG data suggests that the frequency of decorated sherds (as opposed to total sherds) in late Mississippian assemblages ranges between 14% and 23% (Table 5.2). Since transmission is measured by frequency of stylistic types, the finer the differences between frequencies that can be statistically distinguished, the more resolution that can be obtained in mapping transmission.

Fine distinctions, however, have a steep price in terms of sample size. We can calculate the minimum sample size needed to distinguish the percentages of types in two populations at a particular significance level (α) using the equation:

Equation 5.1.

$$z = \frac{(p_1 - p_2)}{\sqrt{p_1(1-p_1)/n_1 + p_2(1-p_2)/n_2}}$$

where $\hat{p}_1$ = the proportion of a class in sample one, $\hat{p}_2$ = the proportion of the same class in sample two, z is the z-score at the specified significance level (α) and n_1 and n_2 are the sizes of each sample. To simplify the example, if we assume the collected samples are of equal size (*i.e.,* $n = n_1 = n_2$) we can solve for n, the sample size needed for any particular significance value (α) and at any particular difference in percentages between the two samples. This solution is shown in Equation 5.2:

Equation 5.2.

$$n = \frac{(p_1 - p_1^2 + p_2 - p_2^2)z^2}{p_1^2 - 2p_1p_2 + p_2^2}$$

Where p_1 and p_2 are the proportions of a class in each sample (samples 1 and 2) and z is the significance value (α). Table 5.3 solves this equation for a variety of p_1 and p_2 values. The table demonstrates that large sample sizes are required for distinguishing differences in the percentage of classes in two populations with relatively fine distinctions (*e.g.,* 1.0%).[1] For example, in order to make 1.0% distinctions in the proportion of a type in two samples (*e.g.,* determining whether a sample with 6% of one type is *statistically* different from another sample with 5%) one must have a sample of between 2,566 and 18,547 total sherds *in each sample.*

My goal was to generate data that permit precision in detecting differences in frequencies between assemblages. In order to address the issues of error here, the ability to detect at least a 2.0% differences in frequencies is adequate. Based on an average of 14% decorated sherds in Mississippian assemblages in the study area (Table 5.2), it was estimated that a representative sample of at least 20,000 total sherds (between any two assemblages) provides sufficient basis to distinguish class frequencies of 2.0% at a significance of α = 0.005. Thus, my target sample size was at least 10,000 sherds *per* assemblage from surface collection so that 1400 to 2300 decorated sherds would be available from each assemblage for analysis.

---

[1] It is possible to solve Equation 5.1 for cases in which one sample is known (n_1) but the other sample (n_2) is unknown. Equation 5.3 shows this calculation:

Equation 5.3.

$$n_1 = -\frac{n_2(-1 + p_2)p_2z^2}{n_2p_1^2 - 2n_2p_1p_2^2 - p_1z^2 + p_1^2z^2}$$

Table 5.2. Frequency of Decorated Sherds from Late Mississippian Assemblages in the Memphis and St. Francis Areas.

Name	Site Designation	Number of PFG Types	PFG Mississippian Ceramics Sample	Number of Decorated Mississippian Ceramics Collected by PFG	Percentage of Decorated Ceramics
Barrett	13-O-1	6	286	39	13.64%
Beck	13-O-7	8	265	87	32.83%
Belle Meade	13-O-5	10	804	260	32.34%
Brackenseik	12-O-6	7	84	8	9.52%
Castile Landing	13-N-21	11	2037	402	19.73%
Cranor	12-O-5	8	1016	146	14.37%
Lakeside	13-N-18	4	236	7	2.97%
Manly	12-N-2	6	249	41	16.47%
Mound Place	12-P-1	12	3201	191	5.97%
Nickel C,D,E	13-N-15	10	739	165	22.33%
Nickel B,	13-N-15	7	120	21	17.50%
Nickey	13-O-4	5	524	12	2.29%
Pouncey	12-O-2	7	445	66	14.83%
Turner	12-O-1	7	458	29	6.33%

Table 5.3. Sample Sizes Required to Make Distinctions in the Proportion of a Type in Samples of the Same Size.[ab]

Proportion of Type in Sample 1 (p_1)	Proportion of Type in Sample 2 (p_2)	($p_1 - p_2$)	Sample Size Required
0.05	0.06	0.01	3949
0.05	0.07	0.02	1058
0.05	0.08	0.03	500
0.25	0.26	0.01	14398
0.25	0.27	0.02	3594
0.25	0.28	0.03	1593
0.55	0.56	0.01	18547
0.55	0.57	0.02	4516
0.55	0.58	0.03	1952
0.75	0.76	0.01	13630
0.75	0.77	0.02	3210
0.75	0.78	0.03	1337
0.95	0.96	0.01	2566
0.95	0.97	0.02	367
0.95	0.98	0.03	39

[a]Calculations made with a 99.95% confidence level ($\alpha = 0.005$).
[b]Table 5.3 solves Equation 5.2 for a variety of p_1 and p_2 values, where p_1 and p_2 differ by 1, 2 and 3 percent.

FIELD METHODS

The goal of the project is to generate assemblages that are representative samples of the deposits from which they are derived. The means by which sherds are collected, therefore, is a critical part of this research. The distribution of decorative types across the deposit and consistency in procedures, individual collectors, and conditions can introduce different ages and/or different functions to the collections. PFG were keenly aware of this possiblitiy in the compositions of the deposits. Although there is no mention of how they made the distinctions in any particular case, PFG routinely divided the collection of sites into smaller "areas" in order to minimize variability. PFG (1951:43) describe that:

> Separate collections were made on different areas of a site wherever there was the slightest indication that they might differ in content. If, at the time of sorting, they were found to be the same, within the normal limits of variability, they were thrown together again to make a larger count for the site.

In order to evaluate the assemblages for variability in content, all collections in this study were made in 10x10-meter blocks. PFG's procedure depended on field "indications" of content differences and a wholly intuitive assessment of similarity. While my units are large enough to allow the collection of large samples across the sites efficiently, they are too gross to describe spatial structure. The null hypothesis, that between block variance is less than or equal to within unit variance, was evaluated during subsequent laboratory analyses (see Chapter 6). During these analyses, only the material from units whose variances were statistically the same was combined to form a single assemblage. When the null hypothesis was rejected, the collections were partitioned into smaller homogenous sub-units.

To create a set of collection units at a particular location, I placed a baseline consisting of a single 100-m tape along one edge of the area to be collected. I recorded information about the location and orientation of this baseline in a field notebook and on field sketches. Using the first tape as a guide, I then placed two parallel 100-m tapes perpendicular to the first tape and ten meters apart in the fields along rows. As the purpose of collecting in 10x10-m units was not to generate absolute information about spatial composition, but rather to collect relative information about how the composition of the deposit changes across space, it was not necessary to record absolute information about the location of the 10x10-m units. However, the locations of the baseline and row were maintained through the use of stakes for multi-day collections. In order to establish parallel lines, I used the crop furrows to initially align and lay out the tapes. After stretching the two 100-meter tapes and placing them in the furrows, I checked the distance between their ends and at 10-meter intervals to insure that the tapes were parallel. Although this procedure does not insure perfectly square units, it does insure units of the same area. Flags were placed at 10-m intervals along the parallel tapes creating 10x10-m areas for collection. As collectors entered each 10x10-m unit, they stretched tapes perpendicular to the two parallel tapes at the 10-meter marks. These tapes bound each side of the collection unit. Once the 10x10-m unit was in place, collectors walked back and forth in the unit, adjacent to one another and following crop rows. Each unit was numbered sequentially and plotted on a sketch of the location in a field notebook.

CONTROLLED COLLECTIONS

In field surveys, ceramics must be collected in a consistent manner and surface visibility calibrated to account for differences caused by variability in collection conditions. All field assistants were instructed to pick up and bag any object larger than 1-cm that resembled a sherd (*i.e.,* has the shape, color and texture of ceramics). Collectors were also instructed attempt to move at the same rate as the other collectors in the unit. In addition, groups of collectors were periodically rotated (after every unit was collected) in order to randomize individual collector bias across units. During the collection process, collectors maintained separate bags for the ceramics they recovered. Bags were labeled with the location, unit number, date, and collector initials. Analysis of the variance of collections made between collectors permit one to determine if individuals biased their collections toward particular kinds of ceramics. Finally, notes about variability in the surface conditions, visibility, and vegetation for each unit were kept in a field notebook as these conditions changed from unit to unit.

FIELD WORK

This research protocol was used to make surface collections at Beck, Belle Meade, Cranor Place, Holden Lake, Nickel, and Rose Mound. The surface collection work was conducted over the spring and summer of 1996 and 1997. During my fieldwork, I was aided by a number of individuals. In the early parts of the summer of 1996, I made collections at the Holden Lake and Nickel sites with John House of the Arkansas Archeological Survey. During the summer of 1996, at the Holden Lake and Castile Landing sites, a number of field school staff and students from the University of Washington Field School in Southeast Missouri worked on the collections. Kris Wilhelmsen and John Dudgeon of the University of Washington spent several weeks with me in the spring of 1997 and helped collect the bulk of my assemblages.

In the next section, I describe the overall results of the fieldwork at each of the locations, including the specific procedures used at each site, the conditions encountered, and the number of ceramics collected.

BELLE MEADE (13-O-5)

Location

Figure 5.5 shows the Belle Meade (Arkansas Archeological Survey [hereafter, AAS] Number: 3CT30) location two kilometers east of Horseshoe Lake in Crittenden County, Arkansas (Hughes, Arkansas 7.5 minute USGS quadrangle, UTM coordinates 3867450 north, 738200 east, zone 15). The site is just north of Highway 38 on a natural levee that runs east to west and is about three meters higher than the adjacent terrain. The area of the archaeological deposit covers approximately nine acres (Dye and Moore 1989:1; Pugh 1990). The current owner of the land on which Belle Meade is situated is Frank Barton of Marion, Arkansas, and the property is farmed by Lan Letner of Hughes, Arkansas. A USGS digital orthophoto quadrangle (Figure 5.6) shows the position of the Belle Meade deposit in the farm structures.

Description and Previous Investigations

The first recorded investigation of Belle Meade came with the visit by Philip Phillips and James Ford on March 15, 1940 and consisted of a surface collection, producing 268 sherds, most of which were from the Middle Mississippian Period. At the time, Phillips and Ford (PFG 1951:49) described Belle Meade as being a "large, St. Francis style site with large platform mounds and small mounds in a plaza arrangement" (Figure 5.7, Figure 5.8).

During the late 1970s and early 1980s, Gerald Smith (Chucalissa Museum, Memphis) with the assistance of Terry Childs, a local enthusiast, conducted excavations at Belle Meade. The purpose of these investigations was to create a detailed map of the site, explore the deposits within the mound, and to help rescue several burials that were washing out along the edges of the mound (Smith 1990). Several test pits were excavated indicated that the depth of the deposit was between 180 and 200 centimeters (Pugh 1990).

Between 1987 and 1991, David Dye of Memphis State University made a series of excavations and surface collections at Belle Meade. The subsurface work uncovered a domestic structure that held the charred remains of persimmon seeds. These seeds were subsequently radiocarbon dated to AD 1288 (Dye and Moore 1989, cited in Pugh 1990 [calibrated at the one-sigma age range to AD 1160 – 1421, Stuiver *et al.* 1998]). Intensive surface collections on the plowed area of the site recovered 1,530 sherds of which 268 were decorated. Table 5.4 presents the results of these surface collections.

Since Dye's excavations, no additional work had been conducted at Belle Meade. Looting for pottery, however, has clearly continued. In fact, much of what is currently known about Belle Meade comes from whole pots removed in backhoe trenches or with probe devices. Unfortunately, this activity appears to be continuing. During the present study, we noted a backhoe trench adjacent to the westernmost barn. When we investigated closer, we found the sides of this trench pincushioned with holes from the metal probes of collectors searching for the telltale feel of whole pots.

New Fieldwork

On May 7-8, 1997, Kris Wilhelmsen, John Dudgeon, and I established a series of eleven 10x10-m collection units that extended east to west (MN) across the area of the deposit that was exposed by plowing (Figure 5.9, 5.10). The southwest corner of Unit 1 was placed 10.2 meters from the easternmost telephone pole on the edge of the plowed field on an azimuth of 140° (MN).

Table 5.4. Frequency of Decorated PFG Types Recovered at Belle Meade From The 1940 PFG Survey and during 1987-91 Memphis State University Collections.

	Dye 1987-1991 (Pugh 1990)		PFG 1951	
PFG Type	Count	Proportion	Count	Proportion
Parkin Punctated	141	0.528	128	0.504
Barton Incised	108	0.404	77	0.303
Old Town Red	7	0.026	8	0.031
Rhodes Incised	6	0.022	2	0.008
Walls Engraved	4	0.015	2	0.008
Hull Engraved	1	0.004	0	0.012
Ranch Incised	0	0.000	3	0.126
Kent Incised	0	0.000	32	0.008
Avenue Polychrome	0	0.000	2	0.504
Σ	267	1.000	254	1.000

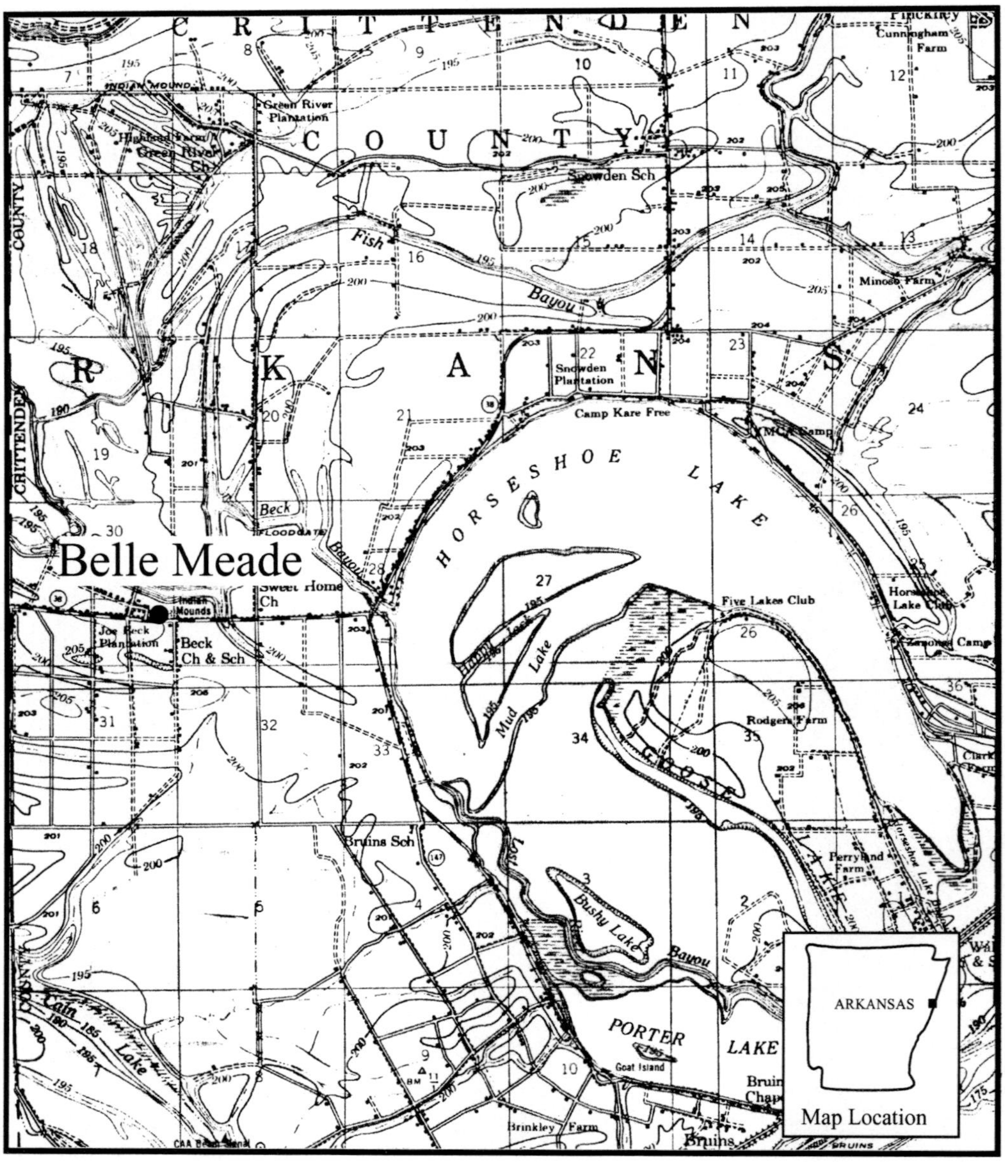

Figure 5.5. Situation of Belle Meade.

Map taken from Horseshoe Lake (1960) 15' USGS Quadrangle.

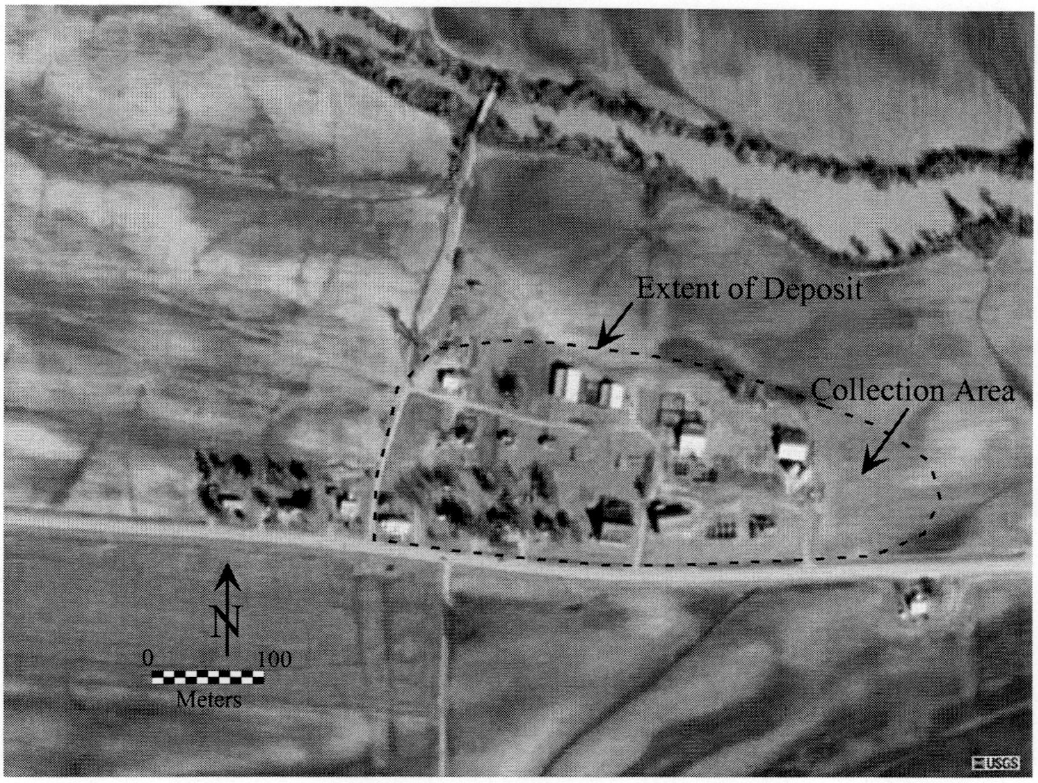

Figure 5.6. Aerial Photograph of the Area around Belle Meade.

This image was extracted from a Color Infra-Red (CIR) digital orthophoto quadrangle provided by the USGS. The image was photographed on February 12, 1996 by the National Aerial Photography Program and digitized by the USGS DOQ Program on July 1, 1997. Each pixel represents 1 by 1 meter.

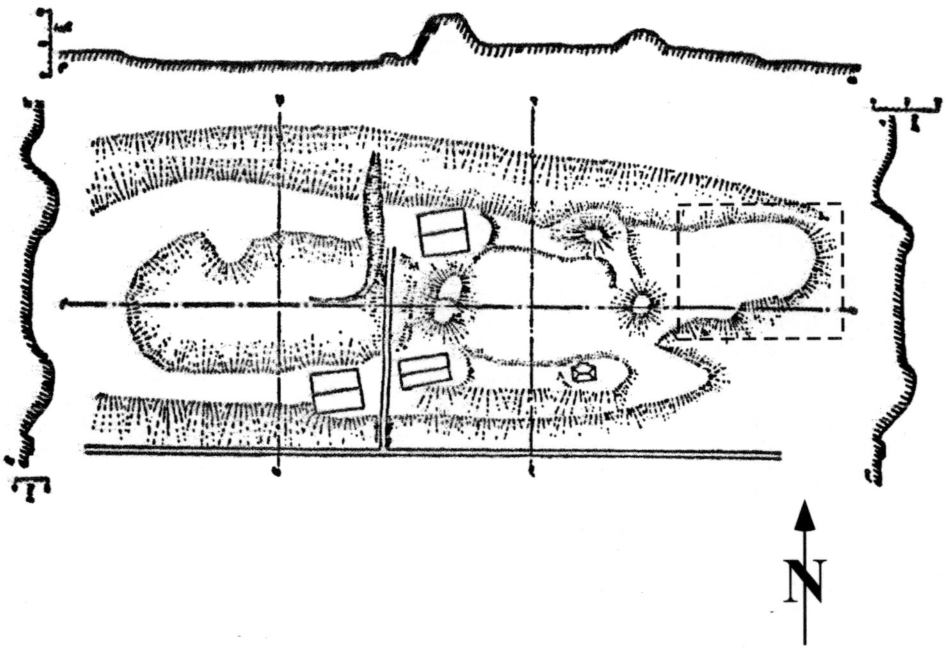

Figure 5.7. PFG Map of Belle Meade.

Dashed lines show area of subsequent collections by Smith and Childs in the late 1970s and early 1980s. This area is presently in cultivation. Figure from Philip Phillips, James A. Ford and James B. Griffin, Archaeological Survey in the Lower Mississippi Alluvial Valley, 1940-1947, Papers of the Peabody Museum of American Archaeology and Ethnology, vol. 25, 1951, p. 331. Reprinted courtesy of the Peabody Museum of Archaeology and Ethnology, Harvard University.

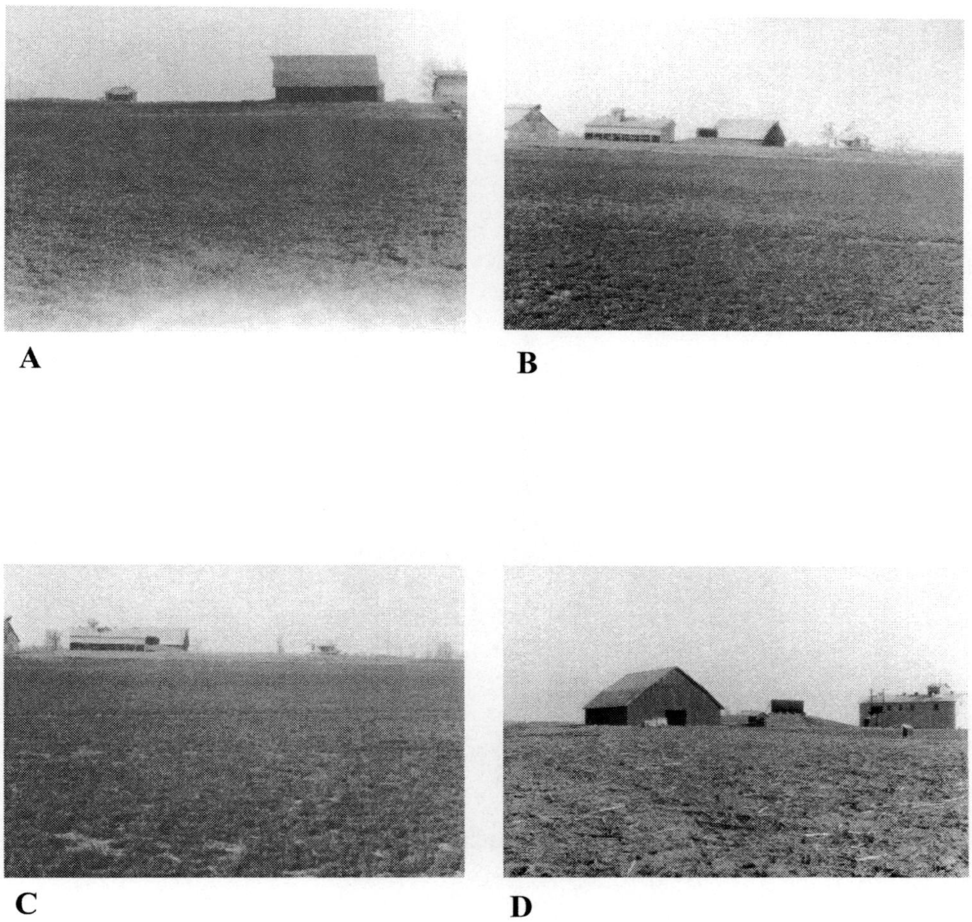

Figure 5.8. Belle Meade at Time of PFG Collections.

Photos are of the site from the north (A), site from the southeast (B), site from the south (C) and site from the west (D). Belle Meade (13-O-5), Box 28, LMS Records 993-23, Peabody Museum Archives, Harvard University. Reprinted courtesy of the Peabody Museum of Archaeology and Ethnology, Harvard University.

Figure 5.9. Belle Meade during 1997 Field Visit.

This photo is of the northeast corner of the site. The raised area to the left of the structures (A) is the exposed and cultivated midden.

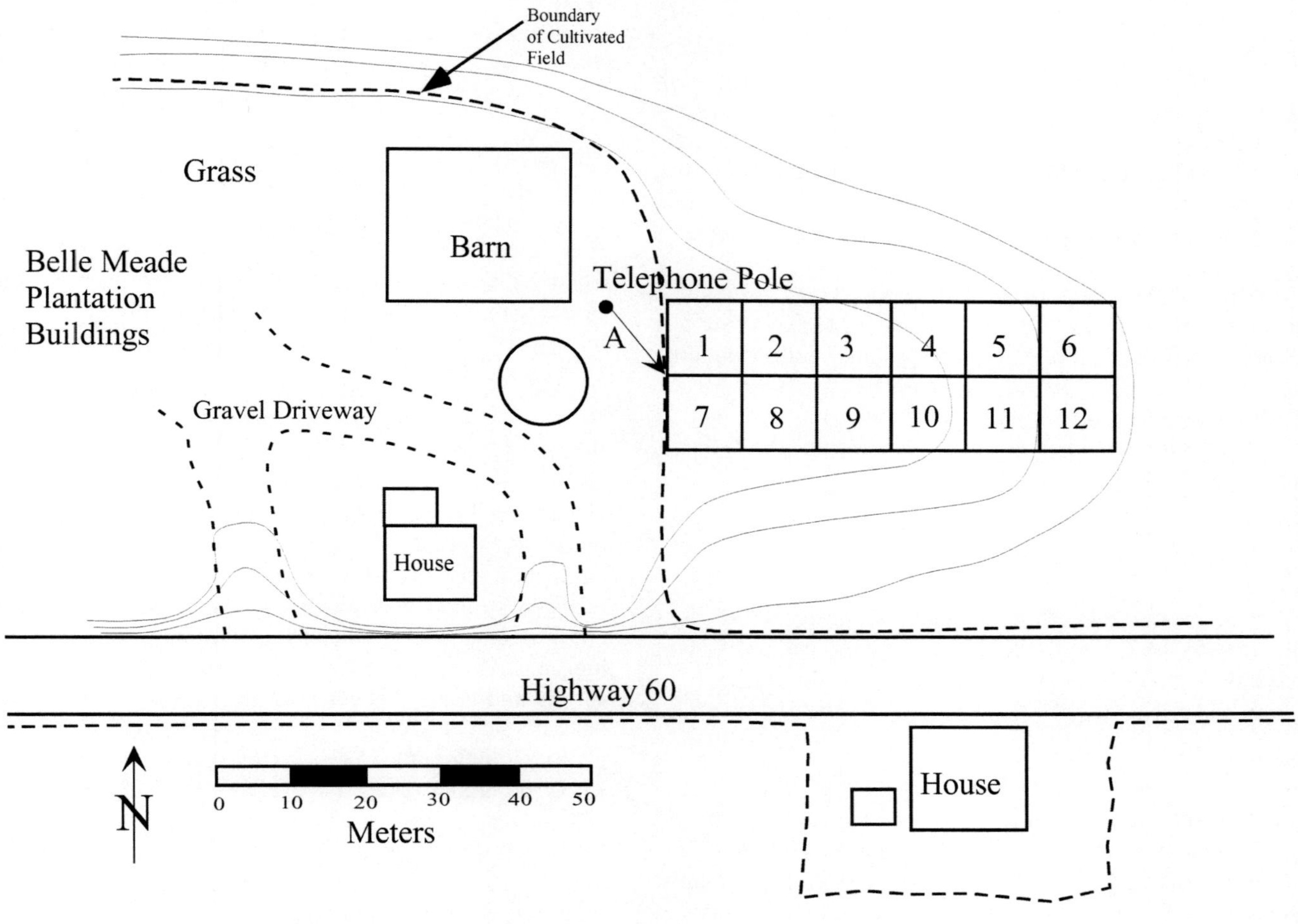

Figure 5.10. 1997 Belle Meade Collection Units

A Southwestern corner of Unit 1 is 140 degrees (MN) and 10 meters from telephone pole. Contour lines depict approximate topography. Contour intervals are not quantitative.

Potential for Current Research

I first visited Belle Meade on June 20, 1995. Since the visit by PFG, however, the mounds have been destroyed or covered by structures (Figure 5.8). Except for a small area on the east edge of the levee, most of the deposit is covered by a ten to twenty centimeters of gravel and is not accessible through surface investigation. Walking along the top of the levee amongst the houses produced several small sherds, but little evidence of the archaeological deposit is visible. Indeed, when asked, one of the inhabitants of the structures was unaware that she was living on top of a very large archaeological deposit. A small area, however, on the north end of the levee is presently under cultivation (see Figure 5.7). The area consists of a low ridge about fifty meters north to south extending about seventy meters from the farm buildings out into a cultivated field. The field slopes towards the west and is regularly planted with cotton and soybeans. According to the current farmer, the field in which the deposit is located is tilled with a chisel plow that creates furrows forty to fifty centimeters in depth. The chisel plow routinely disturbs deep areas of the deposit and makes this location, though small, very productive in terms of surface collection. Thus, the potential for efficiently collecting large numbers of ceramics from this small area of Belle Meade appeared very high. It was clear that even a modest collection effort could produce the large number of sherds necessary for this study.

Conditions

Collection conditions at Belle Meade were excellent. The surface was well washed and had virtually no vegetation. Cultural materials were easily identifiable on the surface.

Results

Although the area of this part of the deposit is relatively small, the density of cultural material on the surface here was extraordinarily high. In the eleven 10x10-m collection units, we gathered 8,157 sherds larger than 1.0-cm. In addition to the ceramics, we also collected numerous adze fragments (broken and whole), Nodena projectile points, decorative elements from vessels (*i.e.,* rim riders), and even a tiny, complete shell-tempered vessel.

BECK (13-O-7)

Location

Beck (AAS Number: 3CT8) is located in Crittenden County, Arkansas less than one kilometer from, and within sight of, Belle Meade (Figure 5.11). It is situated approximately a kilometer south of Highway 38 (Hughes, Arkansas 7.5-minute USGS quadrangle, UTM coordinates: 3866850 north, 738580 east and zone 15). The mound is located in the center of a field that is currently farmed. Like Belle Meade, the property upon which the deposit sits is owned by Frank Barton of Marion, Arkansas and is farmed by Lan Letner of Hughes, Arkansas. A USGS digital orthophoto quadrangle (Figure 5.12) clearly shows the Beck deposit in the center of a cultivated field.

Description and Previous Investigations

The first description of Beck comes from the survey by Phillips and Ford on March 14, 1940. Phillips and Ford describe Beck as a St. Francis-type site that is smaller than Belle Meade, about 180 meters E/W and 90 meters N/S. At the time of the PFG visit, the site consisted of eight mounds and abundant midden debris along the edges of a plaza (*i.e.,* lower) area and over the mounds (Figure 5.13). Phillips and Ford collected 260 sherds during their field visit.

The work at Belle Meade by Childs and Smith in the early 1980s also included some work at Beck (Smith 1990). This work consisted of a surface collection and several deep excavation pits. One of these deep excavations was expanded to expose several post houses that were situated in overlapping shallow pits. The structures were 4.8 to 6 meters on a side with rounded corners. This work has not been published.

Potential for Current Research

I first visited Beck on May 6, 1997. The deposit was located in an agricultural field and has been heavily impacted by cultivation activity (Figure 5.14). None of the mounds reported by PFG were visible and the entire deposit has the appearance of a low, almost square mound, about 1.5 meters high, that is slightly higher along the edges (particularly the eastern and western edges). As the deposit is situated on the edge of Fish Bayou, it slopes steeply to the north but more gently to the south. The surface of the mound, especially along the higher edges, is covered with daub, bones, lithics, and sherds. Surface inspection showed that the highest concentration of cultural materials is distributed along the high edges of the deposit: a pattern characteristic of St. Francis-type sites (PFG 1951:329). Given the high density of the cultural material on the surface and the fact that the entire deposit is under cultivation, it was clear that collections at Beck would produce the large number of ceramics needed in the present analyses.

New Fieldwork

A collection was made at the Beck site by three people who worked for sixteen hours over two days (May 6-7, 1997). We set up two series of units (Figure 5.15). The first series of units is along the eastern edge of the deposit where we noted a particularly high density of daub, bones, and ceramics. The northeast corner of the first collection unit was located on an azimuth of 40°(MN) and 206 meters from the center of a wellhead situated southwest of the midden. The corner point is also located 204° (MN) from the large water tower located east of Hughes and visible on the horizon. We placed parallel tapes along a north-south axis (MN) to establish fifteen 10x10-m collection units. We also

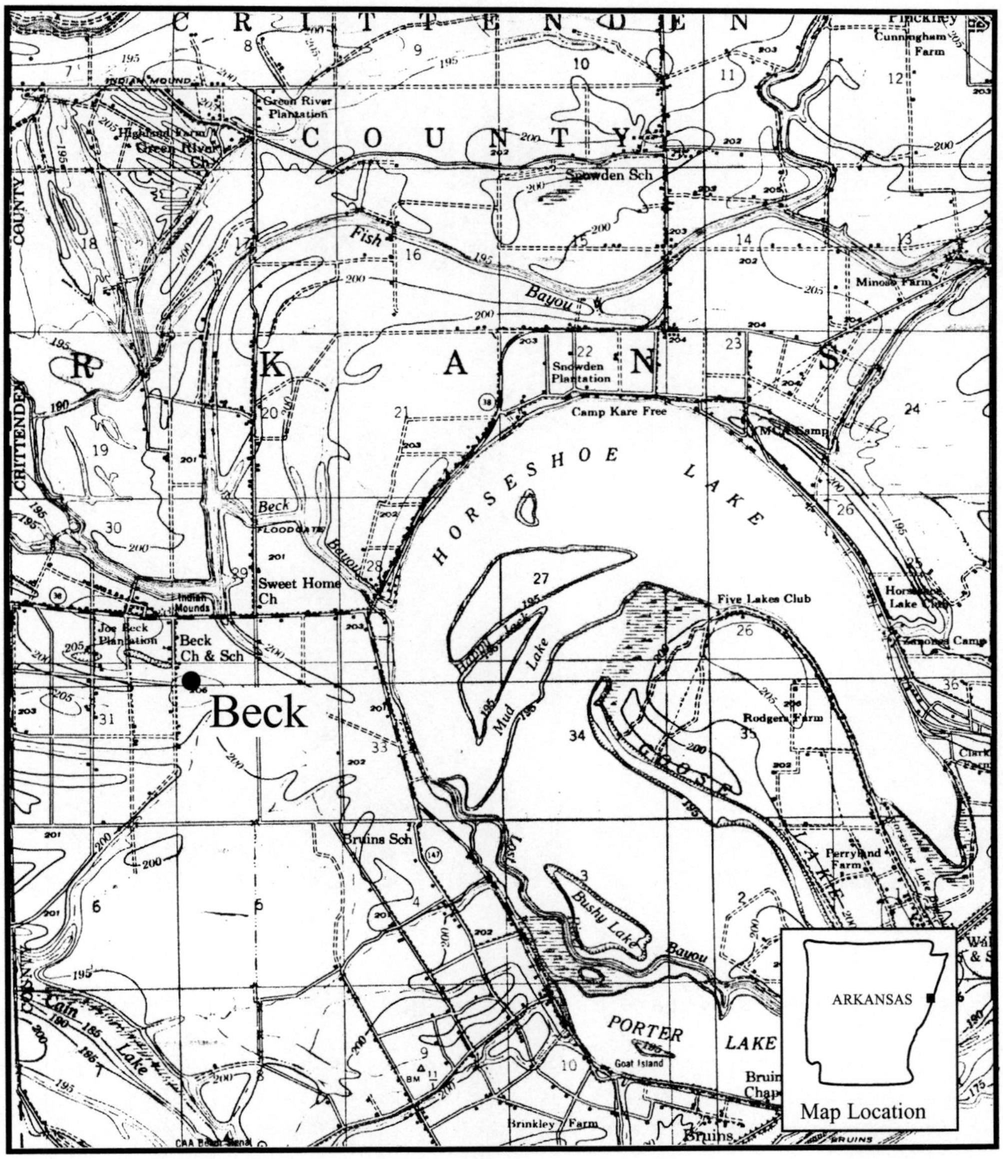

Figure 5.11. Situation of Beck.

Map taken from Horseshoe Lake (1960) 15' USGS Quadrangle.

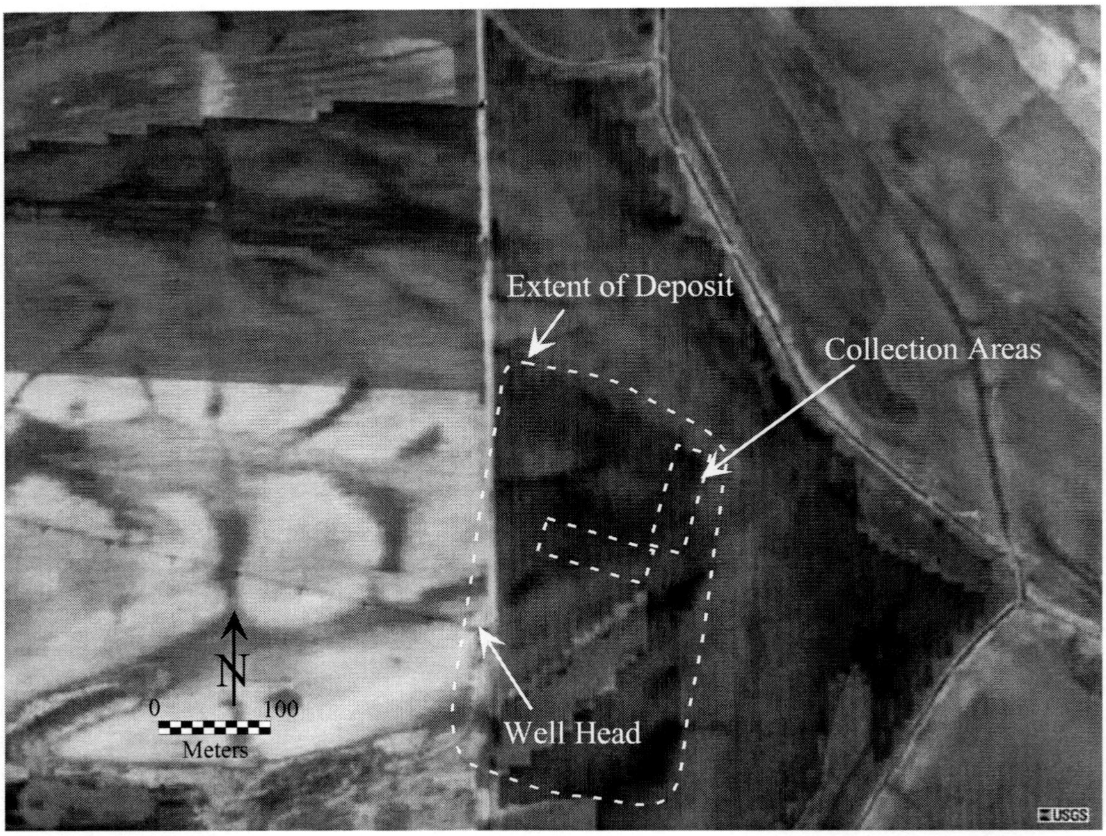

Figure 5.12. Aerial Photograph of the Area Around Beck.

This image was extracted from a Color Infra-Red (CIR) digital orthophoto quadrangle provided by the USGS. The image was photographed on February 12, 1996 by the National Aerial Photography Program and digitized by the USGS DOQ Program on July 1, 1997. Each pixel represents 1 by 1 meter.

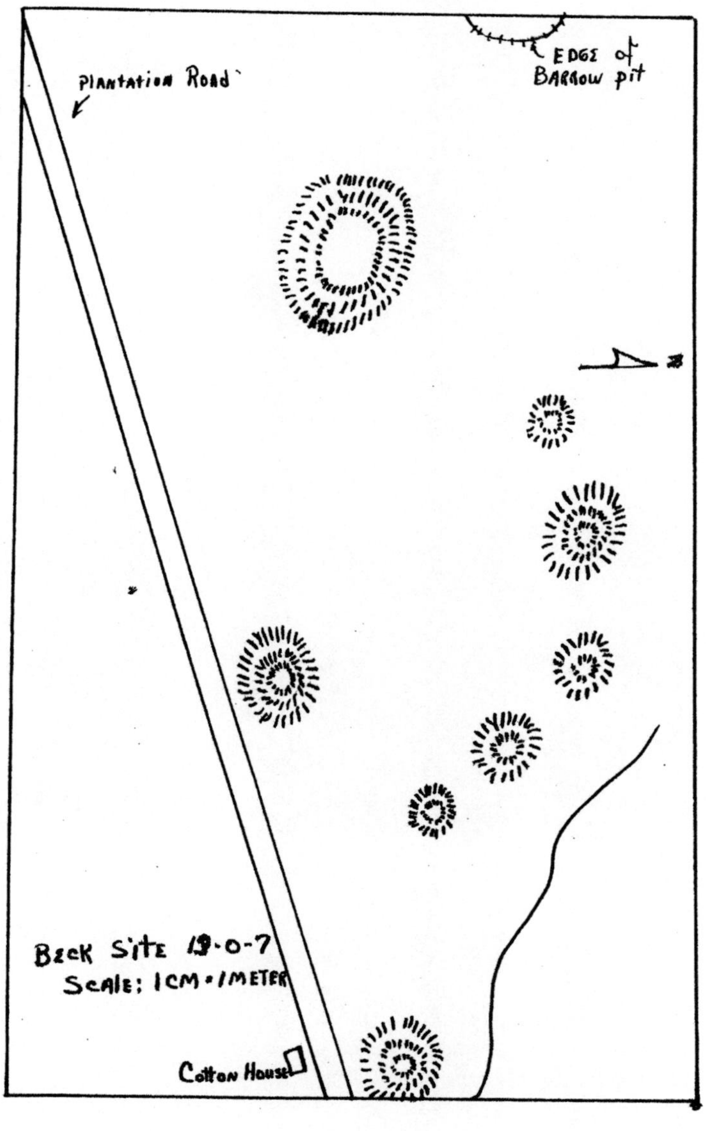

Figure 5.13. Field Drawing of Beck Made by PFG.

Map shows eight mounds none of which are discernible today. The "plaza" area discussed by PFG is probably the area in the center of the mounds and the habitation midden debris is located amongst the mound group. Scale no longer applies. Beck (13-O-7), Box 28, LMS Records 993-23, Peabody Museum Archives, Harvard University. Reprinted courtesy of the Peabody Museum of Archaeology and Ethnology, Harvard University.

Figure 5.14. Beck Taken during 1997 Field Collection

This photo is of the eastern edge of the site looking south. The deposit is the low elevated surface in the foreground. The 50-meter tape in the center of the photo was used as a baseline for ceramic collections.

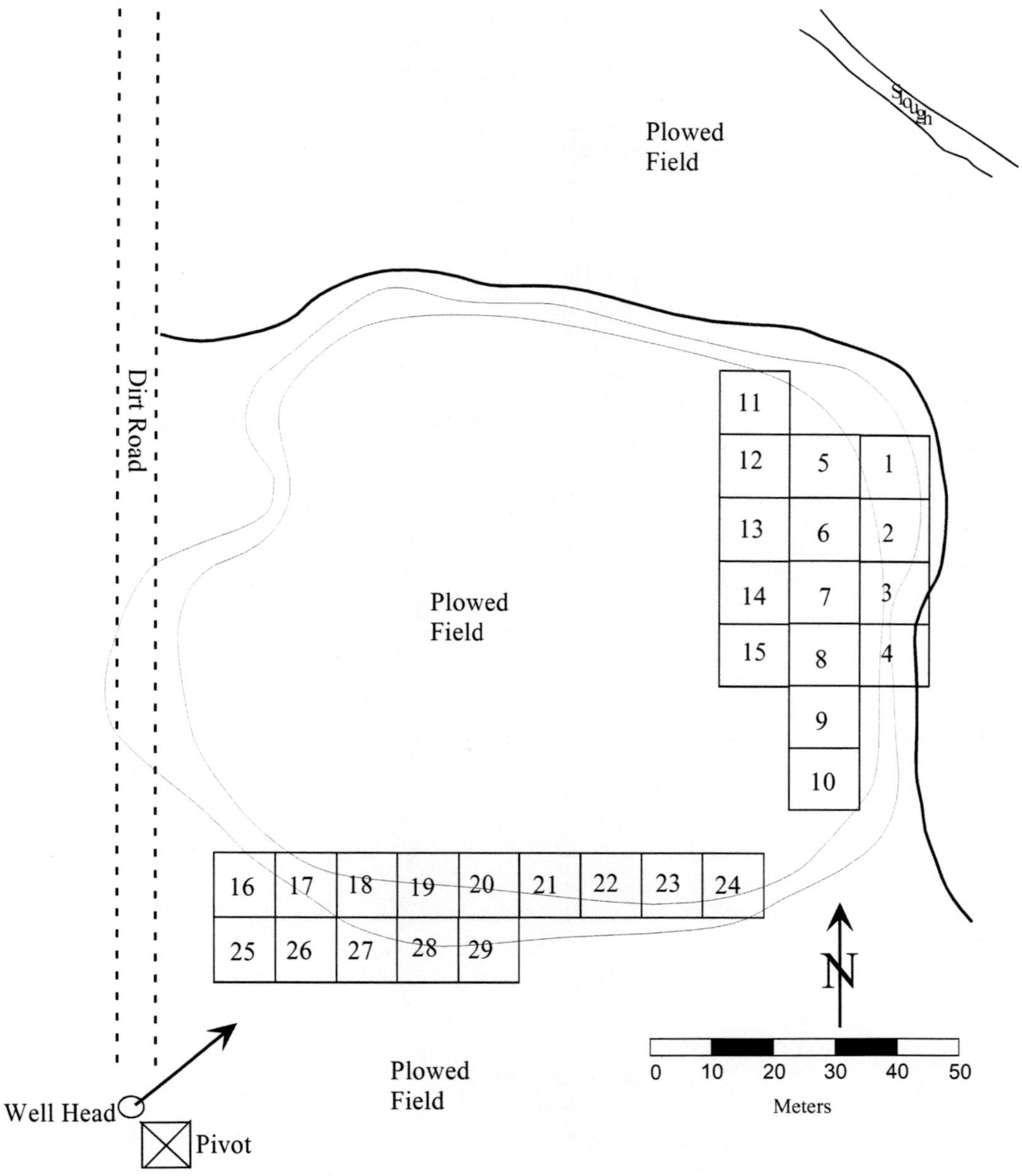

Figure 5.15. 1997 Beck Collection Units.

Contour lines are not quantitative but only illustrate topographic features. Northeast corner of Unit 1 is 40 degrees (MN) and 206 meters from well head.

established a second series of thirteen collection units (16-29) along the southern edge of the deposit (Figure 5.15). The southwest corner of Unit 16 is located at an azimuth of 19 (MN) and 85 meters from the wellhead in the center of the field.

Conditions

Surface conditions during collection were excellent (Figure 5.14). The surface had little vegetation and had been thoroughly washed by months of rain prior to our visit. The field had been plowed several months earlier but due to the rain, the farmers were unable to enter the field to begin planting. Cultural material was readily visible on the surface.

Results

We collected 6,569 sherds from twenty-eight 10x10-m units. In these units, we recovered several fragments of adzes and Nodena projectile points. We also noted the presence of human bone on the surface.

CRANOR (12-O-5)

Location

Cranor (AAS Number: 3SF38; Figure 5.16) is located approximately 0.8 km north of Fifteenmile Bayou and 4.0 km west of Greasy Corner (Chatfield, Arkansas 7.5-minute USGS quadrangle, UTM coordinates north 3876440, east 729450, zone 15) in St. Francis County, Arkansas. Currently, Cranor is the site of farm buildings and an abandoned farmhouse.

Description and Previous Investigations

Ford visited Cranor Place (known to him as "Cramor Place") on April 15, 1940. He collected 1,027 sherds and made a plane table map (Figure 5.17). Ford describes Cranor Place as a St. Francis-type site with two to four mounds situated on top of a levee about eighty meters by sixty meters in width and length. Ford (cited in Phillips 1940) reports that even then the mounds were only identifiable as "small rises on surface" and that the site was "badly cut up by farm buildings." He also notes that the surface material was not particularly abundant.

Other than PFG, Cranor Place does not appear in the archaeological literature except for Smith (1990:159). Smith defined a "Cramor Phase" based on similarity in the proportion of culture-historical types and the geographic propinquity of sites in the area around Cranor Place.

Potential for Current Research

My first visit to "Cramor/Cranor Place" was on May 3, 1997. Local residents now call the mound Wildwood Mound but also report that the name "Cramor" was incorrectly applied by PFG. At the time of the PFG survey, the farm located on the mound was inhabited by the Cranor family. Consequently, I corrected the name throughout this discussion. Most of the site is now covered by a set of farm buildings, silos, and areas for farming equipment (Figure 5.18). Very little of the surface is visible as it is covered by sod, gravel, or buildings. The very western edge of the site, however, is currently in cultivation and the deposit is well exposed. While not a large area, the surface here had dense bone, daub, lithics, and ceramics and the color of the sediment was a very dark gray/brown; a color typical of midden deposits. This brief visual inspection suggested that an intensive survey would produce large numbers of ceramics.

New Fieldwork

With two assistants, I made intensive collections at the Cranor location on May 3, 1997. The plowed westernmost edge of the deposit was the focus of our work (Figure 5.19). We established a north-south base line along the edge of the field. We then placed 100-meter tapes perpendicular to the baseline and along the east-west crop furrows. The area consists of two low ridges that extend from the levee to the west, separated by a lower area. From surface inspection, the highest density of cultural material (daub, ceramics, bone, shell, and grayish brown sediment) appeared to be on the top and edges of these ridges. 10x10-m units, therefore, were then placed on top of the ridges. The NE corner of Unit 1 was placed at an azimuth of 130 degrees (MN) and 12.72 meters from a telephone pole on the southwest edge of the levee. Thirteen 10x10-m units were collected.

Conditions

Surface conditions at the time of collection were excellent. The surface was well washed and the crops were not yet sprouting. Ceramics and other cultural materials were easily identifiable.

Results

It is likely that this area of the Cranor deposit is relatively shallow. The surface sherds are small and look as though they have been heavily abraded by the action of plowing (Dunnell and Simek 1995). Nonetheless, the collection of the thirteen 10x10-meter units produced a sample of 3,612 sherds.

NICKEL (13-N-15)

Location

The Nickel site (AAS Number: 3SF4; Figure 5.20) is located 400 meters east of Fifteenmile Bayou, about 6.4 km west of Hughes, Arkansas in Crittenden County (Mud Lake, Arkansas 7.5-minute USGS quadrangle, UTM coordinates north 3872002, east 724028 and zone 15). The site is located

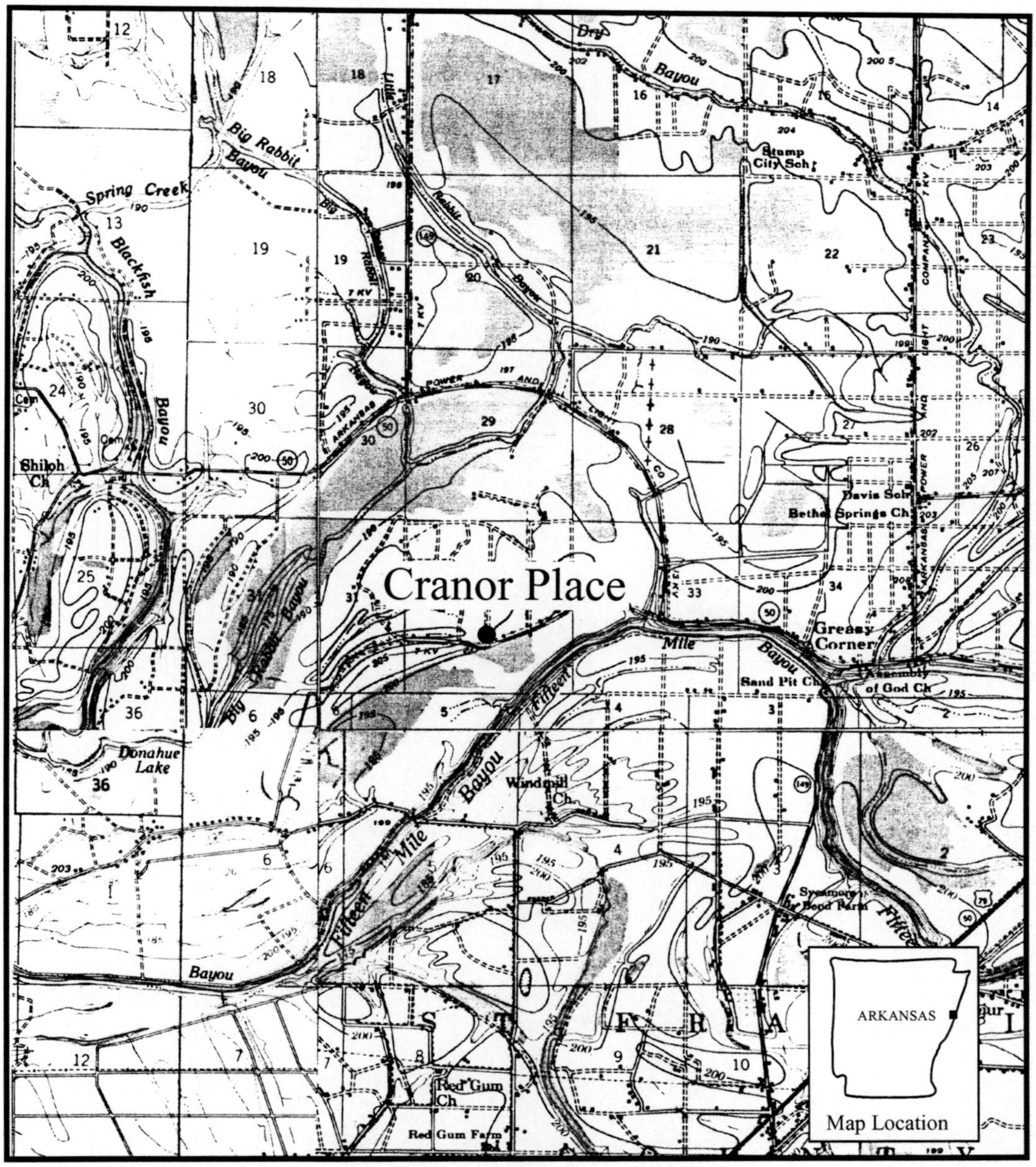

Figure 5.16. Situation of Cranor Place.

Map taken from Whitmore (1960), Edmondsun (1954), Park Place (1962) and Horseshoe Lake (1960) 15' USGS quadrangles.

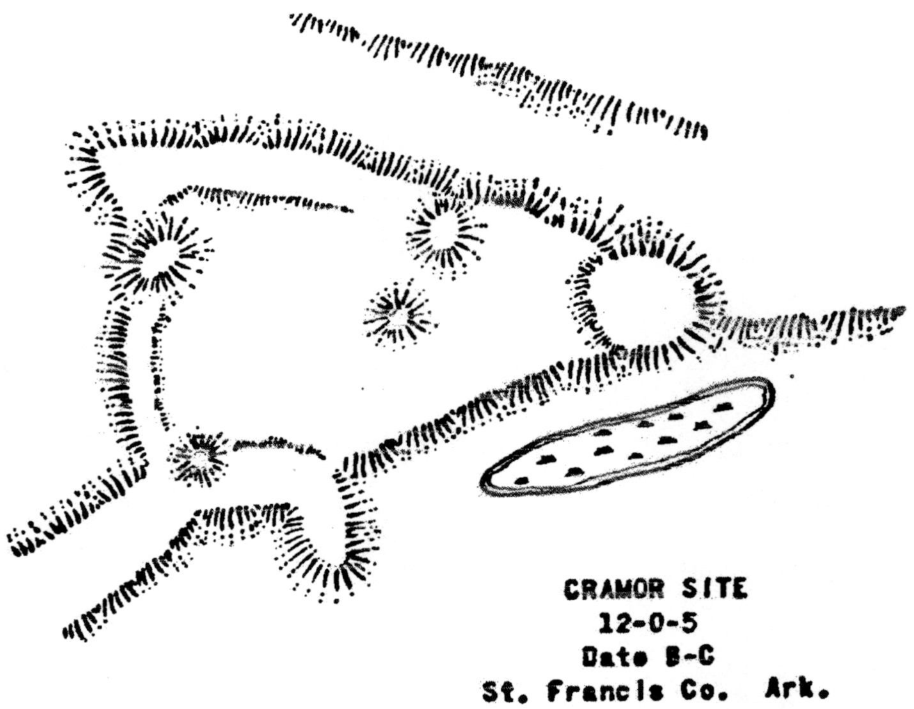

Figure 5.17. PFG Plane Table Map of Cranor.

Figure from Philip Phillips, James A. Ford and James B. Griffin, *Archaeological Survey in the Lower Mississippi Alluvial Valley, 1940-1947*, Papers of the Peabody Museum of American Archaeology and Ethnology, vol. 25, 1951, p. 317, Figure 61. Reprinted courtesy of the Peabody Museum of Archaeology and Ethnology, Harvard University.

Figure 5.18. Cranor Place during 1997 Field Visit.

Photo was taken looking north at the southern edge of the deposit. The deposit is located on the raised areas in the mid-ground of the photo (A). The area that was exposed by cultivation (B) is located to the left of the telephone poles on the left edge of the photograph.

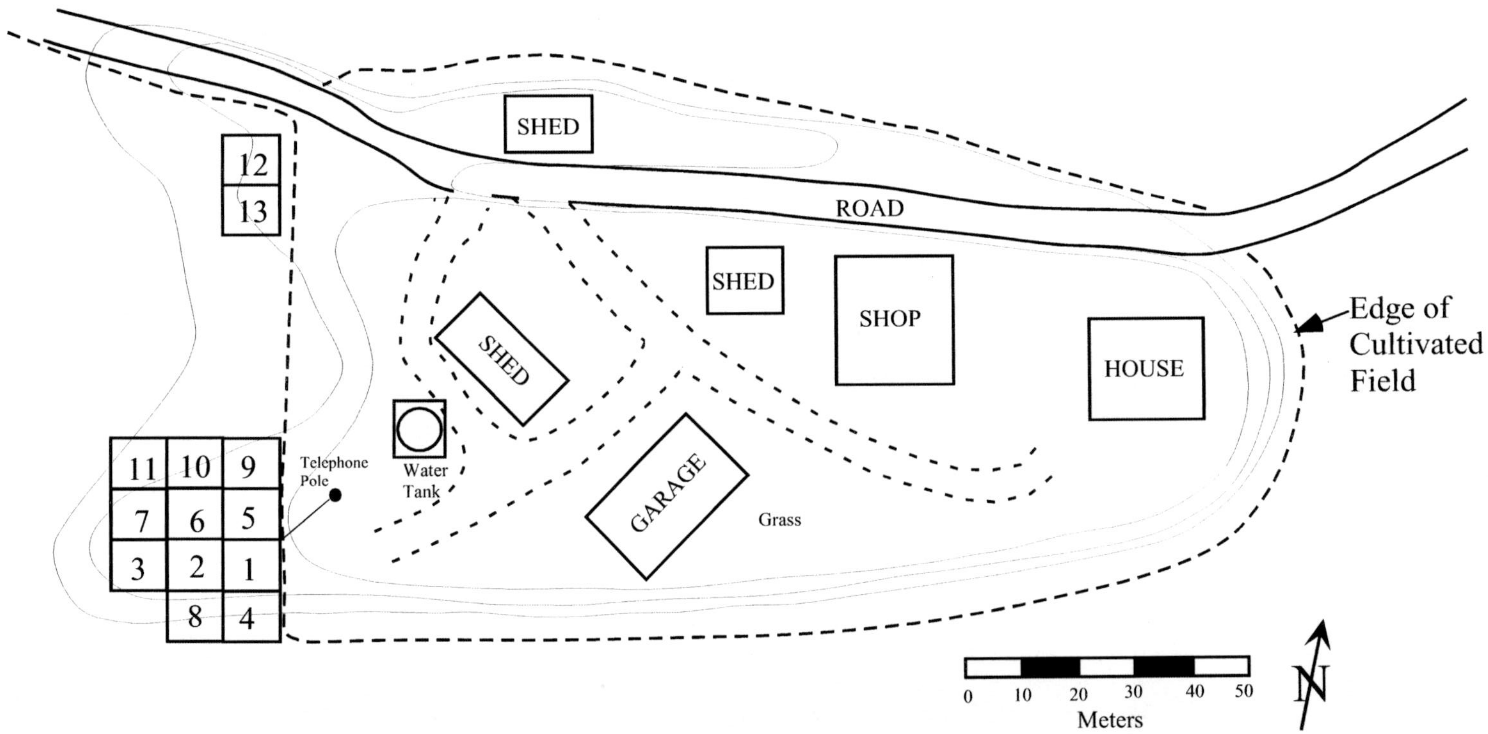

Figure 5.19. 1996 Cranor Place Collection Units.

Northeast corner of Unit 1 is located at an azimuth of 130 degrees (MN) and 12.72m from telephone pole. Contour lines only illustrate topography and not elevation intervals.

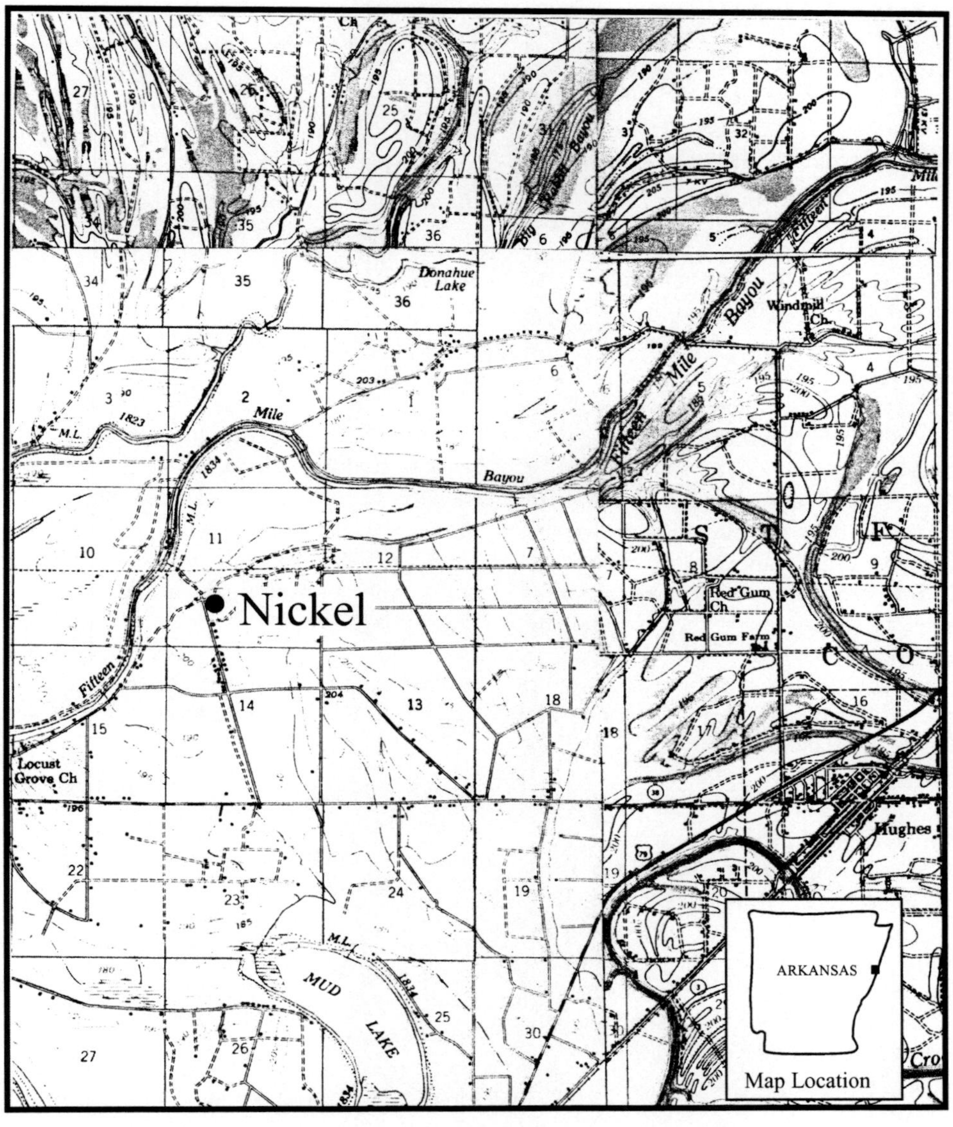

Figure 5.20. Situation of Nickel.

Map taken from Edmundson (1954), Whitmore (1960), Horseshoe Lake (1960), Park Place (1960) 15' USGS quadrangles.

on a north-facing levee bank, several meters above Fifteenmile Bayou. Nickel is located on property owned by Amanda Wise of Hughes, Arkansas and farmed by Ed Cupples of Hughes, Arkansas.

Description and Previous Investigations

The first professional notice of the Nickel site comes from Ford and Griffin's visit to the site on March 19, 1940 during the LMV survey. They describe the site as a "small ceremonial center" consisting of a village area with two relatively large mounds, one rectangular with a flat top (42 meters north/south by 36 meters east/west and 3.5 meters high) and the other plowed-down, about 30 meters in diameter and 1.5 meters high (PFG 1951:321). Figure 5.21 shows their sketch map of the site at the time of their visit.

During the survey, Ford and Griffin noted several large concentrations of daub and ceramics on the surface of the site. Following the PFG field protocol that "separate collections were made on different areas of a site where ever there was the slightest indication that they might differ in content" (PFG 1951:43), Ford and Griffin divided the site into a series of areas labeled B, C, D, and E. They then made separate ceramic collections from each of the four different areas. Area B was located around the low, round mound and had nothing but clay-tempered Baytown pottery. Areas D and E were located around the areas of daub and were found to have only shell-tempered ceramics. Area C was found to have a mix of clay and shell-tempered ceramics as well as daub.

During analysis, however, the ceramics from these collections were lumped into two groups: ceramics from Area B (770 sherds) and the ceramics from C, D, and E (808 sherds). Although PFG make no mention of their rationale, the lumping of collections was probably the result of an intuitive evaluation that determined that the composition of the ceramic types in areas C, D and E were indistinguishable from each other yet distinct from those of collection B.

Other than PFG, the only mention of the Nickel site is by Smith (1990:159) who places it within his "Cramor" phase. Local collectors, however, have not ignored the site. PFG note that local residents had conducted excavations prior to their visit. My discussions with both the farmer and current owner suggest that, collectors have intensively probed the mounds in search of burial pottery. This activity was particularly heavy during the late 1970s and early 1980s.

My first visit to the Nickel site on March 21, 1996 confirmed that significant excavation had taken place in the past by local collectors and that the site had seen extensive alterations since the visit by Ford and Griffin. Small concentrations of human remains on the surface of the mound indicated that the site had seen at least some excavations for burial pottery. Evidence of backhoe trenches into the mound are still visible from the bottom of the bayou channel that abuts the Nickel deposit. In addition, the site has been severely impacted by the farm buildings that cut into the mound and the edge of the site (Figure 5.22). A silo slices east/west through mound A, creating a large trench through the southern end. According to the property owner, in the 1950s the northwestern corner of the deposit was deeply excavated to form the base of a barn. This barn was torn down in the late 1970s or early 1980s but the cement foundation is still extant. In addition to the construction of these structures, a road has been bulldozed that leads to the north through the center of deposit and immediately adjacent to the large mound. The structures that PFG noted as existing on the site are no longer standing. These appear to have been torn down and their remains plowed into the deposit. Much of the deposit to the south is in currently plowed fields and the edge along the levee is in high grass on the west and trees to the east.

Potential for Current Research

My visit also indicated that the Nickel site might provide an excellent sample of late prehistoric ceramics. While not identified by PFG as a "St. Francis-type site," the deposit appears to be managed into an artificial platform along at least one edge. Along the northwest and western edges of the site, there is midden deposit that is mounded to form an elevated rectangular corner. In addition, the areas to the south of the mound appear slightly higher than the surrounding plowed fields, are dark gray-brown in color, and display large amounts of ceramics and other cultural debris. From this examination, it seemed likely that a surface survey would be able to recover large numbers of ceramics from conditions closely resembling St. Francis-type sites.

New Fieldwork

I made collections at Nickel over two field seasons with several assistants. The first set of collections made on March 21-22, 1996, consisted of a series of five units on the northwestern slope of the Mound A (Figure 5.23). The vegetation on this mound had been burned off and the surface was well suited for collections. I established these 10x10-m units along a grid skewed on an azimuth of 45° south from magnetic north and by placing the corners of Units 1 and 2 at the upper most point of the mound. I also placed 3 irregular units (Units 7, 10, 11) around the foundations of the structure embedded in the northwestern corner of the deposit, a unit (Unit 12) on the portion of Mound A separated by the foundation of the feed silo and an irregular unit (Unit 13) to the east of Mound A. Finally, I placed ten 10x10-m units (6, 8, 9, 14-20) in the plowed field to the south of the mound and along the north edge of the field. Generally, the collection conditions along the edges of the mound were excellent since all vegetation had been recently burned. The fields, though planted in cotton, had little vegetation and visibility was good.

With John House of the Arkansas Archeological Survey, I revisited the Nickel site on June 10, 1996 and collected six additional 10x10-m units (Units 14-19). All of these units were in the field southeast of Mound A. The alignment of the units was north to south following the field rows. Conditions remained good for this set of collections, as the cotton had not yet grown substantially.

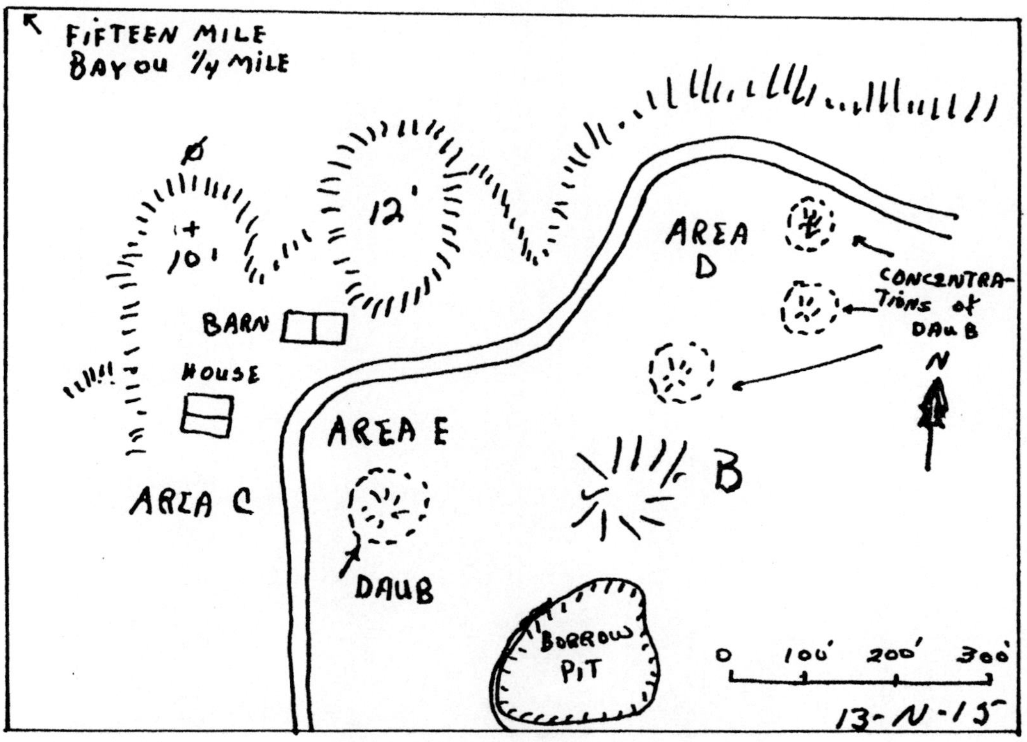

Figure 5.21. PFG Field Map of Nickel.

The PFG map shows the four collection areas, labelled B, C, D, and E. In addition, three mounds are depicted, two labelled as 10', 12' and the third unlabeled near area 'B.' Nickel (13-N-15), Box 27, LMS Records 993-23, Peabody Museum Archives, Harvard University. Reprinted courtesy of the Peabody Museum of Archaeology and Ethnology, Harvard University.

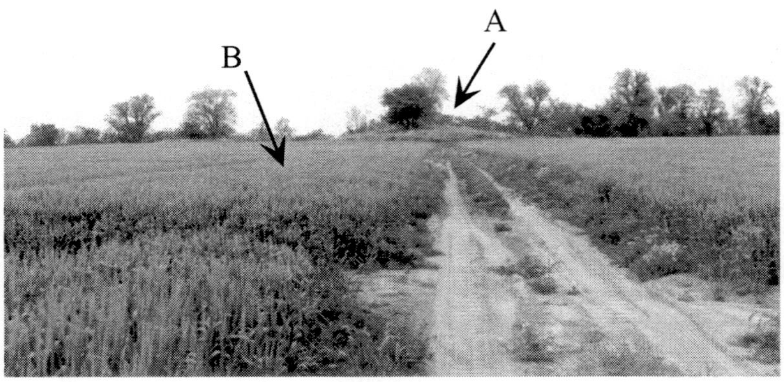

Figure 5.22. Nickel during 1997 Field Visit.

This photo was taken facing north from the south side of the Nickel deposit. In the center of the photograph is a large mound (A). Midden is located in the mid-ground of the photo (B) on either side of the field road.

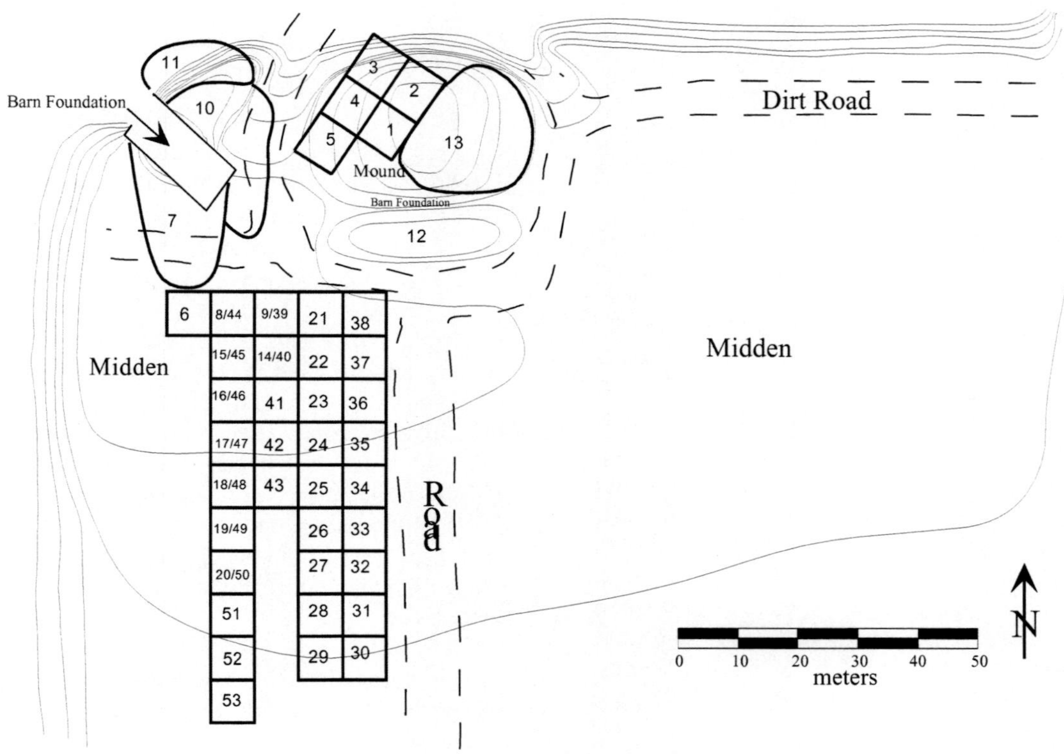

Figure 5.23. 1997 Nickel Collection Units.

Contour lines illustrate topography but intervals are not quantitative.

Finally, with John House and a group of students from the University of Washington Field School in SE Missouri, I collected thirty-three additional 10x10-m units on June 15, 1996 (Units 21-53). Although the field had been plowed, there had been substantial rain since my first collections. Consequently, the collections were very productive; there were substantial cultural materials in this area of the deposit. Nine of these units overlapped my previous set of units and provided resamples of the same area of the deposit that was previously collected. At this time, visibility was still good, though the size of the cotton plants made it clear that further collections that season would be impossible.

Results

The collections at Nickel were productive. Over the course of three different periods, I gathered 9,604 sherds from 53 10x10-meter collection units.

ROSE MOUND (12-N-3)

Location

Rose Mound (AAS Number 3CS27; Figure 5.24) is located about 1.2 km south of the St. Francis River, 7.5 km overland from the town of Parkin (and the Parkin site) and substantially north of most of the sites in this study (Gieseck, Arkansas 7.5-minute USGS quadrangle, UTM coordinates north 3901650, east 717460 and zone 15). The site is situated among a series of old channels of the St. Francis and isolated from the surrounding banks by depressions and swales. Rose Mound is located on property that is currently owned and farmed by Kenn E. Thompson of Wynne, Arkansas.

Description and Previous Investigations

Rose Mound is a deposit so large it serves as a separate field for farmers. It is a St. Francis-type site three hundred meters long by one hundred and fifty meters wide and about four meters higher than the surrounding landscape. Rose Mound is not so much a "mound," but a huge refuse pile of house debris, daub, sherds, bones, lithics, and, at one time, burials and pottery. Figure 5.4, taken by PFG during their survey, gives some indication as to the extent and magnitude of this deposit. Early visitors to Rose Mound noted a circular mound that was 13 meters in diameter and about 1.5 meters high on top of the deposit (Morse 1981:50). From the ground, no evidence of this mound currently exists, having been long since leveled by tillage (PFG 1951:285). A satellite photo (Figure 5.25), however, of the area clearly shows the remains of this mound in the center of the large field.

Rose Mound is bounded by old river channels, on the east and west, and appears to have been cut off from adjoining high ground by artificial ditches (PFG 1951:284). As PFG (1951: 284) report "in effect, the site occupies a sort of rectangular island surrounded on all four sides by depressions of slight depth which increase the mound-like effect the site gives when viewed from any direction." PFG (1951:285) also state that Rose Mound is:

> in fact a huge refuse mound, so sharply defined by either its 'island' situation or by a former stockade – of which no evidences are at present visible – as to present the appearance of a deliberate artificial construction, which of course, it is not. Like other large St. Francis sites of the Mississippi Period, it is simply a result of the accumulation of house remains and refuse in a circumscribed space.

Early Investigators

Given its prominence, Rose Mound has long attracted investigators. Rose was first excavated in 1879 by Edwin Curtis who was traveling in northeastern Arkansas collecting pottery for the Peabody Museum at Harvard University (Morse 1978). Curtis (Morse 1978:7) described "Roses Mounds" as a fifteen and a half-acre field and reported that:

> this field is 15 feet higher than the surrounding country and the field is covered by smaller mounds or ridges. They are full of human remains, burnt brick-like material, ashes, chard wood, fragments of pottery, flints fragments, celts, petrified wood, hickory principally, and fragments of discoid stones. In the mounds explored by me I find worked shells, shell beads, beads of copper and pottery with animal shaped handles, some in the shape of fish, owls and other birds.

During his excavations in Arkansas, Curtis recovered 933 pottery vessels, including one hundred and six effigy vessels, eighty-one painted vessels, and one head pot (Putnam 1881:18). In his diggings, Curtis noted that the deposit was extremely rich with cultural materials and contained "chard human remains, fish bones, turtle shells or fragments, animal bones, mustle shells, pomegranates and all that kind of material that is found in these ancient dwelling places usually" (Morse 1978:8). The surface of the mound, Curtis noted, was "covered with fragments of pottery" (Morse 1978:8).

During his explorations of the mounds of the St. Francis in his private steamboat, the *Gopher* of Philadelphia, Clarence B. Moore made a visit to Rose Mound in 1910. Moore reported that since the initial probings by Curtis, the mound had been well probed by pottery collectors. Moore (1910:277) writes that:

> the entire surface of the great mound is scarred with remains of holes made by seekers after pottery, who have so thoroughly dug the burial places of St. Francis river. It was difficult to find an area of even a few square feet on any part of the great mound, which did not show traces of the spade.

Despite the degree of this previous collecting activity, in twelve days of excavation and with the help of thirteen

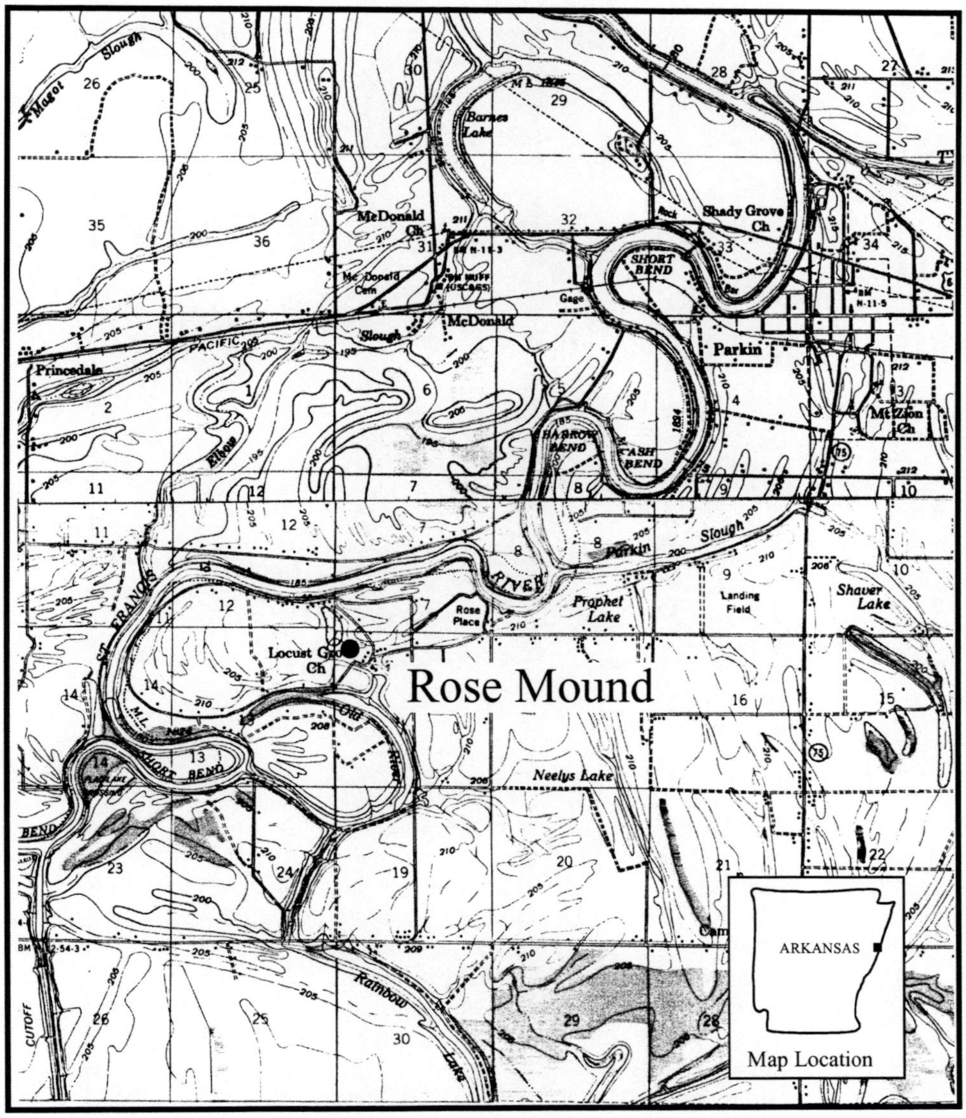

Figure 5.24. Situation of Rose Mound.

Map taken from Princedale (1962) and Whitmore (1960) 15' USGS quadrangles.

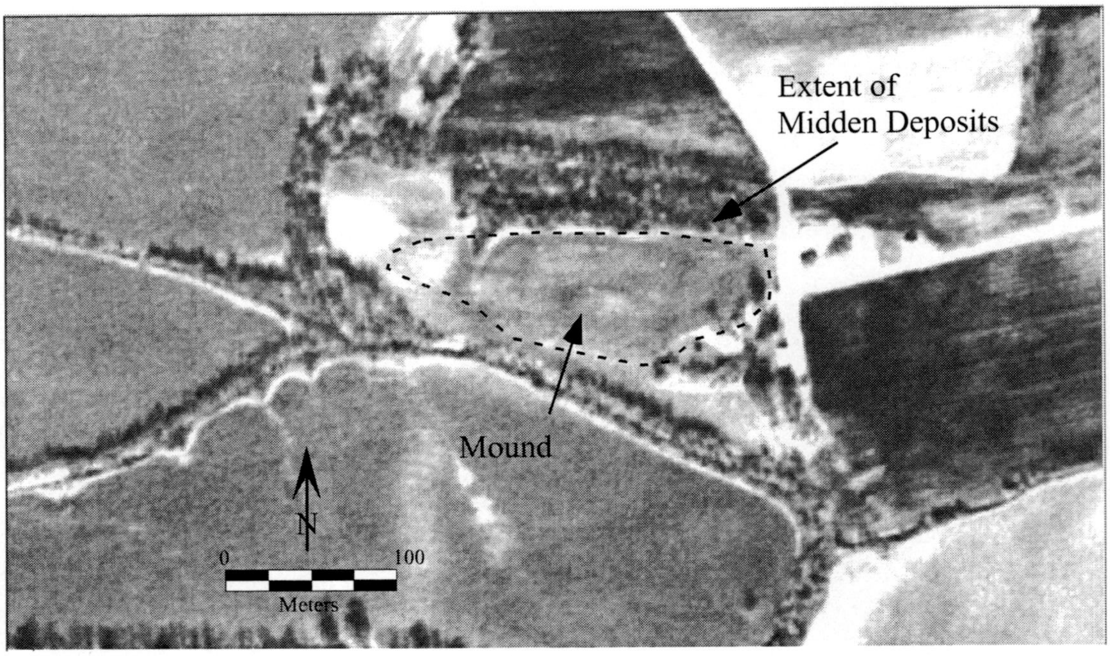

Figure 5.25. Satellite Photo of the Area around Rose Mound.

In the center of the midden area, the oval lighter area records the original position of a mound that is no longer visible from the surface. SPIN-2 imagery consists of digital, panchromatic, 2-meter resolution images that have been geo-references. SPIN-2 imagery is the result of a joint venture of SOVINFORMSPUTNIK (the Interbranch Association of Russian space agencies), based in Moscow; Aerial Images, Inc. of Raleigh, North Carolina; and Central Trading Systems, Inc. of Huntington Bay, New York. SPIN-2 satellite imagery is available via http://www.terraserver.com, a site that operates as an online catalog for SPIN-2 images. This image was taken on January 4, 1989 at 17:40 GMT from an altitude of 218.6 kilometers, with a sun angle of 32.5 degrees and with a KVR-1000 camera system. The image resolution is 1.56 meters per pixel. Satellite photo used by permission of SOVINFORMSPUTNIK

excavators, Moore recovered 587 vessels in 207 burials (Moore 1910:278). Along with many of these vessels, a variety of the burial artifacts (*e.g.*, ear plugs, copper objects, pipes, shell ornaments, bone combs, shell gorgets) were exquisitely depicted in full color illustrations in Moore's 1910 publication in the *Journal of the Academy of Natural Sciences in Philadelphia.*

PFG Fieldwork

Phillips and Ford made an initial visit to Rose Mound during the PFG survey on March 28, 1940 (Figure 26). Figure 5.27 (a, b, c, d, e) depicts photos taken by PFG during the time of their survey. The surveyors made separate ceramic collections from three areas of the site. Area A is located on the eastern end of the midden. Area B is situated toward the western end of the mound. Area C consists of an area of high ground that is located just west of the midden area. Subsequent studies on the collections showed that sherds gathered from areas A and B are all shell-tempered. Consequently, PFG lumped these sherds together for further analyses. Sherds from Area C, however, are a mix of clay and shell-tempered ceramics and thus Ford kept these sherds separate during his seriation studies. During the survey, Phillips and Ford collected 1,406 sherds from Areas A and B and 723 sherds from Area C.

In addition to making a very large surface collection, Phillips conducted limited stratigraphic excavations at Rose Mound in order to obtain chronological information on the relationship between earlier clay-tempered (Baytown) and later shell-tempered (Neeley's Ferry Plain) ceramics. The results of these excavations led Phillips to argue that there was a relatively abrupt change between the earlier and later periods (PFG 1951:284-292). These conclusions, however, were in marked contrast with the seriation results by Ford who argued that the change in ceramics was gradual and continuous. The difference of opinion on this point exposed a deep rift in the perspectives of the PFG researchers as to the nature of culture and culture change (Dunnell 1985a; Lyman *et al.*1997:144-145). While Phillips and Griffin held an essentialist view that treated culture as an empirical entity and change as transformational, Ford took a materialist stance that treated culture as a continuous changing stream of shared ideas. No reconciliation was possible and the final publication of PFG contains both sets of arguments.

While there was considerable debate among the PFG researchers about the nature of change, there was certainly no disagreement about the potential for Rose Mound for future archaeological research. PFG (1951:290) write that:

> the Rose mound is one over which many tears have been shed by archaeologists on account of the activities of several generations of indefatigable pot-hunters. Our small excavation was undoubtedly the first time the site was dug by any one who was not looking for pots. If it accomplishes nothing else, it proved that the important phases of the history of the site are still intact, safe in the deeper levels below the reach of sounding rod and casual shovel. Notwithstanding all the crimes that have been perpetrated here and the righteous, if slightly envious, indignation they have evoked, the Rose Mound would still repay excavation, in fact from the point of view of Mississippi culture beginnings, in the Survey Area it is the most promising site we have so far encountered.

Surprisingly, since the efforts of PFG, relatively little professional attention has been paid to Rose Mound. In the 1950s, the University of Arkansas held a field school there but the results have not been published. There have been numerous other, less formal excavations by local collectors in search of vessels. Many of these vessels are currently in private collections. No other research has been done here.

Potential for Current Research

My first visit to Rose Mound was on June 19, 1995 with Robert C. Dunnell (University of Washington) and Timothy Hunt (University of Washington). Although the field conditions were poor during this visit (*i.e.,* the field crops consisted of burned wheat), it was clear that that the deposit had a substantial surface component. At subsequent visits, however, conditions of the deposit for making large surface collections of ceramics were ideal (Figure 5.28). The entire top surface of the deposit is still a regularly farmed field and one expects that the action of the plow is still bringing up new material to the surface. A quick walkover confirmed this as the surface is literally covered with daub, bones, lithics, and ceramics. The density of ceramics on the surface was very high, and the collectable area enormous, so it was apparent that even a modest surface collection would efficiently generate thousands of sherds.

New Fieldwork

With John Dudgeon and Kris Wilhelmsen, I made collections at Rose Mound on May 4-5, 1997. Since the highest density of artifacts on Rose Mound, like other St. Francis-type sites, is located along the edges of the midden, we established two groups of collection units along the eastern and southern sides of the mound (Figure 5.29). The first set, on the east edge, was established by stretching parallel 100-m tapes north-south (MN) and ten meters from each other and measuring off 10-meter increments. The northeast corner of Unit 1 was placed at an azimuth of 36° (MN) and 52.70 meters from the telephone pole closest to the southeastern corner of the mound. We placed the second set of units on the southern edges of the midden with the northeastern corner of Unit 11 at an azimuth of 155° (MN) and 160 meters from the telephone pole. In order to make collection easier, we aligned this set of units along crop rows on a $110^{o}/290^{o}$ (MN) line.

At the time of these investigations, the surface of Rose Mound was in excellent condition for collections. The field was planted in milo (sorghum) but the growth of these crops was minimal. Since the planting, the ground was washed by

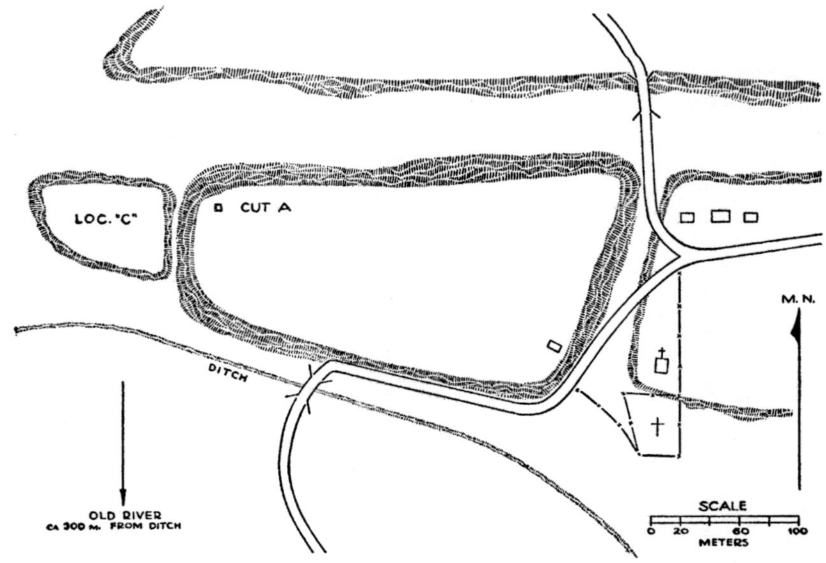

Figure 5.26. PFG Plane Table Map of Rose Mound

Figure from Philip Phillips, James A. Ford and James B. Griffin, *Archaeological Survey in the Lower Mississippi Alluvial Valley, 1940-1947*, Papers of the Peabody Museum of American Archaeology and Ethnology, vol. 25, 1951, p. 52, Figure 52. Reprinted courtesy of the Peabody Museum of Archaeology and Ethnology, Harvard University.

A

B

C

D

E

Figure 5.27. Rose Mound at Time of PFG Collections.

Photos are (A) northwest corner of mound showing artificial cut, (B) southwest corner of mound from location "C", (C) mound from the northeast, (D) east end of mound from the north, (e) Rose Mound from southwest. Rose Mound (12-N-3), Box 27, LMS Records 993-23, Peabody Museum Archives, Harvard University. Reprinted courtesy of the Peabody Museum of Archaeology and Ethnology, Harvard University.

Figure 5.28. Rose Mound during 1997 Field Collections.

Photo was is facing west and is of the east side of the deposit. The deposit comprises the broad sloping area in the mid-ground of the photo. As seen here, the field conditions were excellent for surface collections; the surface was well-washed and there is little vegetation.

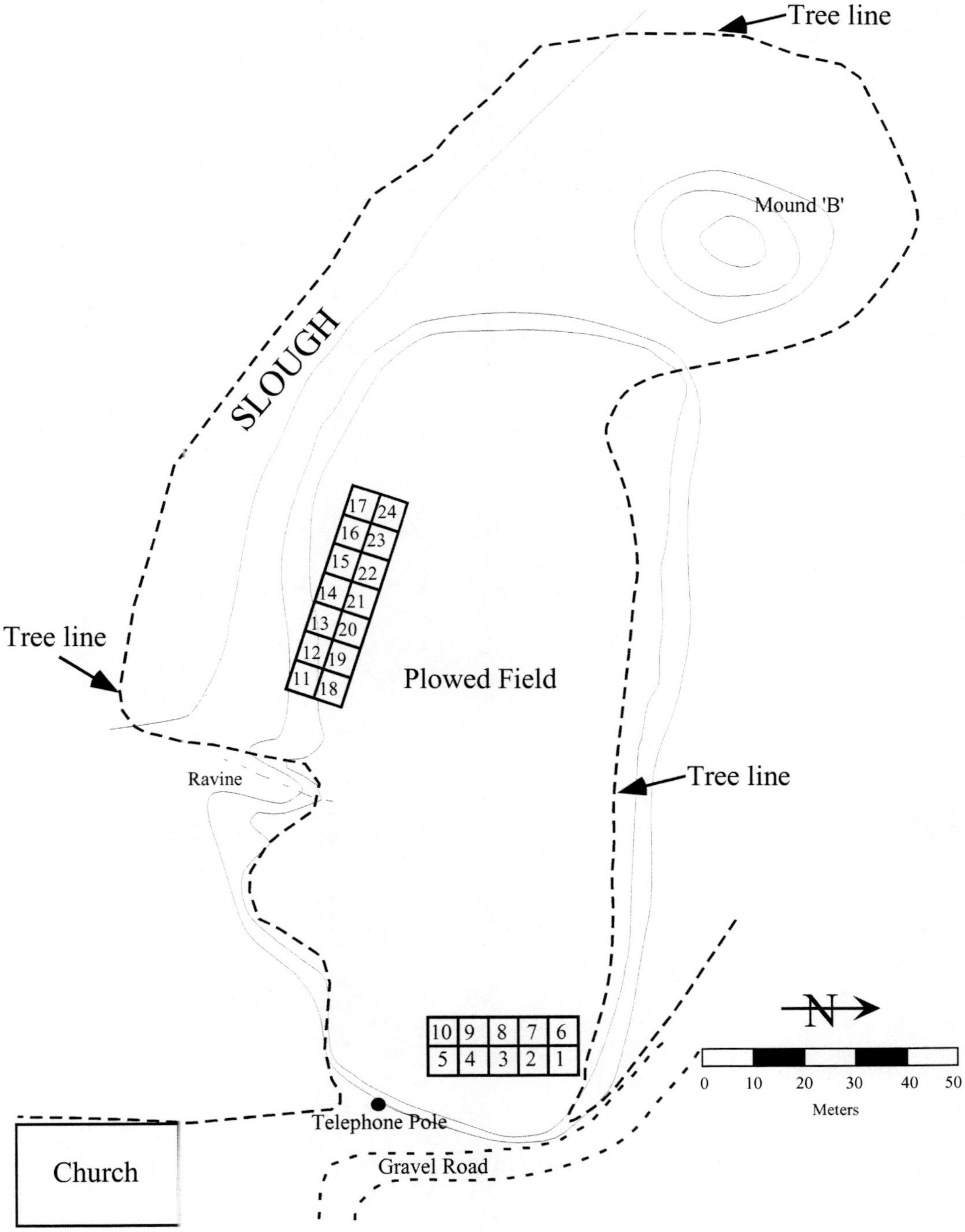

Figure 5.29. 1997 Rose Mound Collection Units.

Contour lines only approximately indicate location of topography.

several weeks of rain. This rain resulted in field conditions in which cultural material was easy to identify.

Results

In two days, we gathered 6,139 sherds from twenty-four 10x10-m collection units. We also recovered a small, hand-sized lump of what appears to be unfired, shell-tempered clay. R.C. Dunnell (University of Washington), D. Bush (University of Washington), H. Neff (University of Missouri-Columbia) and S. Toyoda (University of Osaka) are currently studying this item.

HOLDEN LAKE

Location

The Holden Lake deposit (AAS Number: 3CT240; Figure 5.30) is located about 6.4-km southeast of the town of Earle in Crittenden County, Arkansas (Jeanette, Arkansas 7.5-minute USGS quadrangle, UTM coordinates north 734346, east 3902979 and zone 15). The site is located on the banks of an old river channel, and on the edge of Holden Lake, the water-filled remains of an old river channel, to the east. Holden Lake is located on land that is owned by David Wallace of Crawfordsville, Arkansas. The location is currently farmed by Buster Briggs of Crawfordsville, Arkansas.

Description and Previous Investigations

Unlike the other six sites in this study, Holden Lake was not examined by PFG. The first mention of the site comes from an Arkansas Archeological Survey site record. Guided to the site by a local collector, Charles McGimsey of the Arkansas Archeological Survey described the site in July of 1966 as a Mississippian village site, two hundred by eighty meters in dimensions with shell-tempered ceramics (Figure 5.31). Since that visit, no other investigations have been made at Holden Lake.

I was first directed to the Holden Lake site by E. Rush Harris of Forrest City, Arkansas. Using his directions and the AAS site record, I visited the site in March 1996 with Tim Mulvihill of the Arkansas Archeological Survey Station at Parkin. The situation of the deposit is somewhat similar to Nickel in that it is comprised of a single mound situated on the edge of a levee, adjacent to an old river channel and surrounded by a village midden (Figure 5.32). From surface inspection, the northeast corner of the site appears to consist of a midden that has been deposited to raise and flatten the area. Core samples made by John House of the Arkansas Archeological Survey (House, personal communication) later confirmed this impression of a deep, managed midden deposit. House placed these cores along the northern edge of the site. Their contents indicate that the depth of the midden exceeds two meters in this area.

This midden and the general outlines of the deposit are visible from the satellite image of the site shown in Figure 5.32. This satellite image shows the edge of the midden as a rectangular feature with a western boundary of Holden Lake, trees along the edge of the channel to the north, and shallow depressions along the east and south.

Except for a north-south road that bisects the area, the entire surface of the deposit is in cultivation. On the surface, the densest distribution of cultural material is along the margins of the area, particularly the northeastern and eastern edges. These materials consist of bone, lithics, daub, and shell-tempered ceramics.

Potential for Current Research

Although nearly nothing is known about the Holden Lake site, the presence of the deep midden deposits and the density of the surface remains in cultivated fields suggests that an intensive surface collection would produce large number of ceramics comparable to the other collections.

New Fieldwork

On June 11, 1996 and June 15, 1997, I made my collections at Holden Lake with the assistance of John House and members of the University of Washington Field School in Southeast Missouri. We made collections in twenty 10x10-meter units along the eastern edge of the deposit (Figure 5.33). To establish these units, we stretched two 100-meter tapes parallel to each other and 10 meters apart along a north-south (MN) line. The northern end of these tapes was at the northern base of a large walnut tree that sits on the northeast edge of the deposit. This area of the deposit, south of the tree line, was in cultivation.

Conditions

At the time of the collection, there was very little vegetation and the surface was well washed from months of prior rainfall. Visibility was excellent.

Results

The work at the Holden Lake was extraordinarily productive. In the two days of collections and using just twenty 10x10-m units, we gathered 12,341 sherds.

CASTILE LANDING (13-N-21)

Location

Castile Landing (AAS Number 3SF12; Figure 5.34) is located on the St. Francis River in St. Francis County, Arkansas about 2 miles south of the small town of Widener (Dansby, Arkansas 7.5-minute USGS quadrangle, UTM coordinates, north 3872820 and east 712000, zone 15). The archaeological deposit is situated on a high southern bank of

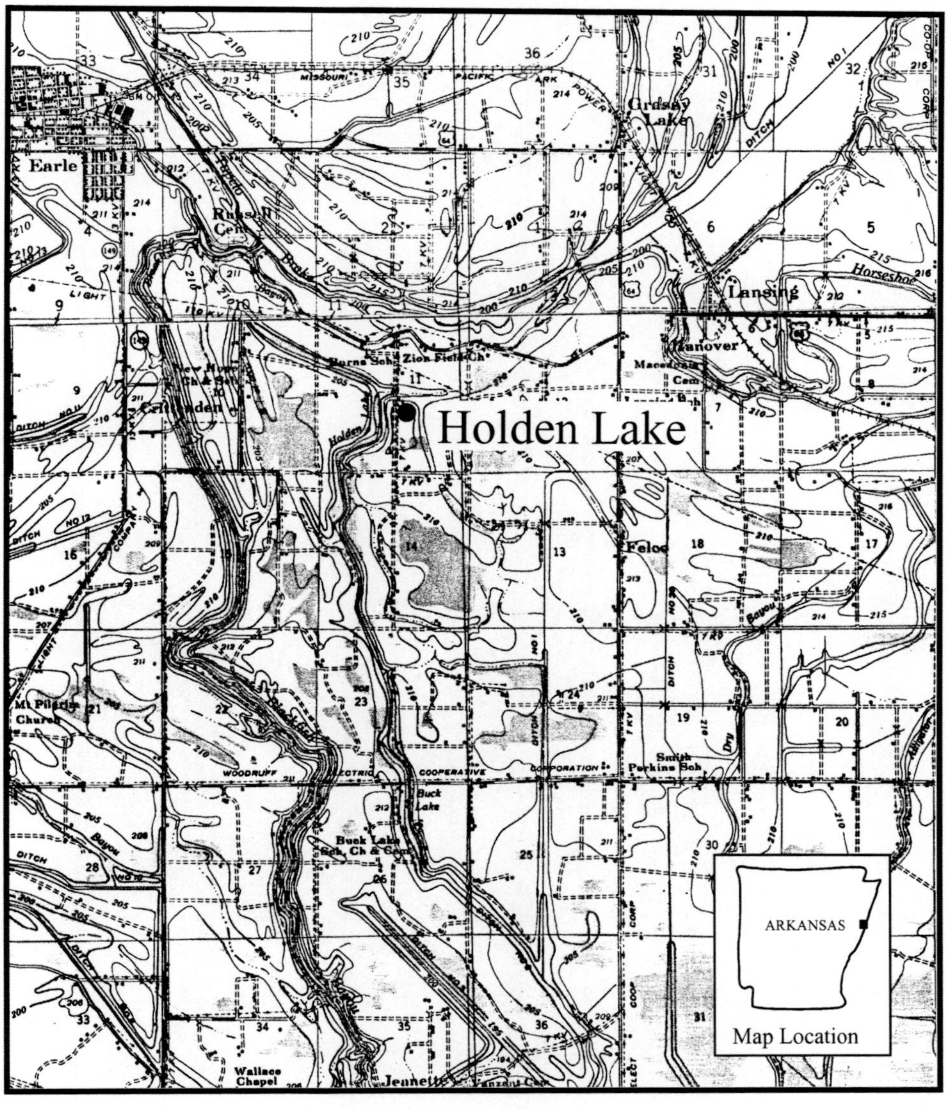

Figure 5.30. Situation of Holden Lake.

Map taken from Deckerville (1957) and Edmundson (1954) 15' USGS quadrangles.

Figure 5.31. Holden Lake during 1996 Field Visit.

Photo is of the western edge of the deposit facing northeast.

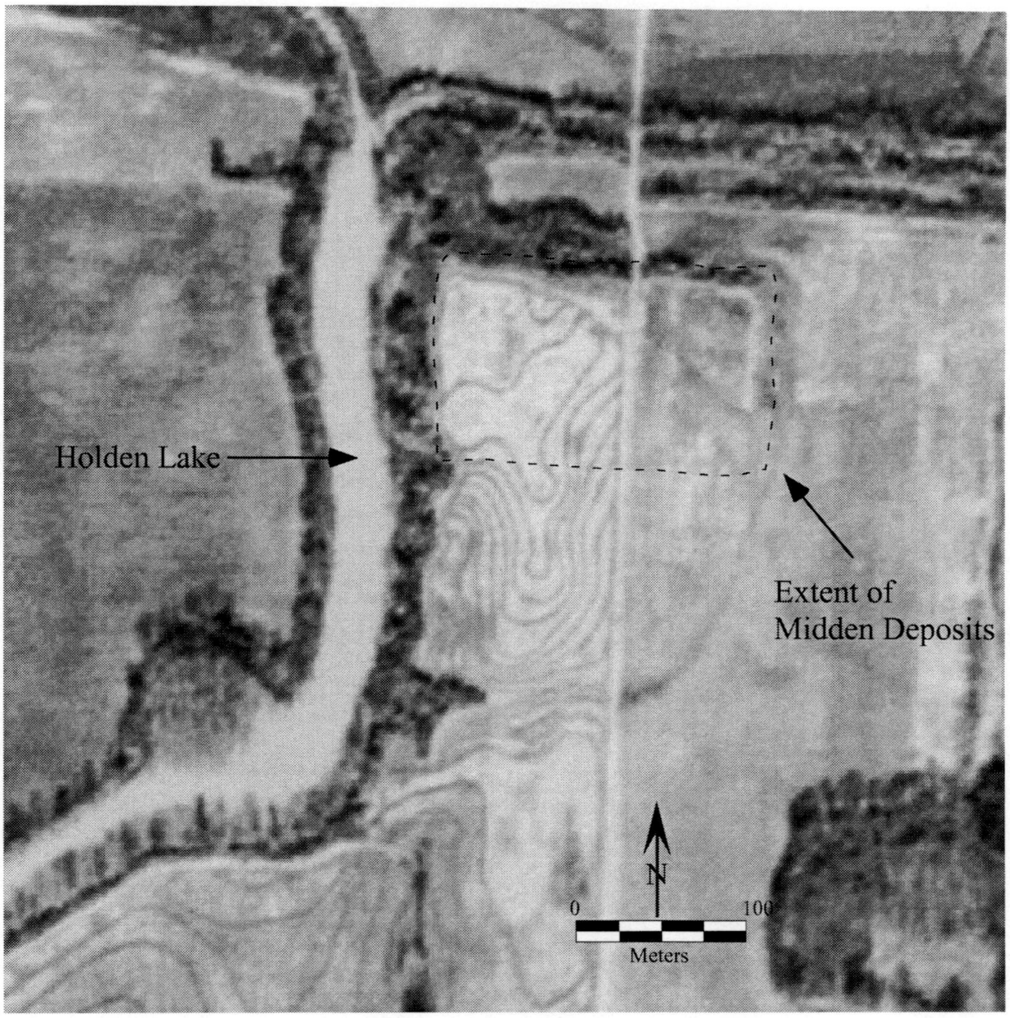

Figure 5.32. Satellite Photo of the Area Around Holden Lake.

The extent of the Holden Lake deposit is clearly visible on this satellite image. The contour-like lines are the results of the preparation of the field for rice farming in which a series of dikes along level contours is created to hold water. The SPIN-2 imagery is the result of a joint venture of SOVINFORMSPUTNIK (the Interbranch Association of Russian space agencies), based in Moscow; Aerial Images, Inc. of Raleigh, North Carolina; and Central Trading Systems, Inc. of Huntington Bay, New York. SPIN-2 satellite imagery is available via http://www.terraserver.com, a site that operates as an online catalog for SPIN-2 images. This image was taken on January 4, 1989 at 17:40 GMT from an altitude of 218.6 kilometers, with a sun angle of 32.5 degrees and with a KVR-1000 camera system. The image resolution is 1.56 meters per pixel. Satellite photo used by permission of SOVINFORMSPUTNIK.

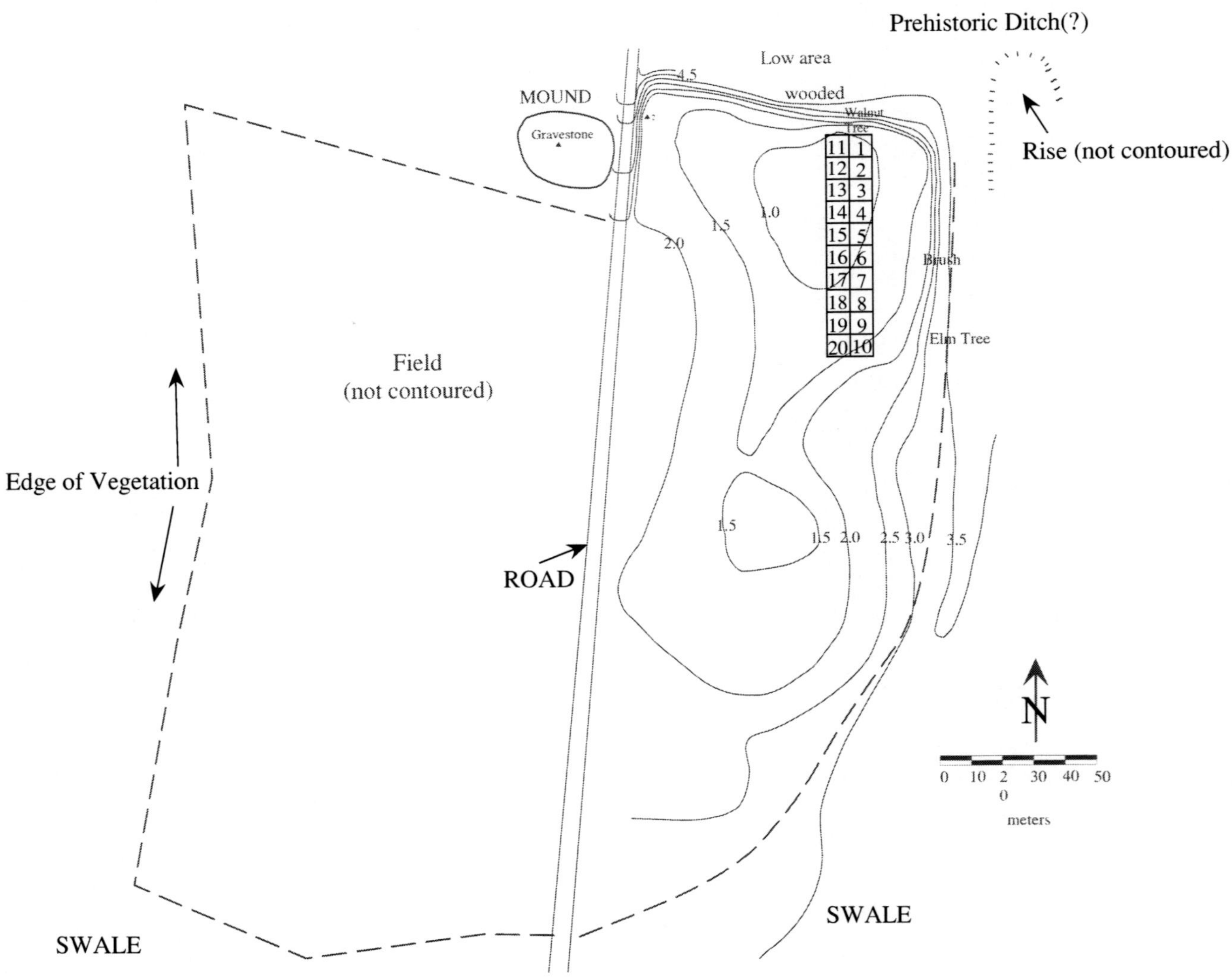

Figure 5.33. Holden Lake Collection Units

Contours were mapped by J. House of the Arkansas Archeological Survey.

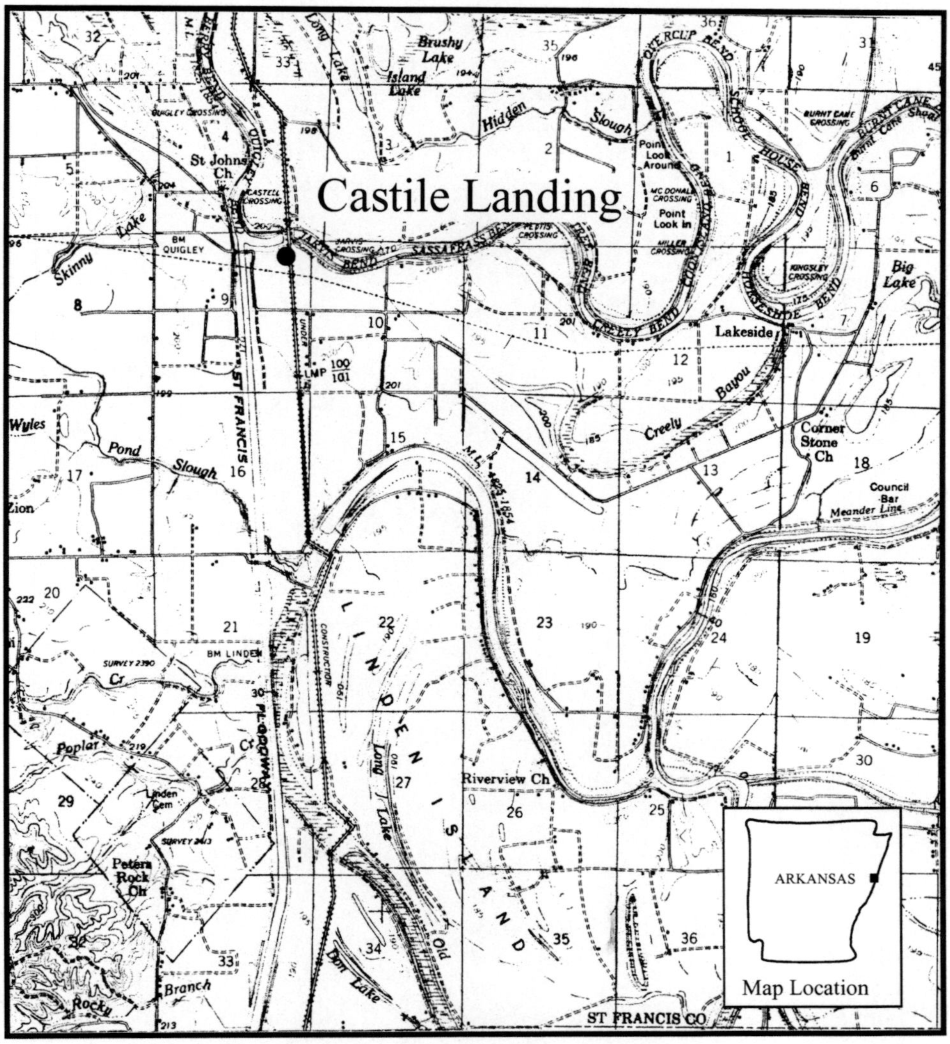

Figure 5.34. Situation of Castile Landing.

Map taken from Park Place (1962) 15' USGS quadrangle.

the St. Francis, just south and west of the Arkansas Game and Fish Commission Burnt Cane river access point. Castile Landing is located on property managed by the St. Francis County Levee District.

Description and Previous Investigations

Castile Landing was originally described by C.B. Moore (1910) during his travels along the St. Francis in the *Gopher*. When Moore visited Castile ("The Castile Place"), it comprised a quadrangular mound about 300 by 350 feet and approximately 7.5 feet high. The surface of the mound was under cultivation and covered with "much debris, consisting of bits of flint, fragments of pottery, and small masses of baked clay" (Moore 1910:267). Moore, interested mostly in recovering decorated pots, excavated twenty-three burials, which produced sixty-four whole vessels. While discovering whole pots was the goal of the expedition, Moore (1910:268) was impressed with the quantity of broken pottery at the site, noting that "although we have given the number of vessels found by us at the Castile Place as sixty-four, this figure by no means represents the tally of those preserved by us, many of vessels having been merely disintegrating fragments, while others had considerable parts missing and were in other ways undesirable."

PFG Fieldwork

Phillips, Ford, and Griffin visited Castile Landing on April 3, 1940 and described the site as a "rectangular town site of St. Francis type" (Phillips 1940; Figures 5.35, 5.36). Ford's plane table map (Figure 5.35) shows the dimensions of the deposit to be about three hundred and fifty feet east/west by two hundred and thirty feet north/south and nine feet high. They also describe the midden mound as being an area about two feet higher on the western edge (Phillips 1940). Characteristic of many St. Francis-type sites, daub, and sherds were "abundant around the periphery, relatively scarce in plaza" (Phillips 1940). During the visit, PFG collected 2,030 sherds.

Today, the Castile Landing deposit is entirely unrecognizable from topography. During the 1950s, a large Army Corps of Engineers levee and floodway construction project nearly obliterated the archaeological deposit. The St. Francis river was rerouted into a new north/south flowing channel (the St. Francis Floodway), several hundred meters to the west of the site (Figure 5.34). Massive amounts of backfill from the construction of this channel were deposited on the Castile location. In addition, the Army Corp of Engineers excavated a large trench parallel to the new channel. This trench was subsequently filled with clay to form the waterproof foundation for the levee. According to a local informant (Margaret Lynn, St. Francis County Museum), the trench sliced directly through the deposit, exposing burials, ceramics, and habitation remains. Finally, the current levee was constructed directly over what remains of the deposit. This levee consists of a clay core covered with packed earth to raise its height. A gravel road runs north/south along its top. The deposit at Castile Landing has been dramatically altered; there is little in the appearance of the area to tell one that it once held remains of a massive St. Francis-type deposit.

No professional work has been conducted at Castile Landing since PFG. The only mention of Castile Landing comes from Phyllis Morse who, during her 1978-79 research at Parkin, argued that Castile Landing belonged to the Parkin phase but that a planned visit to the site was not possible due to flooding (Morse 1981). Undoubtedly, the lack of further investigations is due to the general perception that the site no longer exists.

Potential for Current Research

A visit to the site on June 17, 1995 with John House of the Arkansas Archeological Survey revealed that, while Castile Landing no longer remotely resembles the location that PFG visited, a significant deposit remains intact and accessible. We noted prehistoric cultural materials in two locations on either side of the levee. On the east side, we observed scattered daub, Mississippian sherds and chert flakes along the crest of the levee ridge, and on the steep slope that leads down to the river to the north. Patches of dark soil at the top of the slope suggested the presence of midden. On the west side, we observed dark soils, daub, and some sparse Mississippian sherds. The results of this field visit suggested that it is possible that at least some of the deposit is still extant. Systematic coring, however, is necessary to reveal the extent of the intact midden beneath the levee spoil.

New Fieldwork

Castile Landing (3SF12) required special consideration in this research since it has been extensively modified by the construction of the St. Francis floodway. Castile Landing was chosen for this study because it was identified as being part of the "Parkin" phase by Morse (1981). My seriation analyses showed Castile Landing as more similar to Nickel and Cranor Place than Rose Mound and deposits to the north. Thus, the generation of new data from Castile Landing was particularly important to evaluate the degree to which sampling played a role in determining the composition of culture-historical types and thus similarity of this deposit to others in the study.

Due to the effects of the levee construction the original location and landscape of the deposit was not immediately identifiable. The results of several non-systematic surface collections indicated that the bulk of the deposit forms part of the St. Francis levee and is covered with grass and, in the west, trees. Despite the ground vegetation, however, surface inspection revealed sherds, bone, and lithics along the eastern slope of the levee, particularly in the area facing the old channel of the St. Francis and the boat landing to the northeast. I also noted cultural material on the west side of the levee road along the slope facing the old St. Francis channel (Figures 5.37, 5.38).

The results of the surface collection suggest that at least some of the Castile Landing deposit remains beneath the levee fill. Most cultural material is found on the upper

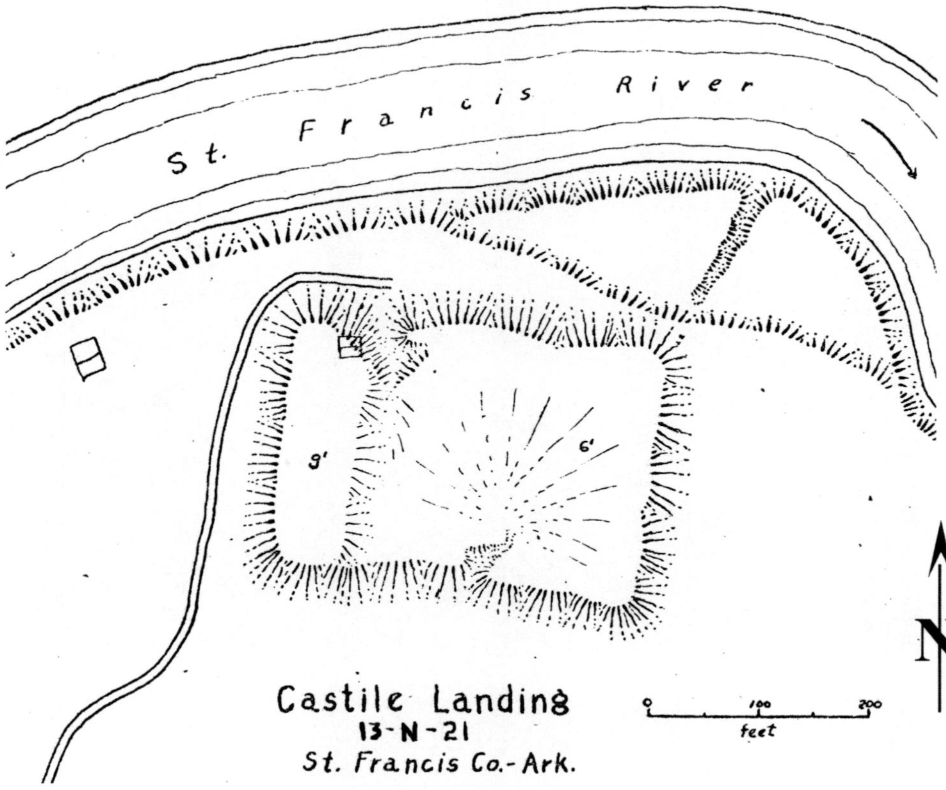

Figure 5.35. PFG Field Map of Castile Landing.

Castile Landing (13-N-21), Box 27, LMS Records 993-23, Peabody Museum Archives, Harvard University. Reprinted courtesy of the Peabody Museum of Archaeology and Ethnology, Harvard University. North arrow added.

Figure 5.36. Castile Landing during PFG Collections.

This photo (probably taken facing the southwest) clearly shows a house situated on a St-Francis type midden deposit. At the present day, this form is no longer recognizable in due to the construction of a levee in the 1950s. Castile Landing (13-N-21), Box 27, LMS Records 993-23, Peabody Museum Archives, Harvard University. Reprinted courtesy of the Peabody Museum of Archaeology and Ethnology, Harvard University.

Figure 5.37. Castile Landing during the 1996 University of Washington Field School Excavations.

Photo was taken from the levee road facing east across the site. The slope on the left hand side of the photo leads down to the old channel of the St. Francis River.

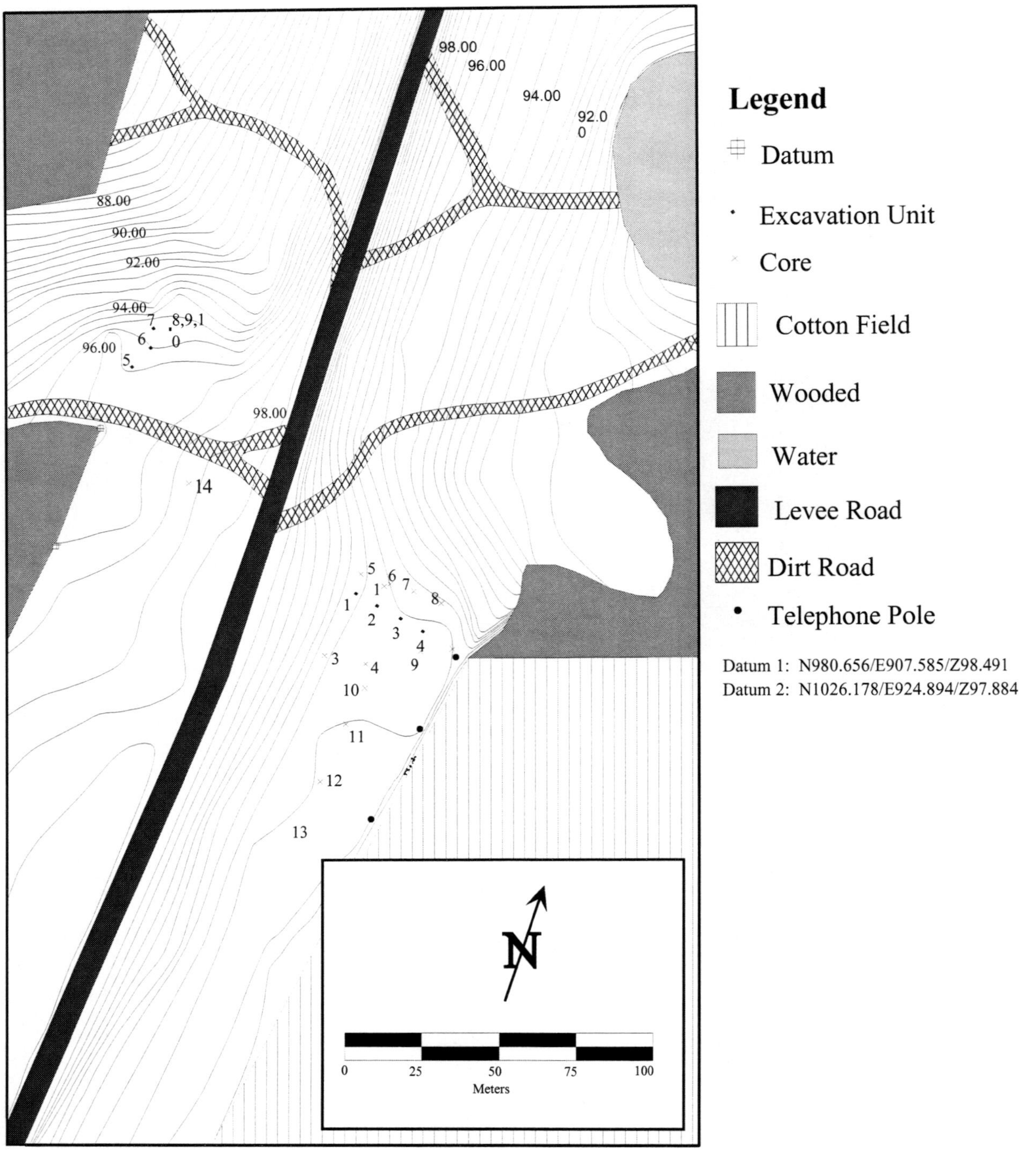

Figure 5.38. 1997 Castile Landing Coring Locations and Excavation Areas.

slopes of the levee indicating that, since the bank of the St. Francis was particularly high in this location, only a meter or so of fill material was added to the mound.

Coring

The next task was delineating the extent of the deposit. On July 4, 1996, I initiated a coring project to examine the subsurface composition of the area. Initially, I was interested in the east side of the levee since the steepness of the levee slope suggested that this area of the deposit might be covered by the least amount of fill. Thirteen cores were taken with a 6" bucket auger across the east side of the levee. I also included one core on the west side of the levee road.

For each core, when possible, augering was continued until sediment with no cultural material was reached. Midden deposit was recognized as dark organic soil with a high density of cultural material. Table 5.5 describes what was found in each core and whether the core indicated there was a buried midden deposit in that location.

Initially, the coring results suggested that there is cultural material on the east side of the levee road and that some of this area contains what appears to be the remains of a St. Francis-type deposit. I was particularly interested in the rectangular lobe that extends from underneath the levee as it looked to be a promising area to conduct further sub-surface exploration and collection. The overburden in this area is shallow and the cultural material was relatively dense. Cores number 4, 5, 6, 7, and 8 produced a relatively large number of sherds given the small diameter of core holes. Table 5.6 provides a summary of the material found in each of the cores. Given the abundance of material in the set of cores, my initial excavation work focused primarily on the east side of the road.

Spatial Coordinates

Prior to further work, I established a spatial coordinate system at Castile Landing so that the locations of the cores could be tied into future excavations and so that a topographic map could be generated for the site. To do this, I aligned an arbitrary grid along magnetic north-south and placed two concrete markers in the tree line to the west of the levee road (Figure 5.37). Monument 1 is located at position N980.655 ± 0.0010, E907.584 ± 0.0022, elevation 98.491 ± 0.0005. Monument 2 is located at position N1026.179 ± 0.0006, E924.892 ± 0.0026, elevation 97.884 ± 0.005. Both monuments are approximately 50-cm by 50-cm in size and have a large, embedded nail to mark the center of their location. I marked the top faces of monuments with the year and the name of my project (Figure 5.39).

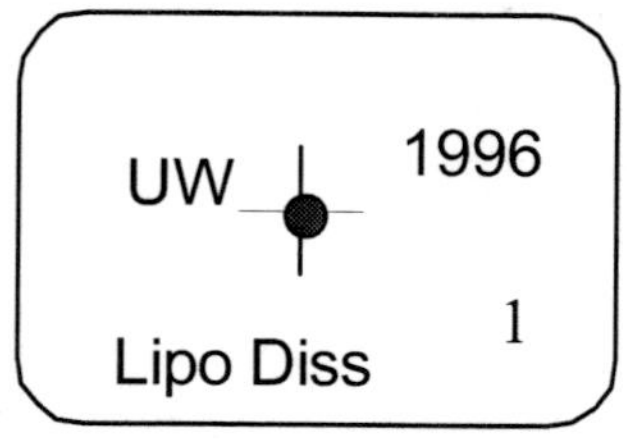

Figure 5.39. Concrete Monuments Established at Castile Landing.

The monuments are flush with the ground. The monuments were placed such that once they were located, it would be possible to reestablish the excavation grid using simple resectioning procedures.

Excavations

In all of the other locations included in this analysis, I used surface techniques to collect large representative samples of the deposit. At Castile Landing, however, systematic surface collection could not be used as the deposit is partly covered with fill from the levee, and is in tall grass. Excavations, therefore, were necessary at Castile Landing.

At all other locations, I had relied on (1) the mixed nature of St. Francis-type sites, and (2) homogenizing effects of plowing that would correct for the effects of spatial autocorrelation present in intact deposits. Although surface collections could not be made at Castile Landing, targeting buried plowzone deposits with excavations, meets the same assumptions that guide the surface work at the other locations in this study.

Potential error does differ between techniques. *Large-scale* spatial autocorrelation is the primary problem in surface collections where autocorrelation is caused, for example, by stylistic spatial structure. In excavation, the primary problem is spatial autocorrelation caused by *breakage*. Thus, at Castile Landing, the sherd assemblages had to be refitted. Since my analyses are restricted to decorated ceramics the combination of sherd shape and decorative pattern could be used to match sherd fragments, this process was neither difficult nor excessively time consuming.

Initial excavations were made in one-meter square units placed at least five meters apart from one another in the areas of the site suspected to have the highest density of ceramics and plowed midden deposits. Since the goal of the excavations is to collect large numbers of ceramics from midden and plowzone deposits, I conducted excavations at arbitrary 20-cm levels and only in deposits that appeared to be disturbed parts of the plowzone or midden. I ceased excavations when cultural features such as pits, house foundations, and hearths were encountered. For efficiency in excavation, and due to the high clay content and the dryness of the deposit, all material was collected in six-liter plastic bags and transported to a field camp for subsequent water screening. At the field camp, each bag was soaked in water to break down the soil peds and washed through a 4-mm screen. All material collected in the 4-mm screen was labeled and bagged by unit number and excavation level.

Since I collected the excavated material by using screens (with 4-mm openings), there was a potential that this material would be systematically smaller than surface collected sherds since surface assemblages were made using a minimum sherd size of 10-mm criterion. If sherd size plays a role the counting of particular types, this discrepancy could potentially bias identifications toward collections with larger sherds (*i.e.,* the surface collections). However, the risk of identification error due to size differences was minimized

Table 5.5. Description of Cores Excavated at Castile Landing

Core	Depth	Description	Midden Deposit?
1	0-60 cm	Brown/black clayey sand fill with clods of clay and historic brick fragments	No
	60-95 cm	Increase in clay. Iron staining in soil. Mottled grayish brown to reddish brown. Chert flake fragment found at 60-62 cm.	No
2	0-40 cm	Dark brownish/black silty sand. Brick fragments, charcoal specks. Shell-tempered pottery fragment and chert flake at 25-30 cm.	Yes
3	0-52 cm	Dark brown sediment. Brick fragment and clay clods.	No
4	0-22 cm	Brown to black sediment. Chert flake at 22 cm.	Yes
	22-57 cm	Shell-tempered ceramics found at 22-25 cm.	Yes
5	0-20 cm	Brownish black silt with clay clods and sand. At 20 cm, ceramic fragments and flake were encountered.	No
	20-63 cm	Brownish black silt. Large, painted, shell-tempered rim sherd.	Yes
	63-90 cm	Very black sediment with charcoal, wood, and organic material. More clayey sediment begins at 90 cm.	Yes
	90 – 130 cm	At 110 cm, sediment color changes to light brown with orange clay mottles	No
6	0 – 25 cm	Dark brown sediment with much daub. 2 sherds, 1 flake found.	Yes
	25 – 28	Dark brown sediment with daub.	Yes
	28 – 53	Medium brown sediment with clay. 1 sherd, 1 flake found.	Yes
	53 – 112	Medium brown sediment with clay and iron oxide nodules.	No
7	0 – 5 cm	Dark brown to black sediment. Ceramics. Possibly midden soil.	Yes?
	5 – 25 cm	Dark brown sediment. Sherds, chert flake, bone found.	Yes
	25 – 30 cm	Dark brown sediment and "midden" sediment ends at 30 cm.	Yes
	30 – 59 cm	Brown sediment. At 54 cm, yellowish clay sand with iron oxide nodules.	No
8	0 – 26 cm	Dark brown to black sediment. Daub, sherds, historic glass.	Yes
	26 – 34 cm	Yellowish clayey sediment. No artifacts.	No
9	0 – 68 cm	Dark brown to black sediment. Daub, 1 sherd found.	Yes
	68 – 87 cm	Yellowish clay. No artifacts	No
10	0 – 64 cm	Dark, brown sediment with some daub, sherds, and flake. Very "midden"-like sediment.	Yes
10	64 – 74 cm	Light brown, clayey sediment with iron oxide mottles.	No
11	0 – 54 cm.	Light brown clayey sediment. 2 sherds.	Yes
	54 – 75 cm	Grayish, clayey sediment with charcoal, sherds.	Yes
	74 – 85 cm	Light brown, sandy sediment with iron oxide mottles.	No
12	0 – 54 cm	Light brown silt loam sediment with daub.	No
	54 cm	Clayey sediment with iron oxide mottling	No
13	0 – 62 cm	Brown silty loam. No daub or other midden debris.	No
	62 – 76 cm	Grey clayey sediment. Charcoal.	No
14	0 – 150 cm	Mottled dark brown-light brown. Daub, charcoal, 3 flakes	No
	150	Mottled brown. Charcoal	No

Table 5.6. Contents of Castile Landing Cores.

Core Number	Plain Sherds	Decorated Sherds	Total Sherds	Lithics	Bone	Daub	Historic Material
1	0	0	0	1	0	0	0
2	3	0	3	1	0	0	0
3	0	0	0	0	0	0	0
4	1	1	2	1	0	0	0
5	9	2	11	3	0	3	0
6	6	0	6	2	1	10	0
7	12	4	16	2	1	3	1
8	7	0	7	1	0	1	2
9	1	2	3	2	8	3	2
10	5	3	8	2	0	9	0
11	5	1	6	0	0	0	0
12	0	0	0	0	0	0	0
13	0	0	0	0	0	0	0
14	0	0	0	3	0	0	0
Σ	49	13	62	18	10	29	5

Table 5.7. Description of Castile Landing 1x1-m Excavation Units.

Unit	Level	Description
1	0 – 20 cm	Hard-packed, brown sediment. Unit contained historic, debris, flakes, and sherds. Appeared to be very hard fill from levee construction.
2	0 – 20 cm	Hard-packed brown sediment that appeared to be well reworked. Contained lithics, lots of daub, charcoal flecks, and bone.
3	0 – 20 cm	Dark-brown/black sediment. Sediment is hard, compact, mottled, and heavily bioturbated. Abundant worms/casts. Many roots in the first 10 cm, less so in the lower 10 cm. Charcoal, daub, sherds, lithics, bone. Possible feature in center of unit at 20 cm. Excavation halted at 20cm.
4	0 – 20 cm	Daub, sherd, flakes, charcoal, silty clay.
	20 – 40 cm	Dark brown sediment, lots of daub, sherds, hard packed silty clay.
5	0 – 10 cm	Below O horizon, sediment was extremely hard packed and filled with very low density of cultural material. Light brown sandy sediment.
6	0 – 25 cm	Brown silty loam with daub, sherds. Sediment is dry, hard packed, dark grayish brown in color, and contains abundant fragments of daub. At bottom of unit are several sherds derived from a single red-slipped bowl. Bone also noted.
	25 – 40 cm	Dark gray clayey silt with sand mottles. The composition of sediment is swirled with lighter and darker clay as though the sediment has been deposited by water. Large amounts of daub, charcoal, large sherds, and bones.
	40 – 60 cm (1/4 of unit)	Very compact sediment. Brown clayey sediment with daub, brick fragments, glass, and iron.
7	0 – 20 cm	Packed brown sandy loam. Cultural material present including daub, sherds, bone.
	20 – 40 cm	Dark midden soil. Many sherds. Bone, daub, and charcoal.
8	0 – 20 cm	Dark brown, very soft sediment with bone, daub, sherds, lithics. At 20cm potential features found with dark gray/brown sediment, sherds, charcoal, daub, lithics, and bone. These "features" are potentially lower midden soil.
	20 – 27 cm	Dark gray brown clayey soils with charcoal, daub, sherds, and bone. At 27cm, soil changes to brown sandy layer with yellow-orange mottling and thin, gray/brown, ashy lenses of charcoal, daub, sherds, and bone. Feature evident at 27 cm so excavation is discontinued.
9	0 – 20 cm	Dark-brown with charcoal, daub, sherds.
10	0 – 40 cm	Grey, silty clay with sherds, ash lenses and bone. Excavation limited to just the western half of the unit.

since *all* sherds were passed through a series of mesh sieves for the estimation of sherd area and *only* the sherds that ended up in the smallest screens (-3.5 phi or 11.31 mm) were used in identifications. In this way, comparability between the collections was maintained for subsequent decorative analyses.

Excavations were conducted on July 20-21, 1996 and July 28, 1996 with the assistance of John House of the Arkansas Archeological Survey and the students and staff of the University of Washington Field School in Southeast Missouri. Based on the cores, I initially placed four (1x1-m) units five meters apart across the low lobe on the east side of the levee in an area that appeared to have a high potential for midden deposit. The sediment in these units was extremely hard packed and difficult to excavate. While we recovered some cultural material, we did not encounter the dark grayish brown sediment with high densities of artifacts that is characteristic of other midden deposits. We found the only indication of intact deposits in this area in Unit 3 where we encountered a small (20-cm) circle of slightly darker material that may have been a pit or post-hole. Due to these conditions, we ceased all excavations in units at the 20-cm level.

We moved our excavations to the west side of the road and along the northern edge of the area facing the old river channel. Although I had not extensively cored this portion of the site, I had noted abundant pottery along the upper slope of this area when the ground was visible in high grass. Initially, I began four additional units that I soon expanded to six. In general, the sediment in this area was much easier to excavate and our work produced greater numbers of artifacts. The presence of historic material along with dark gray/brown sediment with abundant charcoal, daub, bone and sherds suggested that while there had been historic disturbance here, at least some of this area contains midden deposit.

In Unit 6, we encountered a large red-slipped bowl fragment at 20-cm. The size of this sherd suggests that at least some of the deposit in this area of the site had not been, or was infrequently, plowed (Dunnell and Simek 1995). This suspicion was confirmed in Unit 8 where we encountered pit features at 40-cm. Excavations, therefore, were halted in these units at 40-cm. Table 5.7 describes the excavated units for Castile Landing.

In addition to the 1x1-m excavations, I also excavated a small exploratory trench into the eroding slope facing the old St. Francis River channel on the west side of the levee road (Figure 5.38). During my surface examination, I had noticed relatively dense amounts of sherds, daub, and bone eroding out of this slope and I suspected that there might be intact midden in this area of the site. To evaluate this suspicion, I excavated a 50-cm wide trench that extended 1.5 meters into the side of the slope. Figure 5.40 depicts the plan and profile of this trench. The profile of the trench revealed mottled gray/brown silty clay sediment with small flecks of charcoal and small pieces of daub extending down to about 60-cm below the upper surface. This material appeared to be mound fill. At 60-cm, I encountered a thick layer of charcoal and red, burned sediment across the unit that was suggestive of a burned surface. In this layer of charcoal, I also found thick deposits of burned grass and sticks as well as heavily burned sherds that lay horizontally on the surface. Although the area exposed by the trench was small, the presence of the burned surface and mottled sediment suggests that my excavations cut through the portion of a mound (based on the sediment deposited on top of the burned area) that included the remains of a burned house.

Summary

In all, we collected 2,629 sherds at Castile Landing. The small size of this assemblage was due to the cost, in terms of time, of excavation. Relative to surface collection, excavation is a slow means for generating large samples. The sample is large enough, however, to evaluate the reliability of the original PFG sample.

One of the main contributions of the research at Castile Landing is information about this poorly known but geographically important late-prehistoric deposit. Due to massive modification of the landscape around and including the Castile Landing site, it was not initially clear if parts of the deposit were extant and accessible. The results of the surface collections, coring, and excavations demonstrate that a substantial portion of the deposit remains. Future research at Castile Landing is clearly possible.

FIELDWORK SUMMARY

Beck, Belle Meade, Cranor Place, Holden Lake, Nickel, and Rose Mound are excellent deposits for evaluating the degree to which the PFG collections reflect the archaeological record or the processes of sampling. Each location consists of a large, late prehistoric midden that is composed of abundant household material, ceramics, and burials. At the time of the field work, each deposit was under cultivation. Although the Castile Landing deposit is not currently tilled, preliminary investigations suggested that disturbed, mixed, midden material is extant and easily accessible through excavation. These seven deposits are comparable and well suited for producing the data required to evaluate the research questions proposed in Chapter 4, including assessing variability due to sampling.

Table 5.8 summarizes the collection efforts. Over the course of the project, I built an assemblage of 50,366 sherds from one hundred and forty-nine 10x10-meter surface collection units and from 2.65 cubic meters of excavated material. The results of the project demonstrate how cost effective surface collection is in terms of the rate of data generation when compared to excavation. The sherds recovered per person-day for the 10x10-meter units are well over an order of magnitude greater than the excavations. These results neatly demonstrate one of the significant advantages of surface work over excavation: when sample size is critical, excavations are not cost effective. In addition, it highlights one of the reasons why the PFG survey has had such a

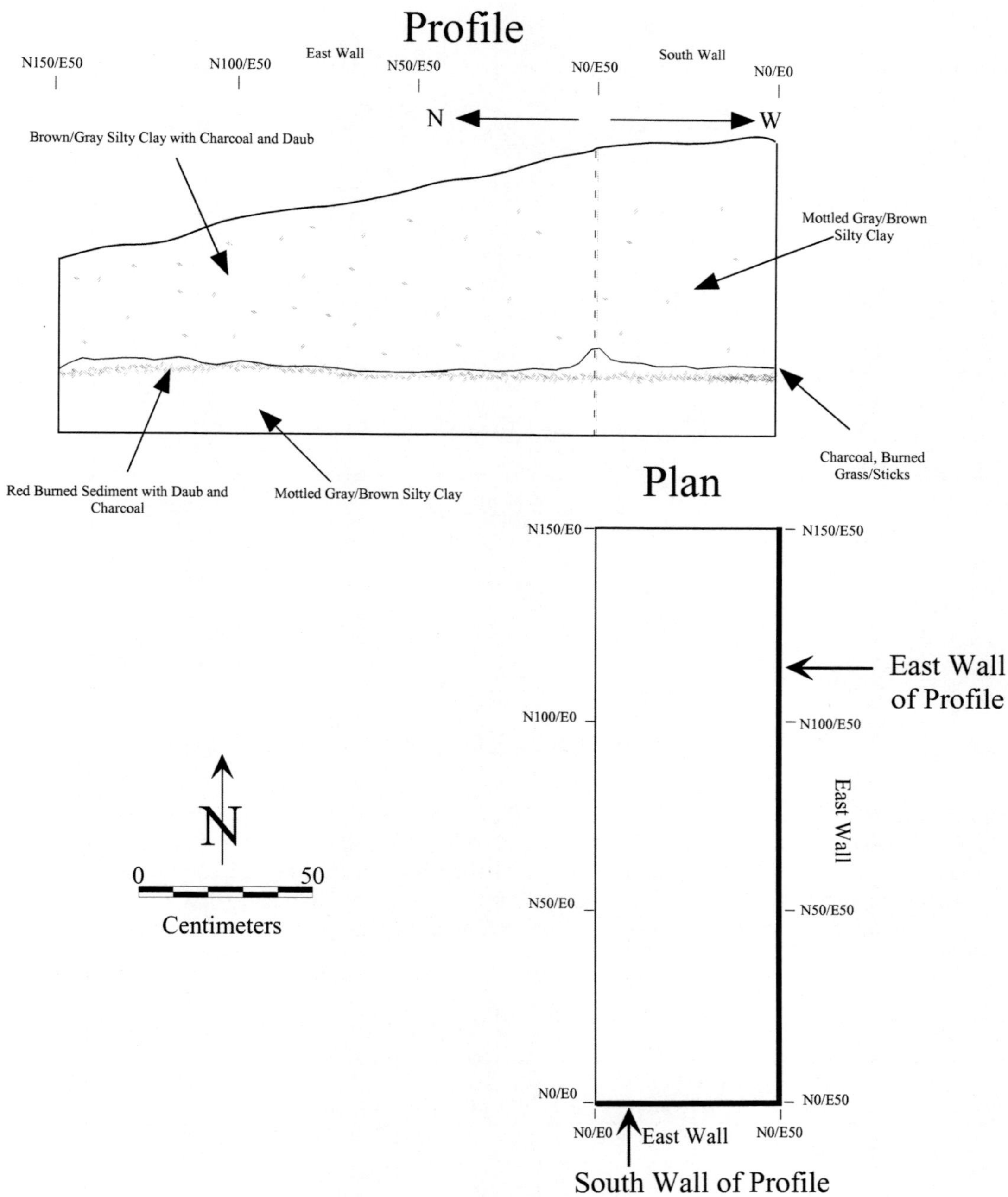

Figure 5.40. Plan and Profile of Exploratory Trench at Castile Landing.

Table 5.8. Summary of Collections.

Location	10x10-m Units Collected	Volume of Excavated Material (m^3)	Sherds	People	Days	Person-Days[a]	Sherds/ Person-Day
Beck	28	--	6569	3	2	6	1095
Belle Meade	11	--	8157	3	2	6	1360
Castile Landing	--	2.65	2629	13	3	39	67
Cranor Place	13	--	3612	3	1	3	1204
Holden Lake	20	--	12341	9	2	18	686
Nickel	53	--	9604	12	5	16[b]	600
Rose Mound	24	--	7454	3	2	6	1242
Σ	149	2.65	50366	46	17	22	6254

[a]Person-days are calculated by multiplying the number of people involved in collections by the number of days worked. Workdays lasted about ten hours.
[b]In the case of the Nickel collections, not all of the individuals worked through the entire period.

Table 5.9. Summary of Sherd Counts for all Project Assemblages.[a]

Assemblage	Σ Plain	Σ Decorated	Σ Total
Beck	5504 ± 133	1443 ± 59	6569 ± 270
Belle Meade	6831 ± 425	2524 ± 107	8157 ± 336
Castile Landing	2383 ± 219	388 ± 16	2629 ± 108
Cranor Place	3326 ± 111	599 ± 25	3612 ± 149
Holden Lake	12043 ± 233	826 ± 34	12341 ± 508
Nickel	8538 ± 234	2036 ± 84	9604 ± 395
Rose Mound	6715 ± 370	1315 ± 54	7454 ± 307
Σ	45340 ± 1727	9131 ± 375	50366 ± 2073

[a]These counts were derived from the initial counting of the collections. The relatively large error terms for the counts of decorated and plain sherds were caused by having several individuals make each of the counts over a period of several days. Each unit bag was counted separately and added together and, as a result, each collection count produced slightly different results when summed together. Ultimately the decorated sherds were narrowed down to "absolute" counts that were used in the seriation analyses. This was accomplished by physically numbering and creating an entry on a form for each decorated sherd.

Table 5.10. Comparison of PFG Samples with Lipo Collections.[a]

Assemblage	PFG Decorated Samples	Lipo Decorated Samples	Proportion of Lipo Samples to PFG Samples (Σ PFG/ Σ Lipo)
Beck	80	1443 ± 59	18.0
Belle Meade	256	2524 ± 104	9.9
Castile Landing	395	388 ± 16	1.0
Cranor Place	152	599 ± 25	3.9
Holden Lake	--	826 ± 34	--
Nickel	201	2036 ± 84	10.1
Rose Mound	245	1315 ± 54	5.4

[a] Samples included all decorated sherds of all temper types.

Table 5.11. Differences in Proportions that Can Be Statistically Distinguished at Confidence Level of α=0.005 Given Sample Size.

Assemblage	Sample Size	p_1=1.0%	p_1=25.0%	p_1=50.0%
Beck	1443	0.006	0.031	0.036
Belle Meade	2524	0.005	0.024	0.028
Castile Landing	388	0.010	0.059	0.070
Cranor Place	599	0.009	0.047	0.057
Holden Lake	826	0.008	0.041	0.048
Nickel	2036	0.005	0.026	0.031
Rose Mound	1315	0.006	0.026	0.048

lasting impact on our knowledge of the Mississippi river valley prehistory. Since PFG focused nearly exclusively on surface collections, their work produced large (for the time), comparable samples.

At each of the selected locations, I was able to collect large numbers of ceramics in a comparable manner. Table 5.9 lists the total counts of plain sherds, decorated sherds, and total sherds for each assemblage made from the surface surveys and excavations. Since counting objects always involves error, I counted each assemblage three times to calculate a mean and a standard deviation for the number of plain and decorated sherds in each collection. Decorated sherds were those identified with any kind of modification to their surface. These modifications included punctation, slipping/painting, incising/engraving, appliqué, and noding.

Table 5.10 compares the size of the original PFG samples with the new samples. The samples vary from 1.0 (Castile Landing) to 18.0 times (Beck) as large as the PFG samples. With the exception of Castile Landing, the samples are at least several times larger than the original PFG samples.

Given these sample sizes, it is now possible to determine the size of the discriminations that can statistically be made between proportions of types in pairs of assemblages. We can determine the differences in proportions that can be statistically distinguished by setting $\Delta p = p_1 - p_2$, where p_1 and p_2 are the proportions being compared and substituting this variable into Equation 5.1. This algebraic manipulation results in:

Equation 5.4.

$$z = -\frac{\Delta p}{\sqrt{\frac{(1-p)p}{n} + \frac{(1-\Delta p - p)(\Delta p + p)}{n}}}$$

where p is the proportion of a type in an assemblage, Δp is the difference between p for one assemblage and another, n is total sample size for the smaller sample, and z is the z-score statistical value. If we solve Equation 5.4 for Δp, we find that:

Equation 5.5.

$$\Delta p = \frac{-(-1+2p)z^2 + \sqrt{(-1+2p)^2 z^4 - 4p(-2+2p)z^2(n+z^2)}}{2(n+z^2)}$$

Using Equation 5.5 we can determine the differences in proportions that can be statistically distinguished using the new samples of decorated ceramics collected in my fieldwork.[1]

Table 5.11 shows the differences in proportions that can be statistically distinguished at confidence level of α=0.005, given the size of the newly collected samples. The differences in proportions that can be statistically resolved depend upon the particular sample size and the proportions being compared. Distinguishing the proportions of a type in two samples when the proportions are small (*e.g.*, < 10.0%) or large (*e.g.*, > 90 %) is statistically easier than distinguishing proportions that are closer to 50% (*e.g.*, 40-60%). For example, if two samples of a type were being compared with proportions (p) of 3.6% and 4.8%, respectively, the sample size requirements would be less than if samples with 48.2% and 50.4% where being compared despite the fact that the absolute differences between the frequencies is the same (i.e., Δp = 2.2%). Conversely, given two samples of particular sizes, one would be able to statistically resolve the smallest differences in frequencies at a specified confidence level (e.g., α=0.005) when the frequencies are either small or large. Thus, since the sample size requirements vary by the magnitude of the proportions, Table 5.11 has three columns to provide example values for Δp at 1.0%, 25.0%, and 50.0%. The values in the table demonstrate that I will be able to discriminate, in the worst-case (when proportions are closer to 50%) differences of 3-7% and in the best-case (when compared proportions are close to 1%) differences of less than 1%.

[1] For example, if we are examining an assemblage, *a*, where n_a = 1,000 and the proportion of a type (p) in the assemblage is 20.0% and we are interested in determining differences in proportions that can be statistically resolved at an α=0.005 confidence level, we set z = 1.96, p = 0.20 and n = 1,000 in Equation 6.2 and solve for Δp. These values result in Δp =0.038 and indicate differences less than 3.8% are not significant at the α=0.005 confidence level. In this case we could confidently report that 23.8% of type p in a second sample was greater than 20% of the type in the first sample assuming that the second assemblage had at least 1,000 sherds.

Several steps are required to compare these new collections with the ones generated by PFG, and to compare new seriations, results to those from PFG assemblages. First, we must describe the collections using observational units comparable to those used by PFG (*i.e.,* historical types). Second, we must determine whether the collection units for each location can be combined to form a single assemblage. I turn to these tasks in Chapter 6.

CHAPTER 6: COLLECTION DESCRIPTIONS AND UNIT AGGREGATION

Employing the protocol described in Chapter 5, I generated seven large ceramic collections in locations previously collected by PFG in the St. Francis river basin. These collections were made in a manner such that that I could estimate sources of variability that impact the frequencies of stylistic type assignments and that potentially influenced the results of the seriation analyses in Chapter 4. Controlling the means by which the ceramics were collected minimizes the contribution of variability in collection technique from the seriation results. In addition, the large assemblage size permits one to estimate the frequency for stylistic types with great precision and minimize the statistical effect of differences between assemblages caused by sample size.

In the first step of the analysis, each decorated sherd was assigned to a culture-historical type using criteria derived from PFG's type descriptions. Second, I evaluated the homogeneity of each collection as a means of assessing the assumption that each ceramic assemblage is composed of samples from a single population and is not the product of the aggregation material from chronologically and/or functionally distinct deposits.

CULTURE-HISTORICAL TYPE ASSIGNMENTS

Differences in counting produces results that are unrelated to the structure of the archaeological record (Orton 1993). Consequently, in order to make unbiased comparisons between the new collections and the PFG assemblages, it is necessary to describe the sherds using the same culture-historical types as PFG. Although PFG are relatively explicit about their type definitions, they include a combination of extensional descriptions with intensional defining criteria. As a result, single set of attributes cannot always be used to distinguish types. For example, the type Wallace Incised is described by PFG as consisting of shell-tempered sherds with wide linear incisions that are both recti- and curvilinear in pattern. This definition, however, overlaps the definition of the Rhodes Incised since it is also defined by wide, curvilinear incisions. Often, from the descriptions a sherd might meet the definitions for several types. Part of the problem in the PFG classification is that while it is applied to sherds, it occasionally makes use of information that is only detectable for whole vessels. PFG (1951:64) admit this problem and state, "let us not forget that we are classifying vessels, though we have for most types only sherds to do it with." This focus on vessels as counting units and types built at the scale of whole pots produces inconsistencies when actual type assignments occur with sherds. Parkin Punctated, for example, can usually be assigned to a sherd of any size provided it is large enough to bear a recognizable portion of a punctate. Assigning Rhodes Incised and Wallace Incised, on the other hand, involves detailed knowledge of decoration design (*i.e.,* particular combinations of sets of elements) and thus requires sherds that represent larger portions of vessels. Similarly, other types, such as Kent Incised, require information about sherd orientation relative to the complete vessel for assignment.

The confusion between sherd and vessel attributes clearly represents a step towards a future that PFG tried to ward off. As PFG (1951:66) point out, the pottery type "has little, if any, correspondent cultural 'reality,' present or past. Exigencies of language require us to think and talk about pottery types as though they had some sort of independent existence." But when the attributes of sherds become mixed with attributes of whole vessels, PFG (1951:66) move from treating types as "a named abstraction, representing a combination of selected characters," to discussing types as real "kinds" of vessels associated with attributes not involved in their definitions. Thus, while Ford's initial types for the Lower Mississippi river valley (Ford 1936a) approach the paradigmatic ideal (Dunnell 1971; Lyman *et al.*1997), PFG began to confuse the ontological nature of types in their descriptions. It is possible that Phillips' in-depth knowledge of decorative variability on whole vessels gained from his dissertation research (Phillips 1939) helped move the PFG discussion of types as arbitrary classes towards types as empirical entities. His later work (*i.e.,* Phillips 1970) certainly completes a move in this direction. By 1970, Phillips' type-varieties are based on PFG types but often lack sorting criteria and are explicitly based on impressions of whole vessels.

Although the PFG type descriptions include a combination of descriptive and definitional attributes (*sensu* Dunnell 1971), it is possible to establish a set of attributes that can be used to assign sherds as one or another PFG type. As a first step in making my counts comparable to the PFG data, I simplified the PFG definitions to create an unambiguous set of criteria. This refinement resulted in the key for assigning types as shown in Figure 6.1. The key makes use of pairs of dichotomous, discrete traits so that decision making during assignment is efficient, simple, and reproducible. Although there is no way to know *a priori* whether these decisions accurately reflect the way in which PFG made their assignments, once the assignments are done I will be able to compare my counts with the counts made by PFG for the same assemblage. If the frequencies of types in the new and PFG assemblages are statistically distinguishable, subsequent analyses will have to determine whether differences are caused by counting or by compositional properties of the assemblages. Varying the defining criteria by lumping and splitting types used to make assignments can assist in this determination.

A key component of the decorated type classification key is the identification of material added to the vessel paste as temper. For the purposes of the culture-historical type key, three kinds of tempered paste were defined: shell, sand, and clay. Shell-tempering was identified as any paste with visible shell in broken sherd cross-sections. Shell appears as white or gray-white platy shapes. In the case of leeched

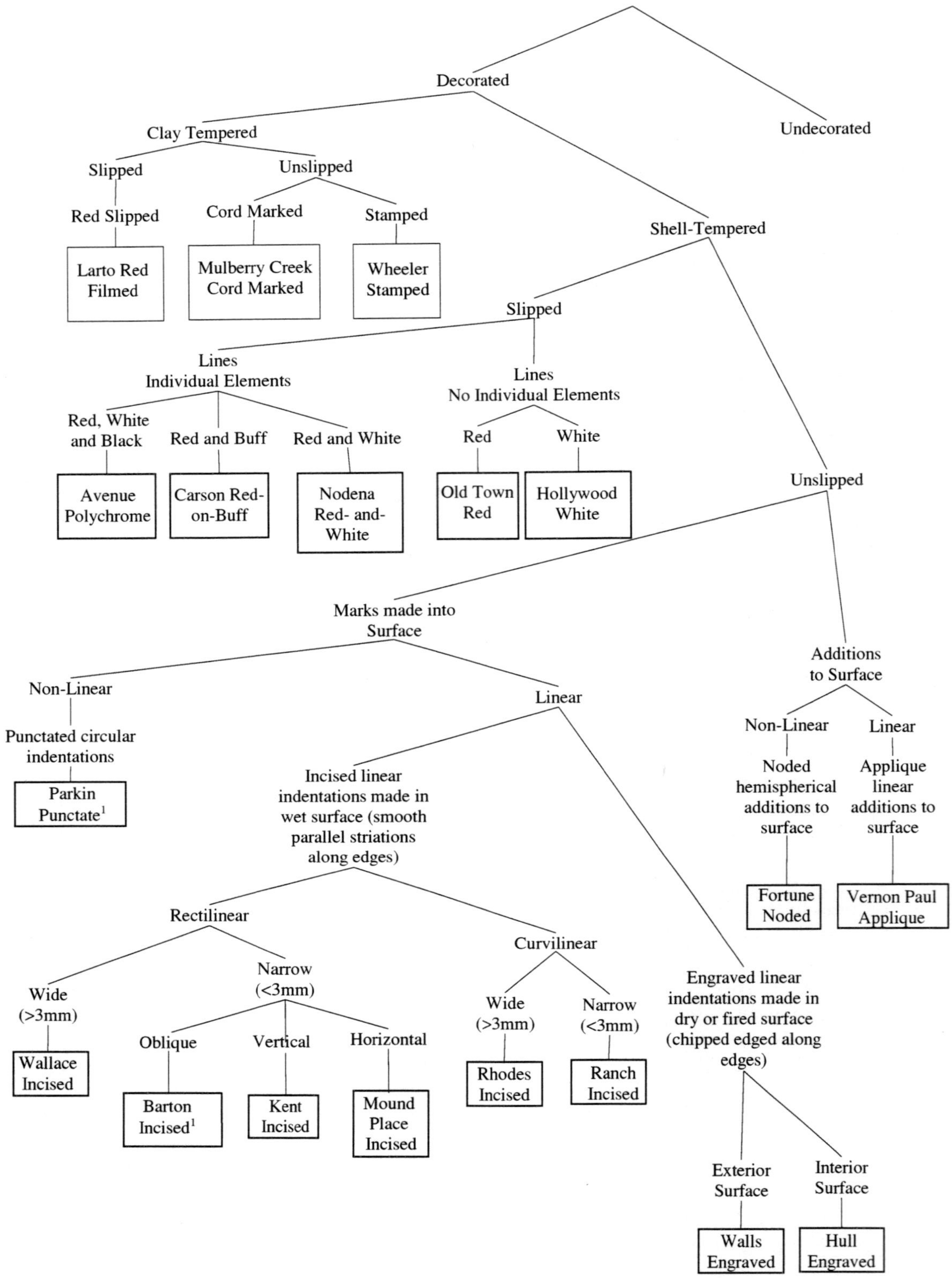

Figure 6.1. Key to PFG Decorate Ceramic Types that Occur in Study Area.

[1]Sherds with incisions and punctation are identified as Barton Incised.

sherds, shell-tempering was identified by the platy voids left by the shell particles.[1] Clay tempering refers to paste tempered with clay instead of shell. This tempering is identified as blocky grains in sherd cross-sections that are the same or different color as the surrounding paste. And posed a problem for PFG (1951) in trying to distinguish sand naturally present in the clay and "temper" that was added. Fortunately, sand was not common in the Lower St. Francis Basin and none occurs in Late Mississippian ceramics.

These definitions of the culture-historical types were easily and efficiently assigned across all sherd size classes. Generally, if a sherd is recognized as decorated then the PFG type is also recognizable since the level of assignment is restricted to decoration technique and form rather than composition and design. This is an important feature for constructing robust, reproducible type frequencies for assemblages of small sherds.

Descriptions are given for Parkin Punctated, Barton Incised, Kent Incised, Wallace Incised, Rhodes Incised, Ranch Incised, Walls Engraved, Hull Engraved, Fortune Noded, Vernon Paul Appliqué, Carson Red-on-Buff, Nodena Red-and-White, Old Town Red, Hollywood White, Avenue Polychrome, Larto Red Filmed, Mulberry Creek Cord Marked, and Wheeler Stamped. Not all of the PFG types could be rigorously defined with dichotomous attributes. Manly Punctated, for example, is described by PFG (1951:147) as a "tentative" type that has "punctates in an incised meander, usually a single line on the upper shoulder area making an arcaded effect around the vessels." PFG recorded instances of Manly Punctated in only two locations in the St. Francis and Memphis collections (Neeley's Ferry [11-N-4] and Barton Ranch [11-O-10]). I did not include Manly here because no set of dichotomous traits visible solely on sherds could be found to clearly distinguish it from Parkin Punctated or Barton Incised in the PFG description (PFG 1951:147). Ultimately, other researchers have found the type to have an inconsistent distribution and not useful as a historical type. Manly Punctated was relegated to a variant of Matthews Incised in the work of Phillips (1970) and eventually abandoned as type by subsequent researchers (*e.g.,* Brown 1998; House 1991; Lyman 1994). Manly, however, is a dominant form on the Malden Plain north of the St. Francis area (R. Dunnell, personal communication).

Below are the shell-tempered types that I used to describe my assemblages and the criteria used for assignments. Following PFG, these descriptions utilize only decoration on body sherds (*e.g.*, non-rim and non-base portion of vessels).

[1] Shell leeching did not turn out to be a problem since the St. Francis-type sites are basic with pH values of ca. 7.0-9.0 (R. Dunnell, personal communication).

PARKIN PUNCTATED

All shell-tempered ceramics bearing only indentations on the exterior surfaces were assigned to Parkin Punctated. Sherds with punctates located only on the rim are not counted as Parkin Punctated. Following PFG, sherds with both punctations and incisions were assigned to Barton Incised (see below).

BARTON INCISED

I assigned all shell-tempered ceramics with rectilinear, parallel, incised lines on the exterior surface and that are obliquely oriented and less than 3-mm in width to Barton Incised. Like all the "incised types," incisions must have smooth striations along their surfaces, indications of lines being drawn onto wet clay (*i.e.,* before the clay was dried or fired). Only sherds that can be oriented in terms of up/down vs. left/right were assigned to Barton Incised. Many sherds, for example, have parallel lines but were not assigned to Barton Incised since orientation was impossible. These sherds were described as "Incised Indeterminate."

KENT INCISED

All shell-tempered sherds with vertically oriented, rectilinear, parallel-incised lines were assigned to Kent Incised. Importantly, *only* sherds that I could orient in terms of up/down vs. left/right were assigned to Kent Incised. Incisions must have parallel striations indicating they were drawn on a wet surface.

MOUND PLACE INCISED

All shell-tempered sherds with horizontally-oriented, rectilinear, parallel-incised lines were assigned to Mound Place Incised. Like Kent Incised, *only* sherds that I could orient in terms of their position on a vessel (*i.e.,* up/down vs. left/right) were described as Mound Place Incised. Incisions must have parallel striations indicating they were drawn on a wet surface.

WALLACE INCISED

I assigned all shell-tempered ceramics with rectilinear incised lines that were at least 3-mm in width to Wallace Incised. In addition, incisions must have parallel striations indicating they were drawn on a wet surface. Some researchers (*e.g.,* O'Brien and Fox 1994) have chosen to include curvilinear elements in their definition of Wallace Incised. O'Brien and Fox (1994:38), for example, argue that "the only other type exhibiting wide, curvilinear elements on the body is Rhodes incised, though sherds of that type are easily distinguishable." To make this distinction, O'Brien and Fox argue that Rhodes Incised has a distinctive spiral that does not appear on Wallace Incised. However, in order to allow a consistent assignment of types, I follow the strategy of Brown (1998:20) and define Wallace as having *only* rectilinear lines with widths greater than 3-mm. All curvilinear incised patterns were assigned to either Rhodes or Ranch Incised.

RHODES INCISED

All shell-tempered ceramics with curvilinear incised lines that were at least 3-mm in width were assigned to Rhodes Incised. Incisions must have parallel striations indicating they were drawn onto a wet surface.

RANCH INCISED

All shell-tempered ceramics with curvilinear incised lines that were less than 3-mm in width were assigned to Ranch Incised. Incisions must have parallel striations indicating they were drawn onto a wet surface.

WALLS ENGRAVED

All shell-tempered or clay-tempered sherds (Bell Plain paste) that exhibited fine lines incised into the exterior surface when the paste was already dry or fired (as evidenced by chipping on the margins of the lines) were assigned to Walls Engraved.

HULL ENGRAVED

Hull Engraved sherds were described as shell-tempered or clay-tempered (Bell Plain paste) with fine lines cut into the interior surface when the paste was already dry or fired (as evidenced by chipping on the margins of the lines and into the interior surface). I identified the interior surfaces of sherds as those with concave curvature in the horizontal axis.

FORTUNE NODED

Shell-tempered sherds with hemispherical or conical nodes of clay added to the exterior surfaces were assigned to Fortune Noded.

VERNON PAUL APPLIQUÉ

I assigned all shell-tempered sherds with strips of clay added to their exterior surface in a vertical orientation as Vernon Paul Appliqué.

OLD TOWN RED

All shell-tempered sherds with red slip covering on one or both surfaces and without any gaps in the coverage of the slip across the surface (*i.e.,* no indication of lines or individual elements) were described as Old Town Red.

CARSON RED-ON-BUFF

All shell-tempered sherds with red slipping not entirely covering one or both sherd surfaces (indicating individual elements or lines) were assigned to Carson-Red-on-Buff.

NODENA RED-AND-WHITE

All shell-tempered sherds with separate areas of red-and-white slipping on one or both sherd surfaces were described as Nodena Red-and-White. This type also includes sherds with red, white, and buff slip.

AVENUE POLYCHROME

All shell-tempered sherds with separate areas of red and white slip as well as black paint on one or both sherd surfaces were assigned to Avenue Polychrome.

HOLLYWOOD WHITE

I assigned all shell-tempered sherds with *only* white slipping on one or both surfaces to Hollywood White.

In order to make the descriptions of the assemblages comparable to the PFG descriptions, in addition to the shell-tempered types several, coarse, clay-tempered, decorated types were also used to describe sherds in the PFG descriptions. Based primarily on the results of PFG's seriations (PFG 1951) and subsequent work (*e.g.*, Chapman 1989; Dunnell and Feathers 1991; Feathers 1990; Ford 1963), clay-tempered types are known to have an earlier temporal distribution than shell-tempered types. Two kinds of decorated, clay-tempered ceramics are described: Larto Red-Filmed and Wheeler Check-Stamped.

LARTO RED-FILMED

All clay-tempered sherds with red slip on one or both sherd surfaces were assigned to Larto Red-Filmed (PFG 1951:102-104).

WHEELER CHECK-STAMPED

Wheeler Checked-stamped sherds were described as those sherds with clay tempering and the impressions of a checkered or grid-like stamp on their surface.

SHERD DESCRIPTION RESULTS

Using these PFG type definitions and following the key in Figure 6.1, I assigned each sherd from my new collections of the PFG locations. If a sherd did not match any of the criteria, it was given a description of "unidentified." Both

Table 6.1. Counts of PFG Culture Historical Types Assigned in Lipo Assemblages.

Collection	Parkin Punctated	Barton Incised	Old Town Red	Fortune Noded	Ranch Incised	Kent Incised	Walls Engraved	Wallace Incised	Rhodes Incised	Mound Place Incised	Vernon Paul Appliqué	Carson Red-On-Buff	Hull Engraved	Hollywood White Slipped	Larto Red Slipped	Wheeler Checkstamped	Σ
Beck	558	346	61	10	20	27	18	8	12	3	7	2	1	2	0	0	1080
Belle Meade	795	395	29	12	30	13	25	15	11	3	5	3	2	0	0	0	1338
Castile Landing	163	20	45	1	0	1	0	1	0	2	0	0	0	0	0	0	233
Cranor Place	121	106	35	6	7	4	4	2	1	4	0	0	0	1	3	0	320
Holden Lake	27	233	3	24	2	3	0	2	1	4	3	0	0	1	0	1	304
Nickel	622	256	136	8	5	14	8	14	1	6	7	17	4	2	1	0	1236
Rose Mound	395	259	55	18	3	2	0	3	1	3	1	3	0	0	0	0	744
Σ	2681	1615	364	79	67	64	55	45	27	25	23	18	12	5	5	3	5255

shell- and clay-tempered decorated type descriptions were used. In all, 5,255 sherds were described across the seven assemblages (Table 6.1). The differences between the numbers of decorated sherds listed in Table 5.9 and Table 6.1 are due to sherds that could not be assigned to a specific culture-historical type. Such a situation occurred when incisions could not be oriented on the sherd as in the case of Barton/Kent/Mound Place Incised), when punctates or incision were found only on the rim, or with peculiar kinds of appliqué (*e.g.,* faces, animal riders, etc.).

AGGREGATION OF UNITS

Before the counts of the culture-historical types can be compared between collections, however, it is necessary to test the assumption that the material from each unit was gathered from a single population or whether the units sampled multiple populations at each location. PFG were well aware of the potential for the collection locations to contain multiple deposits. Consequently, PFG routinely divided the collection of sites into smaller "areas" in order to minimize variability "whenever there was the slightest indication that [the collection areas] might differ in content" (PFG 1951:43). By selecting St. Francis-type settlements, a large component of which is redeposited trash, I sought to minimize spatial variability that could lead to chronologically and/or functionally mixed samples. There is no *a priori* guarantee, however, that this assumption is met in any particular collection.

It was for this reason that I collected all surfaces in 10x10-meter units. Evaluating the assumption of spatial homogeneity consists of testing a hypothesis that each of the units is a sample of a single larger population and that difference between the frequencies of PFG types in each unit is a function of sample size. Falsifying the hypothesis indicates that there are multiple populations that are being sampled by the 10x10-meter collection units. Multiple populations would occur if the collection units covered multiple deposits and if the deposits were chronologically and/or functionally (and therefore compositionally) distinct.

The evaluation of homogeneity is accomplished by a cluster analysis of the units measuring similarity as the abundances of PFG types in each unit. If the units are simply samples of a single population, the dendrogram structures formed should not be spatially contiguous and produce clusters but rather should be randomly distributed. Thus, if the variance in frequencies between spatially contiguous collection units is relatively the same as variance within units (*i.e.,* it is not possible to create spatial clusters with similarity data), then lumping together samples produces samples representative of the entire, single deposit. If significant variance in similarity is measured between units (*i.e.,* spatial clusters can be created), then combining all collections into a single unit for analytic purposes is not warranted. Units that are statistically distinct from other units are either removed, or if the resulting groups are large enough, treated as separate analytic groups. Units are formed by combining the largest number of units that can be statistically placed together. These units become the basis for comparisons with the PFG assemblages.

BELLE MEADE

At Belle Meade, I described 1,338 sherds with PFG culture-historical types. Tables 6.2 and 6.3 gives the counts and the proportion of PFG types across all of the Belle Meade collection units. Inspection of these frequencies does not immediately suggest that particular units or groups of units are compositionally distinctive. Indeed, the frequencies of types for any unit are within one standard deviation of the

Table 6.2. Counts of Culture-Historical Types for Belle Meade Surface Collection Units.

Unit	Parkin Punctated	Barton Incised	Old Town Red	Fortune Noded	Ranch Incised	Kent Incised	Walls Engraved	Wallace Incised	Rhodes Incised	Mound Place Incised	Vernon Paul Appliqué	Carson Red-On-Buff	Hull Engraved	Σ
1	45	22	1	1	0	2	2	0	0	2	0	0	1	76
2	40	28	1	2	2	0	1	2	0	0	1	0	0	77
3	40	27	4	0	4	1	2	0	0	0	1	0	0	79
4	116	63	6	2	2	1	3	6	1	0	1	0	0	201
5	128	48	3	2	6	1	10	1	0	0	0	2	0	201
6	22	4	3	0	1	0	0	0	1	0	0	0	0	31
7	56	36	1	0	3	2	2	3	2	0	1	0	0	106
8	69	33	3	0	0	1	2	2	0	0	1	0	0	111
9	72	34	1	2	7	1	2	1	2	1	0	0	0	123
10	95	51	3	0	1	0	0	0	1	0	0	0	0	151
11	95	34	3	3	4	2	1	0	4	0	0	1	1	148
12	17	15	0	0	0	2	0	0	0	0	0	0	0	34
Σ	795	395	29	12	30	13	25	15	11	3	5	3	2	1338

Table 6.3. Proportions of Culture-Historical Types for Belle Meade Surface Collection Units.

Unit	Parkin Punctated	Barton Incised	Old Town Red	Fortune Noded	Ranch Incised	Kent Incised	Walls Engraved	Wallace Incised	Rhodes Incised	Mound Place Incised	Vernon Paul Appliqué	Carson Red-On-Buff	Hull Engraved	Σ
1	0.592	0.290	0.013	0.013	0.000	0.026	0.026	0.000	0.000	0.026	0.000	0.000	0.013	1.000
2	0.520	0.364	0.013	0.026	0.026	0.000	0.013	0.026	0.000	0.000	0.013	0.000	0.000	1.000
3	0.506	0.342	0.051	0.000	0.051	0.013	0.025	0.000	0.000	0.000	0.013	0.000	0.000	1.000
4	0.577	0.313	0.030	0.010	0.010	0.005	0.015	0.030	0.005	0.000	0.005	0.000	0.000	1.000
5	0.637	0.239	0.015	0.010	0.030	0.005	0.050	0.005	0.000	0.000	0.000	0.010	0.000	1.000
6	0.710	0.129	0.097	0.000	0.032	0.000	0.000	0.000	0.032	0.000	0.000	0.000	0.000	1.000
7	0.528	0.340	0.009	0.000	0.028	0.019	0.019	0.028	0.019	0.000	0.009	0.000	0.000	1.000
8	0.622	0.297	0.027	0.000	0.000	0.009	0.018	0.018	0.000	0.000	0.009	0.000	0.000	1.000
9	0.585	0.276	0.008	0.016	0.057	0.008	0.016	0.008	0.016	0.008	0.000	0.000	0.000	1.000
10	0.629	0.338	0.020	0.000	0.007	0.000	0.000	0.000	0.007	0.000	0.000	0.000	0.000	1.000
11	0.642	0.230	0.020	0.020	0.027	0.014	0.007	0.000	0.027	0.000	0.000	0.007	0.007	1.000
12	0.500	0.441	0.000	0.000	0.000	0.059	0.000	0.000	0.000	0.000	0.000	0.000	0.000	1.000
$\overline{\Sigma}$	0.594	0.295	0.022	0.009	0.022	0.010	0.019	0.011	0.008	0.002	0.004	0.002	0.002	1.000

Table 6.4. Counts of Culture-Historical Types for Beck Surface Collection Units.[ab]

Units	Parkin Punctated	Barton Incised	Old Town Red	Fortune Noded	Ranch Incised	Kent Incised	Walls Engraved	Wallace Incised	Rhodes Incised	Mound Place Incised	Vernon Paul Appliqué	Carson Red On Buff	Hull Engraved	Hollywood White	Σ
1	13	3	1	0	3	1	2	0	0	0	0	0	0	1	14
2	23	10	0	0	1	1	0	0	0	0	1	0	0	0	36
3	23	6	2	0	2	0	2	0	0	0	1	0	0	1	37
4	10	11	0	0	0	0	1	0	0	1	0	0	0	0	23
5	29	12	2	0	1	0	3	2	0	0	0	0	0	0	49
6	9	7	3	0	0	2	0	0	1	0	0	0	0	0	22
7	20	12	2	0	2	0	0	0	1	0	0	0	0	0	37
8	35	15	5	2	1	1	0	1	0	0	0	0	0	0	60
9	34	12	3	1	1	2	0	0	0	0	0	0	0	0	53
10	17	4	1	0	1	1	0	0	0	0	0	1	0	0	25
11	34	9	6	2	0	3	0	0	0	0	0	0	0	0	54
12	21	12	6	0	1	2	0	2	3	0	2	0	0	0	49
13	9	11	1	0	0	0	0	0	0	0	0	0	0	0	21
14	13	7	2	1	0	0	0	0	0	0	0	0	0	0	23
15	15	7	2	1	1	1	4	0	0	0	0	0	0	0	31
16	32	9	8	1	0	0	1	1	0	0	1	0	0	0	53
17	31	33	4	0	0	2	0	0	1	0	0	0	0	0	71
18	22	20	3	0	0	0	1	0	2	0	0	1	0	0	49
19	24	24	2	0	1	3	0	0	0	0	0	0	0	0	54
20	18	17	2	0	1	3	0	0	0	0	0	0	0	0	41
21	25	14	3	1	1	0	0	0	0	1	1	0	0	0	46
22	7	11	0	0	1	1	0	0	2	0	0	0	0	0	22
23	17	28	1	0	0	1	0	1	0	1	0	0	1	0	50
24	4	5	0	0	1	0	0	0	0	0	0	0	0	0	10
25	40	12	2	0	1	1	2	1	1	0	0	0	0	0	60
26	10	12	0	0	0	1	0	0	1	0	0	0	0	0	24
27	9	13	0	1	0	1	1	0	0	0	1	0	0	0	26
28	14	10	0	0	0	0	1	0	0	0	0	0	0	0	25
Σ Units 1-15	305	138	36	7	14	14	12	5	5	1	4	1	0	2	544
Σ Units 16-28	253	208	25	3	6	13	6	3	7	2	3	1	1	0	531
Σ All Units	558	346	61	10	20	27	18	8	12	3	7	2	1	2	1075

[a] No decorated sherds were collected from Unit 29.
[b] Units 1-15 and 16-28 are spatially contiguous and thus were evaluated separately.

Table 6.5. Proportions of Culture-Historical Types for Beck Surface Collection Units.[a]

Units	Parkin Punctated	Barton Incised	Old Town Red	Fortune Noded	Ranch Incised	Kent Incised	Walls Engraved	Wallace Incised	Rhodes Incised	Mound Place Incised	Vernon Paul Appliqué	Carson Red On Buff	Hull Engraved	Hollywood White	Σ
1	0.929	0.214	0.071	0.000	0.214	0.071	0.143	0.000	0.000	0.000	0.000	0.000	0.000	0.071	1.000
2	0.639	0.278	0.000	0.000	0.028	0.028	0.000	0.000	0.000	0.000	0.028	0.000	0.000	0.000	1.000
3	0.622	0.162	0.054	0.000	0.054	0.000	0.054	0.000	0.000	0.000	0.027	0.000	0.000	0.027	1.000
4	0.435	0.478	0.000	0.000	0.000	0.000	0.044	0.000	0.000	0.044	0.000	0.000	0.000	0.000	1.000
5	0.592	0.245	0.041	0.000	0.020	0.000	0.061	0.041	0.000	0.000	0.000	0.000	0.000	0.000	1.000
6	0.409	0.318	0.136	0.000	0.000	0.091	0.000	0.000	0.046	0.000	0.000	0.000	0.000	0.000	1.000
7	0.541	0.324	0.054	0.000	0.054	0.000	0.000	0.000	0.027	0.000	0.000	0.000	0.000	0.000	1.000
8	0.583	0.250	0.083	0.033	0.017	0.017	0.000	0.017	0.000	0.000	0.000	0.000	0.000	0.000	1.000
9	0.642	0.226	0.057	0.019	0.019	0.038	0.000	0.000	0.000	0.000	0.000	0.000	0.000	0.000	1.000
10	0.680	0.160	0.040	0.000	0.040	0.040	0.000	0.000	0.000	0.000	0.000	0.040	0.000	0.000	1.000
11	0.630	0.167	0.111	0.037	0.000	0.056	0.000	0.000	0.000	0.000	0.000	0.000	0.000	0.000	1.000
12	0.429	0.245	0.122	0.000	0.020	0.041	0.000	0.041	0.061	0.000	0.041	0.000	0.000	0.000	1.000
13	0.429	0.524	0.048	0.000	0.000	0.000	0.000	0.000	0.000	0.000	0.000	0.000	0.000	0.000	1.000
14	0.565	0.304	0.087	0.044	0.000	0.000	0.000	0.000	0.000	0.000	0.000	0.000	0.000	0.000	1.000
15	0.484	0.226	0.065	0.032	0.032	0.032	0.129	0.000	0.000	0.000	0.000	0.000	0.000	0.000	1.000
16	0.604	0.170	0.151	0.019	0.000	0.000	0.019	0.019	0.000	0.000	0.019	0.000	0.000	0.000	1.000
17	0.574	0.250	0.075	0.014	0.024	0.024	0.022	0.010	0.009	0.002	0.009	0.002	0.000	0.003	1.000
18	0.437	0.465	0.056	0.000	0.000	0.028	0.000	0.000	0.014	0.000	0.000	0.000	0.000	0.000	1.000
19	0.449	0.408	0.061	0.000	0.000	0.000	0.020	0.000	0.041	0.000	0.000	0.020	0.000	0.000	1.000
20	0.444	0.444	0.037	0.000	0.019	0.056	0.000	0.000	0.000	0.000	0.000	0.000	0.000	0.000	1.000
21	0.439	0.415	0.049	0.000	0.024	0.073	0.000	0.000	0.000	0.000	0.000	0.000	0.000	0.000	1.000
22	0.544	0.304	0.065	0.022	0.022	0.000	0.000	0.000	0.000	0.022	0.022	0.000	0.000	0.000	1.000
23	0.318	0.500	0.000	0.000	0.046	0.046	0.000	0.000	0.091	0.000	0.000	0.000	0.000	0.000	1.000
24	0.340	0.560	0.020	0.000	0.000	0.020	0.000	0.020	0.000	0.020	0.000	0.000	0.020	0.000	1.000
25	0.400	0.500	0.000	0.000	0.100	0.000	0.000	0.000	0.000	0.000	0.000	0.000	0.000	0.000	1.000
26	0.667	0.200	0.033	0.000	0.017	0.017	0.033	0.017	0.017	0.000	0.000	0.000	0.000	0.000	1.000
27	0.417	0.500	0.000	0.000	0.000	0.042	0.000	0.000	0.042	0.000	0.000	0.000	0.000	0.000	1.000
28	0.346	0.500	0.000	0.039	0.000	0.039	0.039	0.000	0.000	0.000	0.039	0.000	0.000	0.000	1.000
$\overline{\Sigma}$ Units 1-15	0.561	0.254	0.066	0.013	0.026	0.026	0.022	0.009	0.009	0.002	0.007	0.002	0.000	0.004	1.000
$\overline{\Sigma}$ Units 16-28	0.476	0.392	0.047	0.006	0.011	0.024	0.011	0.006	0.013	0.004	0.006	0.002	0.002	0.000	1.000
$\overline{\Sigma}$ All Units	0.519	0.322	0.057	0.009	0.019	0.025	0.017	0.007	0.011	0.003	0.007	0.002	0.001	0.002	1.000

frequencies of types for the total assemblage. The graphical comparison of the types in each unit shown in Figure 6.2 as proportional horizontal bars with confidence intervals (α=0.05) confirms an impression that the compositions of the units are statistically indistinguishable from one another as the error terms of types for each unit overlap with those of other units. The frequency of culture-historical types is variable between units. We can quickly evaluate whether there is significant patterning to the collection units by conducting a hierarchical cluster analysis of the units. If, on one hand, there are differences in type composition across the deposit, we should be able to statistically cluster sets of collection units that are spatially contiguous through cluster analysis techniques (Thompson 1978). On the other hand, if the units were drawn from a single homogenous deposit, one would expect that the results of a cluster analysis performed on the frequencies of types would not be spatially structured. Instead, the units should be randomly distributed in space.

Figure 6.3 shows the results of a hierarchical cluster analysis for the Belle Meade units. Clustering was conducted by the type counts and by calculating the distances as the chi-square measure of independence of rows and columns on 2-by-n frequency tables, formed by pairs of cases (see Whallon 1972). The chi-square metric is appropriately used when the data are counts of objects. Clusters were formed using Ward's method (Aldenderfer and Blashfield 1984; Ward 1963), which averages all distances between pairs of objects in different clusters, with adjustments for covariances, to decide how far apart the clusters are. Comparing the results of the cluster analysis with the map of the collection units (*i.e.*, Figure 5.10) shows that groups of similar units are not spatially contiguous.

Comparison of the unit composition and the results of the cluster analysis provides strong support that the units are all drawn from a single population. The differences in the frequencies of the types between units, therefore, are indistinguishable from the effects of sample size and other random processes and do not reflect differences in underlying composition. Thus, it is reasonable to aggregate all of the units into a single, larger collection as PFG did.

BECK

At Beck, I described 1,075 sherds with PFG culture-historical types from 28 10x10-m collection units. Counts and proportions of culture-historical types for each collection are given in Tables 6.4 and 6.5. Although the frequencies are relatively similar, the distribution of the type frequencies (Figure 6.4) suggest that there is much more variability in frequencies than seen at Belle Meade, particularly in the Parkin Punctated and Barton Incised frequencies. In order to examine whether this variability is spatially patterned, I used a chi-square metric on counts of sherds to measure similarity between units and Ward's method to form clusters. The results of this cluster analysis (Figure 6.5) suggest that variability in the abundance of types within each unit is structured into two large spatial clusters.

By comparing the clusters in Figure 6.5 with the distribution of units in Figure 5.29, we can see that units are not randomly included within clusters as a group. Instead, the units form two spatial sets that are spatially contiguous that are equivalent to the two collection areas (Units 1-15 and Units 16-29) made at Beck. The probability that this distribution of units across these clusters occurred by chance is low. The tabulation of units within groups and across clusters is shown in Table 6.6. With a chi-square value of 14.30, the configuration of units and clusters has only a 0.25% chance of occurring by chance alone.

Table 6.6. Tabulation of Surface Collection Units across Clusters and Surface Collection Groups Shown in Figure 6.5.[a]

Cluster	Surface Collection Group 1 (Units 1-15)	Surface Collection Group 2 (Units 16-28)
1	2	9
2	13	4

[a]The two clusters of units are significant at $p<0.01$ (X^2=9.123.).

The compositions of culture-historical types within each unit show that the two clusters have relatively distinct compositions of relative type abundances. The proportion of types that make up these two groups of units (Figure 6.4) indicates that the differences in these groups are, for the most part, due to differences in the relative abundances of the three most numerous types: Parkin Punctated, Barton Incised, and Old Town Red. Table 6.7 displays the counts of these PFG types for the two groups of units. The distribution of types is significantly distinct (p<0.0001). Table 6.8 shows the mean proportions and 95% confidence intervals of Parkin Punctated, Barton Incised, and Old Town Red for the two clusters. It is clear from inspection of these proportions that the distribution of the types is distinct.

Table 6.7. Counts of Parkin Punctated, Barton Incised, Old Town Red, and Other PFG Types in the Two Beck Collection Areas.[a]

Cluster	Parkin Punctated	Barton Incised	Old Town Red	Other	Σ
Units 1-15	337	147	44	74	602
Units 16-28	221	199	17	41	478

[a] The two clusters of units are significant at $p<0.0001$ (X^2=39.64.).

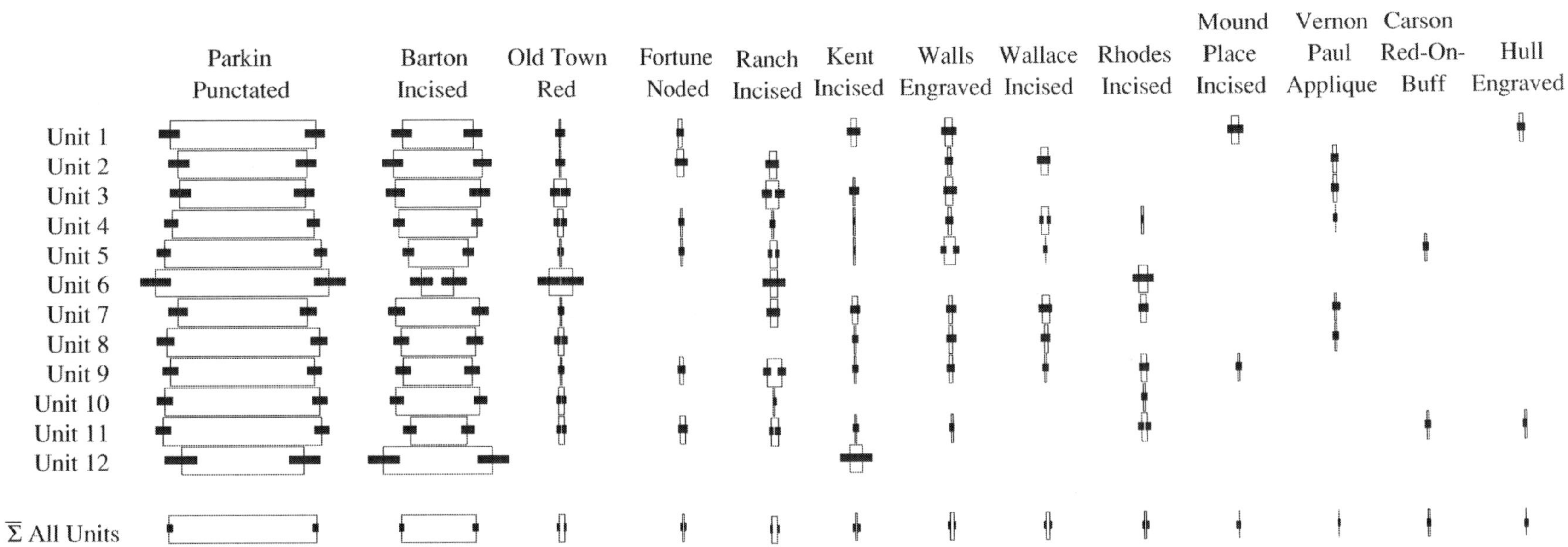

Figure 6.2. Comparison of the Proportions of Culture-Historical Types for the Belle Meade Surface Collection Units.

Confidence intervals of α=0.05 are shown for the relative proportion of each type. For example, for the frequency of any type in an unit there is a 95% chance that the population frequencies lies X% and Y%.

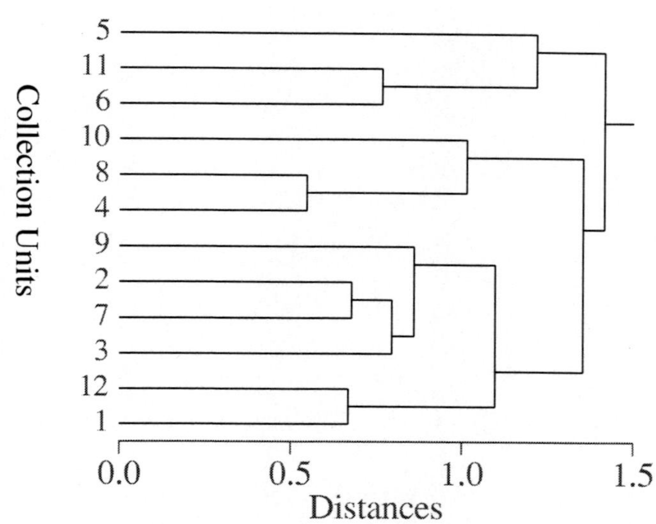

Figure 6.3. Hierarchical Cluster Analysis of Culture-Historical Type Frequencies from Belle Meade Collection Units.

Similarity of units was estimated from sherd counts and by calculating the distances as the chi-square measure of independence of rows and columns on 2-by-n frequency tables, formed by pairs of cases. Dendrogram was generated by Ward's method (Aldenderfer and Blashfield 1984, Ward 1963).

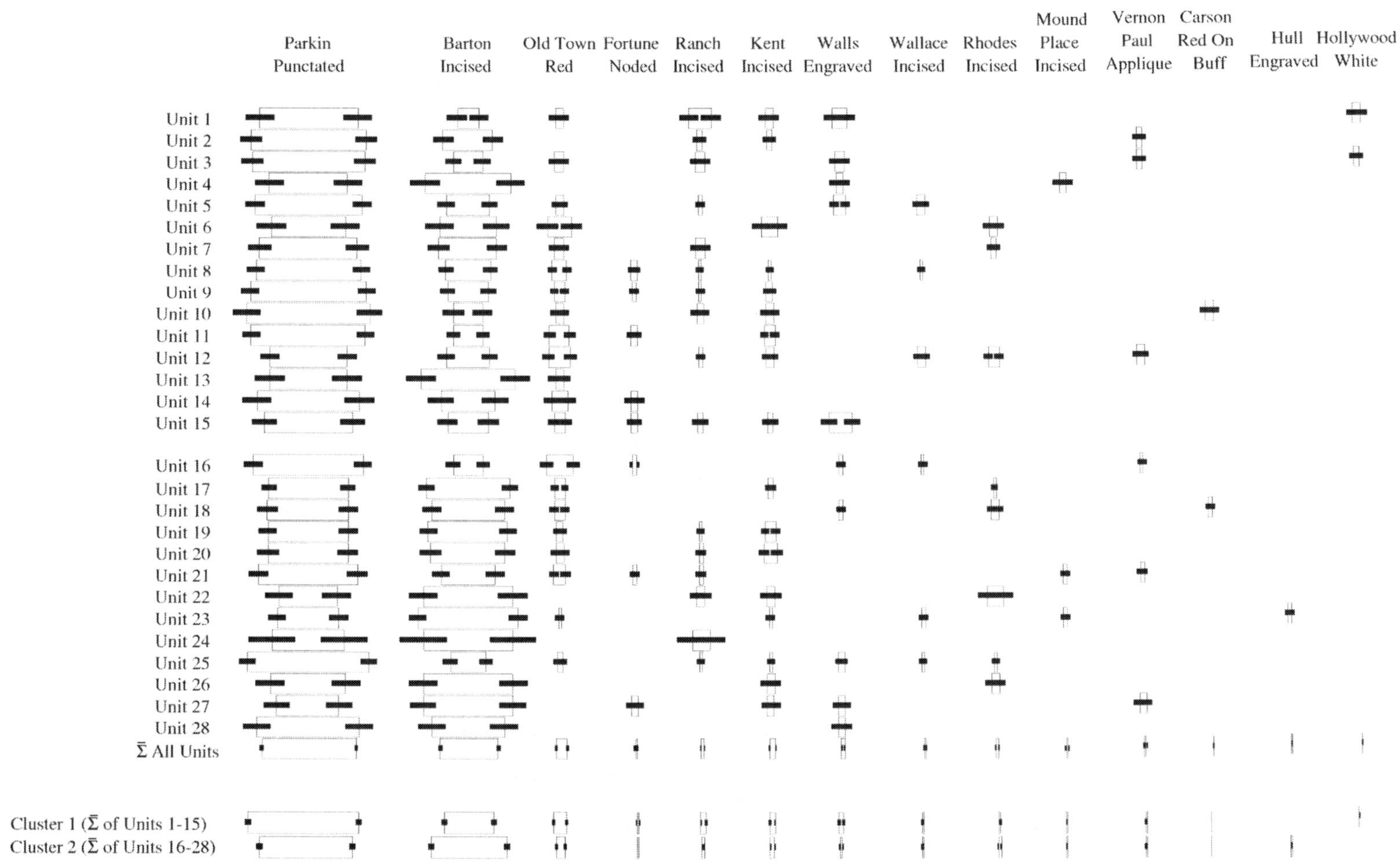

Figure 6.4. Comparison of the Proportions of Culture-Historical Types for the Beck Surface Collection Units.

Confidence intervals of α=0.05 (95%) are shown for the relative proportion of each type.

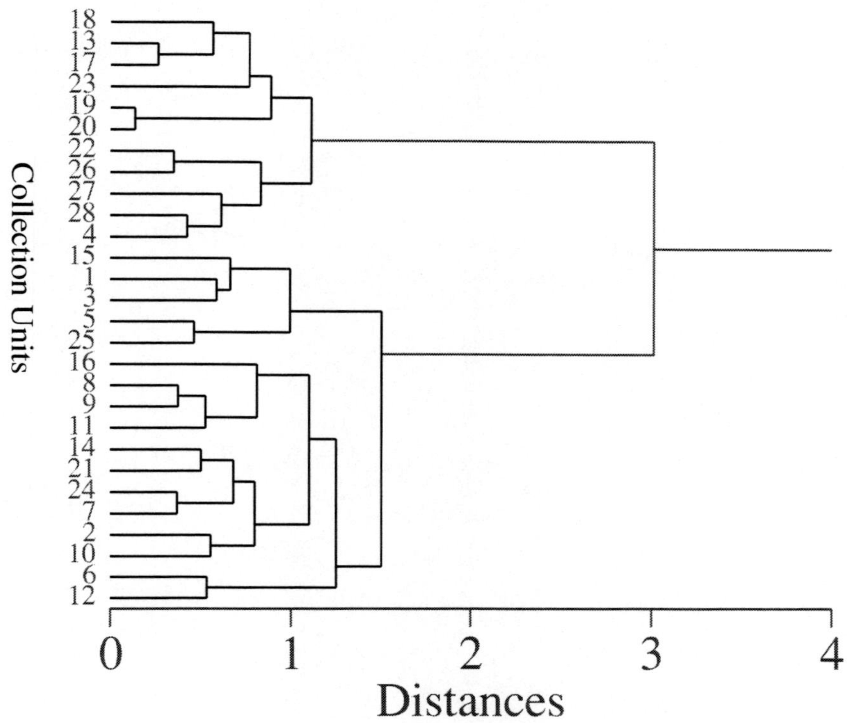

Figure 6.5. Hierarchical Cluster Analysis of Culture-Historical Type Frequencies from Beck Collection Units.

Similarity was estimated from sherd counts and by calculating the distances as the chi-square measure of independence of rows and columns on 2-by-n frequency tables, formed by pairs of cases. Dendrogram was generated using Ward's method.

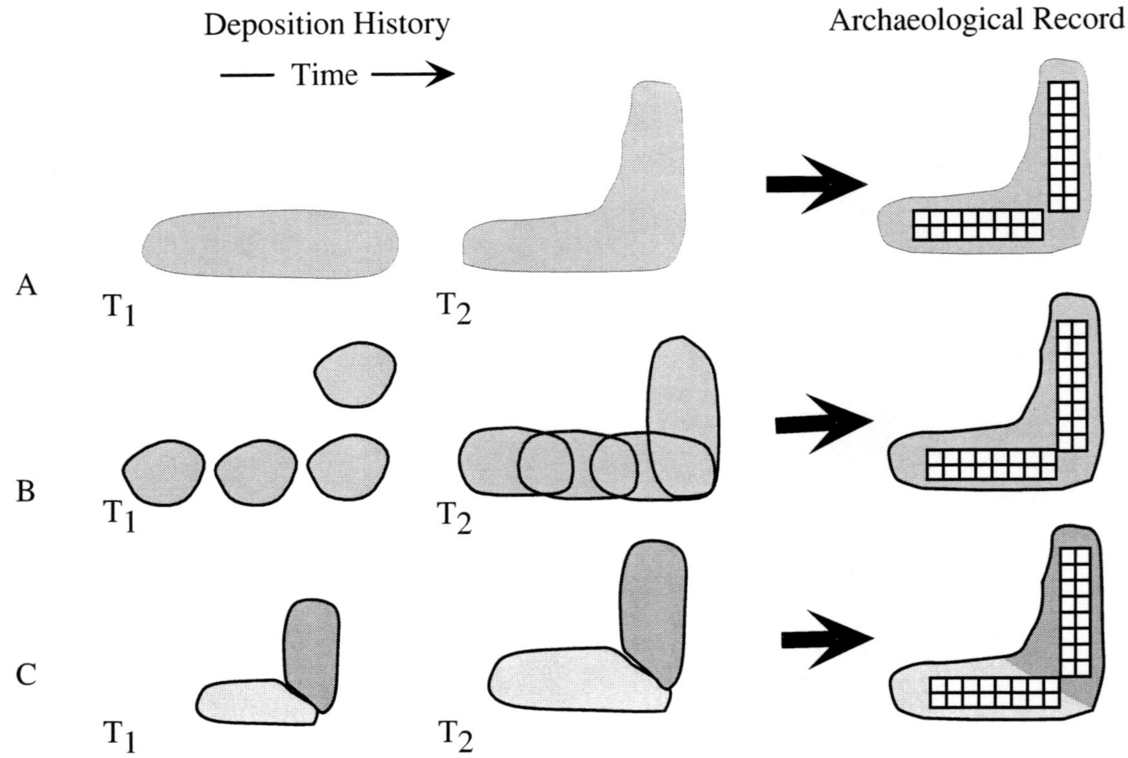

Figure 6.6. Three Hypothetical Depositional Histories for the Beck Deposit.

Three hypothetical depositional histories (A-C) can account for the accretion of the Beck deposit over time (T_1 - T_2). Shapes represent deposits. The grids represent the 1997 collection units. In first case (A), deposition begins in one area and is extended over time to a second location. In this case, the composition in material from the two collection areas reflects temporal differences in deposition. In the second case (B), deposition is distributed over the area but the deposit is formed by independent events. If over time the independent deposits overlap and form a single large midden. and if the populations interact equally over time, there will be minimal differences between the areas due to time or space. In the final case (C), the archaeological deposit is formed by two distinct sets of depositional events in which populations do not interact across the area Here, populations do no differences detected by the collection units will reflect differences in population structure. In summary, deposition history must be carefully considered with evaluating the results of surface collection units.

Table 6.8. Proportions of Parkin Punctated, Barton Incised, Old Town Red, and Other PFG Types in the Two Beck Collection Areas.[a]

Cluster	Parkin Punctated	Barton Incised	Old Town Red	Other
Units 1-15	0.560 ± 0.041	0.244 ± 0.035	0.073 ± 0.021	0.123 ± 0.027
Units 16-28	0.462 ± 0.046	0.416 ± 0.045	0.036 ± 0.018	0.086 ± 0.026

[a] Confidence intervals are calculated to $\alpha = 0.05$ level of significance.

When the units in each cluster are aggregated, they form statistically significant distinct sets. The possibility exists, therefore, that the units are drawn from different deposits. In this case, the history of the deposit must be taken into consideration.

As shown in Figure 6.6, the deposit may have been produced by three depositional scenarios. Two of these can account for the pattern of two distinct clusters. First, it is possible that the two clusters are chronologically distinct. This situation would occur if one set of collection units was located on an area with an earlier mean age than the area from which the other set of units was drawn. Finally, it is possible that the differences in the clusters are produced by the sampling of two contemporaneous but functionally distinct areas of the deposit. In this case, the samples will be unrelated and potentially compositionally unique.

Since it is impossible to tell which of these processes is at work with these data (without absolute dates), I took a conservative approach and treated each cluster as an independent sample. The two collections will be referred to as Beck A (Units 1-15) and Beck B (Units 16-28).

In contrast to the approach taken here, PFG did not make any spatial distinctions in their analysis of the Beck assemblages. This highlights the differences in my analytic methods. With intuition alone to guide decisions PFG were unable to detect the differences in sherd compositions across the deposit. More careful and controlled collections are able to detect spatial variability. Chapter 7 analyses evaluate whether the differences are temporal or caused by functional differences in the community

CRANOR PLACE

The abundance of ceramics is lower at Cranor Place than at Belle Meade or Beck. I assigned 291 sherds to PFG decorated types from thirteen 10x10-meter collection units. The counts and proportion of PFG types for each unit are given in Tables 6.9 and 6.10. Although there appears to be a substantial difference in composition between the units (Figure 6.7), the large error terms (α=0.05) suggest that much of the difference may be due to sample size. The results of a cluster analysis (Figure 6.8) using the chi-square metric and Ward's method for the aggregation of clusters suggests that Units 1,4,5,8 are slightly distinct from Units 2, 3, 6, 7, 9, 10, and 11.

Comparison of the unit composition and the results of the cluster analysis shows that similar units tend to be adjacent to one another. This pattern supports a hypothesis that the deposit is not entirely homogenous and has some structure at a scale smaller than the sampled deposit as a whole. This would be the case, for example, if areas of the deposit were formed from independent piles of unmixed house debris. The similarity of samples taken near local areas in these cases would tend to be more similar than the surrounding areas. However, for the purposes of this study there does not appear to be structure at the scale of the entire, sampled deposit. In other words, though there is local similarity, there is no clear dissimilarity between sets of units. The differences in the frequencies of the types in each unit, therefore, are likely due to the effects of sample size and other random processes rather than reflecting functional or chronological differences in underlying composition.

Since the results obtained by clustering algorithms are sensitive to differences in samples caused by sample size, we can pose two alternative hypotheses for the results of the cluster analysis. First, since the number of sherds in each unit is relatively small (average number = 26.7 ± 11.7 sherds) it is possible that the clusters are just formed by differences in sampling. Alternatively, it is possible that these clusters reflect differences in the deposits from which the samples were generated and that the compositions of the units are a product of chronological or functional differences.

We can test the first hypothesis using a two-sample Kolmogorov-Smirnov (KS) test to compare the composition of each unit with each other unit. If the first hypothesis is falsified the alternative hypothesis will be examined in the subsequent seriation analyses. The two-sample KS test evaluates whether two independent samples come from the same distribution by comparing the two-sample cumulative distribution functions. The test assumes that both samples were independently drawn from exactly the same distribution and is sensitive to differences in sample size. Of interest in the test is not the value of the KS test *per se* but the p-values that indicate the likelihood that two samples are derived from the same distribution. Sets of units that are statistically distinct will consistently have very low p-values when compared with one another. Units that cannot be distinguished will have high p-values in the KS test.

Table 6.11 shows the two-sided probability values calculated for the KS test for each collection unit with each other collection unit in the Cranor Place collection set. In the Cranor collections, the p-values for this unit compared with each other are all greater than 0.05 and the compositions of

Table 6.9. Counts of Culture-Historical Types for Cranor Place Surface Collection Units.[a]

Unit	Parkin Punctated	Barton Incised	Old Town Red	Ranch Incised	Fortune Noded	Kent Incised	Mound Place Incised	Walls Engraved	Wallace Incised	Hollywood White Slipped	Rhodes Incised	Σ
1	15	21	2	2	0	3	3	0	1	0	0	47
2	2	6	5	0	0	0	0	0	0	0	0	13
3	0	0	0	0	0	0	0	0	0	0	0	0
4	2	3	1	0	0	1	1	0	0	0	1	9
5	16	17	0	0	0	0	0	0	0	0	0	33
6	14	10	3	1	1	0	0	1	0	0	0	30
7	14	5	2	2	1	0	0	1	0	0	0	25
8	4	7	1	1	0	0	0	0	0	0	0	13
9	14	9	5	0	0	0	0	1	0	0	0	29
10	17	10	9	1	0	0	0	1	0	0	0	38
11	7	10	5	0	1	0	0	0	0	1	0	24
12	8	2	1	0	2	0	0	0	0	0	0	13
13	8	6	1	0	1	0	0	0	1	0	0	17
Σ All Units	121	106	35	7	6	4	4	4	2	1	1	291

[a] No decorated sherds were identified in Unit 3.

Table 6.10. Proportions of Culture-Historical Types for Cranor Place Surface Collection Units.

Unit	Parkin Punctated	Barton Incised	Old Town Red	Ranch Incised	Fortune Noded	Kent Incised	Mound Place Incised	Walls Engraved	Wallace Incised	Hollywood White Slipped	Rhodes Incised	Σ
1	0.319	0.447	0.043	0.043	0.000	0.064	0.064	0.000	0.021	0.000	0.000	1.000
2	0.154	0.462	0.385	0.000	0.000	0.000	0.000	0.000	0.000	0.000	0.000	1.000
3	0.000	0.000	0.000	0.000	0.000	0.000	0.000	0.000	0.000	0.000	0.000	1.000
4	0.222	0.333	0.111	0.000	0.000	0.111	0.111	0.000	0.000	0.000	0.111	1.000
5	0.485	0.515	0.000	0.000	0.000	0.000	0.000	0.000	0.000	0.000	0.000	1.000
6	0.467	0.333	0.100	0.033	0.033	0.000	0.000	0.033	0.000	0.000	0.000	1.000
7	0.560	0.200	0.080	0.080	0.040	0.000	0.000	0.040	0.000	0.000	0.000	1.000
8	0.308	0.539	0.077	0.077	0.000	0.000	0.000	0.000	0.000	0.000	0.000	1.000
9	0.483	0.310	0.172	0.000	0.000	0.000	0.000	0.035	0.000	0.000	0.000	1.000
10	0.447	0.263	0.237	0.026	0.000	0.000	0.000	0.026	0.000	0.000	0.000	1.000
11	0.292	0.417	0.208	0.000	0.042	0.000	0.000	0.000	0.000	0.042	0.000	1.000
12	0.615	0.154	0.077	0.000	0.154	0.000	0.000	0.000	0.000	0.000	0.000	1.000
13	0.471	0.353	0.059	0.000	0.059	0.000	0.000	0.000	0.059	0.000	0.000	1.000
Σ All Units	0.416	0.364	0.120	0.024	0.021	0.014	0.014	0.014	0.007	0.003	0.003	1.000

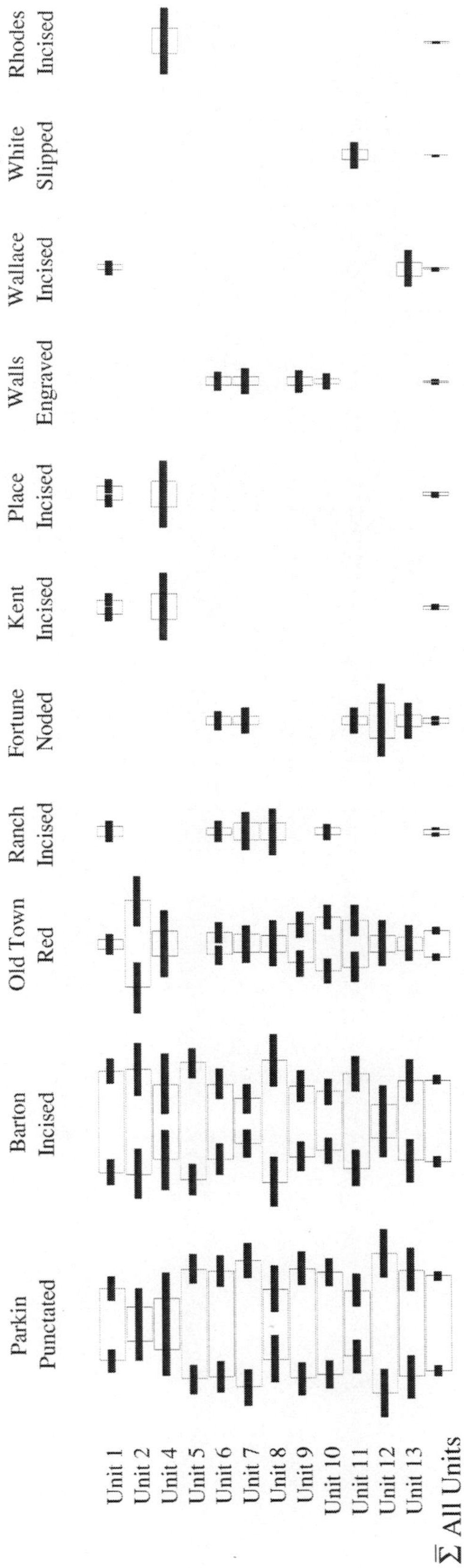

Figure 6.7. Comparison of the Proportion of Culture-Historical Types for the Cranor Place Surface Collection Units.

Confidence intervals of α=0.05 are shown for the relative proportion of each type.

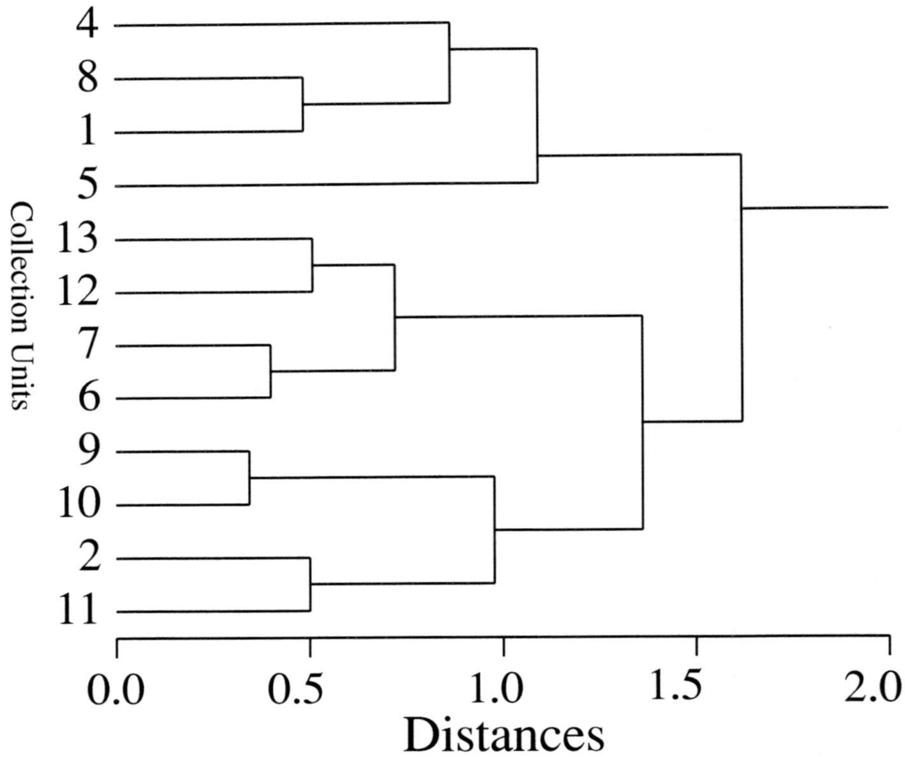

Figure 6.8. Hierarchical Cluster Analysis of Culture-Historical Type Frequencies from Cranor Place Collection Units.

Similarity was estimated from sherd counts and by calculating the distances as the chi-square measure of independence of rows and columns on 2-by-n frequency tables, formed by pairs of cases. Dendrogram was generated using Ward's method.

Table 6.11. Two-sided Probabilities that Pairs of Assemblages are Derived from the Same Distribution: Results of Two-Sample Kolmogorov-Smirnov Test for Cranor Place Surface Collection Units.

Unit	1	2	4	5	6	7	8	9	10	11	12	13
1	.											
2	0.201	.										
4	0.911	1.000	.									
5	0.156	0.104	0.089	.								
6	0.730	0.279	0.792	0.522	.							
7	0.238	0.105	0.408	0.340	0.999	.						
8	1.000	0.843	0.701	0.896	0.951	0.602	.					
9	0.441	0.333	0.847	0.228	0.948	0.908	0.881	.				
10	0.322	0.349	0.800	0.094	0.887	0.801	0.860	1.000	.			
11	0.319	0.992	0.968	0.179	0.706	0.297	0.784	0.843	0.837	.		
12	0.261	0.109	0.327	0.644	0.972	1.000	0.523	0.939	0.926	0.302	.	
13	0.880	0.401	0.785	0.826	1.000	1.000	0.977	0.974	0.948	0.634	0.993	

the remainder of the units are not statistically distinguishable from one another. We cannot falsify the hypothesis, therefore, that the Cranor Place units represent samples of a single deposit. Thus, it is reasonable for the purposes of these analyses, to aggregate all of the units into a single, larger collection.

HOLDEN LAKE

Holden Lake exhibited a low density of decorated ceramics, so sherds that could be described by PFG types were infrequent in relation to the other sites in this study. Out of a total of 12,341 sherds collected at Holden Lake from twenty 10x10-meter units, only 826 of these sherds were decorated and out of these, only 304 were assignable to decorated PFG types. As seen in Tables 6.12 and 6.13 and the graph of frequencies in Figure 6.9, the distribution of decorated types across the collection units is relatively similar. The results of a cluster analysis of the units using the chi-square metric and Ward's method for the aggregation of clusters are given in Figure 6.10. These results indicate that similarity between collection units is not spatially structured and that units in the two largest clusters contain a mix of spatially contiguous and noncontiguous units.

NICKEL

Out of 9,604 sherds collected at the Nickel site from 53 10x10-meter collection units, 2,036 were decorated and 1,101 could be assigned to decorated PFG types. The counts and proportions of culture-historical types for each unit in the Nickel collection are given in Tables 6.14 and 6.15. Figure 6.11 graphically shows the proportion of types among units. While the unit frequencies vary, the relatively small samples have large standard deviations.

Figure 6.12 gives the results of a cluster analyses using chi-square as a distance metric and Ward's method to aggregate clusters. No spatial pattern is evident in the patterns of similarity among the units. A two-sided KS test of all of the units compared with each other (944 comparisons) produces only 11 pairs of unit that have distributions of types that are significantly different from one another ($p<0.05$). This number of significantly distinct pairs is expected by chance alone since they consist of only approximately 1.2% of all pairs. Thus, it is not possible to falsify a hypothesis that the Nickel units are a sample of a single homogenous deposit. We can aggregate the Nickel collections, therefore, into a single set.

The results of KS tests for each unit pair in the collection indicate differences in the frequencies of PFG types across the deposit are not statistically significant. The analyses, therefore, do not suggest that there are compositional differences in the deposit across space. No pair of units had distributions that could be distinguished at a p-value less than 0.27. The frequency distributions for each of the units are indistinguishable from one another for every unit pair. The hypothesis that all of these units represent samples from a single population cannot be falsified. It is reasonable, therefore, to aggregate all of the Holden Lake units into a single group that can be considered representative of a single deposit.

ROSE MOUND

Rose Mound produced 7,454 sherds from 24 10x10-meter collection units. 1,315 sherds were decorated and 743 of these sherds were assignable to PFG types. The tabulation of culture-historical types and their proportions within collection units are given in Tables 6.16 and 6.17, respectively. These tables and the graphical depiction of the proportions shown in Figure 6.13 indicate that the PFG types are unevenly distributed across the collection units. The large error terms on the type frequencies, however, suggest that these differences may be due to variability in sample size rather than reflecting underlying differences in the unit compositions. Figure 6.14 shows the results of a hierarchical cluster analysis on the PFG type frequencies of the collection units. The clustering of the units, however, does not suggest

Table 6.12. Counts of Culture-Historical Types for Holden Lake Surface Collection Units.

Unit	Parkin Punctated	Barton Incised	Old Town Red	Fortune Noded	Ranch Incised	Kent Incised	Wallace Incised	Rhodes Incised	Mound Place Incised	Vernon Paul Appliqué	Hollywood White	Wheeler Checkstamped	Σ
1	1	12	0	1	1	0	0	0	0	0	0	0	15
2	1	3	0	1	0	0	0	0	0	0	0	0	5
3	0	6	0	2	0	0	0	0	0	0	0	0	8
4	1	4	0	0	0	1	0	0	0	0	0	0	6
5	2	16	0	0	0	0	0	0	0	0	0	0	18
6	1	7	1	2	0	1	0	0	0	0	0	0	12
7	0	11	0	1	0	0	0	0	1	1	0	0	14
8	0	7	1	1	0	0	0	0	0	0	0	0	9
9	1	11	0	4	0	0	0	0	0	0	0	0	16
10	2	11	0	2	0	0	0	0	1	0	0	0	16
11	4	25	0	0	0	0	0	0	0	0	0	0	29
12	2	25	0	2	0	0	0	0	0	0	0	0	29
13	4	20	0	2	0	0	0	0	0	2	0	0	28
14	2	13	0	1	0	0	1	0	0	0	0	0	17
15	2	11	0	0	0	0	1	1	2	0	0	0	17
16	2	11	1	1	1	0	0	0	0	0	0	0	16
17	1	6	0	1	0	0	0	0	0	0	1	1	10
18	0	5	0	0	0	0	0	0	0	0	0	0	5
19	0	10	0	2	0	0	0	0	0	0	0	0	12
20	1	19	0	1	0	1	0	0	0	0	0	0	22
Σ All Units	27	233	3	24	2	3	2	1	4	3	1	1	304

Table 6.13. Proportions of Culture-Historical Types for Holden Lake Surface Collection Units.

Unit	Parkin Punctated	Barton Incised	Old Town Red	Fortune Noded	Ranch Incised	Kent Incised	Wallace Incised	Rhodes Incised	Mound Place Incised	Vernon Paul Appliqué	Hollywood White	Wheeler Checkstamped	Σ
1	0.067	0.800	0.000	0.067	0.067	0.000	0.000	0.000	0.000	0.000	0.000	0.000	1.000
2	0.200	0.600	0.000	0.200	0.000	0.000	0.000	0.000	0.000	0.000	0.000	0.000	1.000
3	0.000	0.750	0.000	0.250	0.000	0.000	0.000	0.000	0.000	0.000	0.000	0.000	1.000
4	0.167	0.667	0.000	0.000	0.000	0.167	0.000	0.000	0.000	0.000	0.000	0.000	1.000
5	0.111	0.889	0.000	0.000	0.000	0.000	0.000	0.000	0.000	0.000	0.000	0.000	1.000
6	0.083	0.583	0.083	0.167	0.000	0.083	0.000	0.000	0.000	0.000	0.000	0.000	1.000
7	0.000	0.786	0.000	0.071	0.000	0.000	0.000	0.000	0.071	0.071	0.000	0.000	1.000
8	0.000	0.778	0.111	0.111	0.000	0.000	0.000	0.000	0.000	0.000	0.000	0.000	1.000
9	0.063	0.688	0.000	0.250	0.000	0.000	0.000	0.000	0.000	0.000	0.000	0.000	1.000
10	0.125	0.688	0.000	0.125	0.000	0.000	0.000	0.000	0.063	0.000	0.000	0.000	1.000
11	0.138	0.862	0.000	0.000	0.000	0.000	0.000	0.000	0.000	0.000	0.000	0.000	1.000
12	0.069	0.862	0.000	0.069	0.000	0.000	0.000	0.000	0.000	0.000	0.000	0.000	1.000
13	0.143	0.714	0.000	0.071	0.000	0.000	0.000	0.000	0.000	0.071	0.000	0.000	1.000
14	0.118	0.765	0.000	0.059	0.000	0.000	0.059	0.000	0.000	0.000	0.000	0.000	1.000
15	0.118	0.647	0.000	0.000	0.000	0.000	0.059	0.059	0.118	0.000	0.000	0.000	1.000
16	0.125	0.688	0.063	0.063	0.063	0.000	0.000	0.000	0.000	0.000	0.000	0.000	1.000
17	0.100	0.600	0.000	0.100	0.000	0.000	0.000	0.000	0.000	0.000	0.100	0.100	1.000
18	0.000	1.000	0.000	0.000	0.000	0.000	0.000	0.000	0.000	0.000	0.000	0.000	1.000
19	0.000	0.833	0.000	0.167	0.000	0.000	0.000	0.000	0.000	0.000	0.000	0.000	1.000
20	0.046	0.864	0.000	0.046	0.000	0.046	0.000	0.000	0.000	0.000	0.000	0.000	1.000
$\overline{\Sigma}$ All Units	0.089	0.766	0.010	0.079	0.007	0.010	0.007	0.003	0.013	0.010	0.003	0.003	1.000

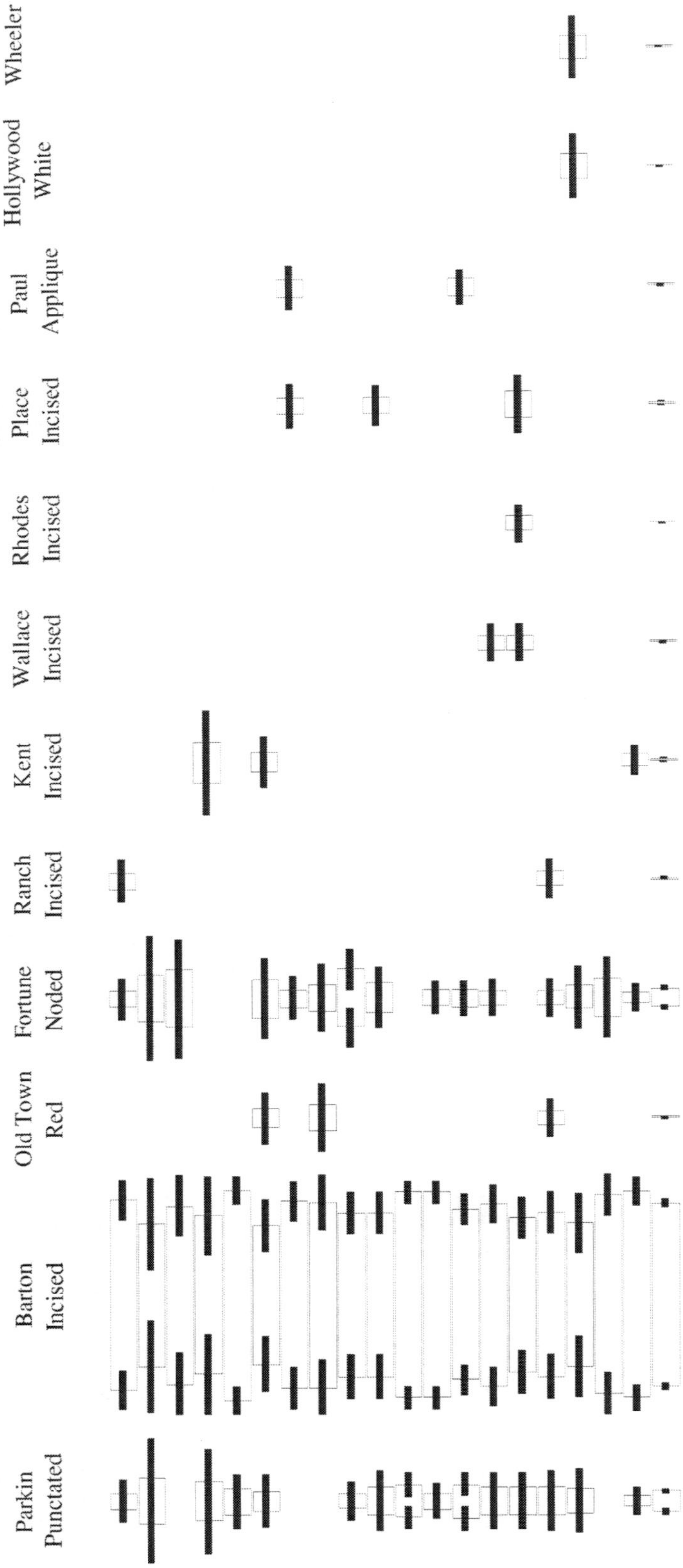

Figure 6.9. Comparison of the Proportion of Culture-Historical Types for the Holden Lake Surface Collection Units.

Confidence intervals of α=0.05 are shown for the relative proportion of each type.

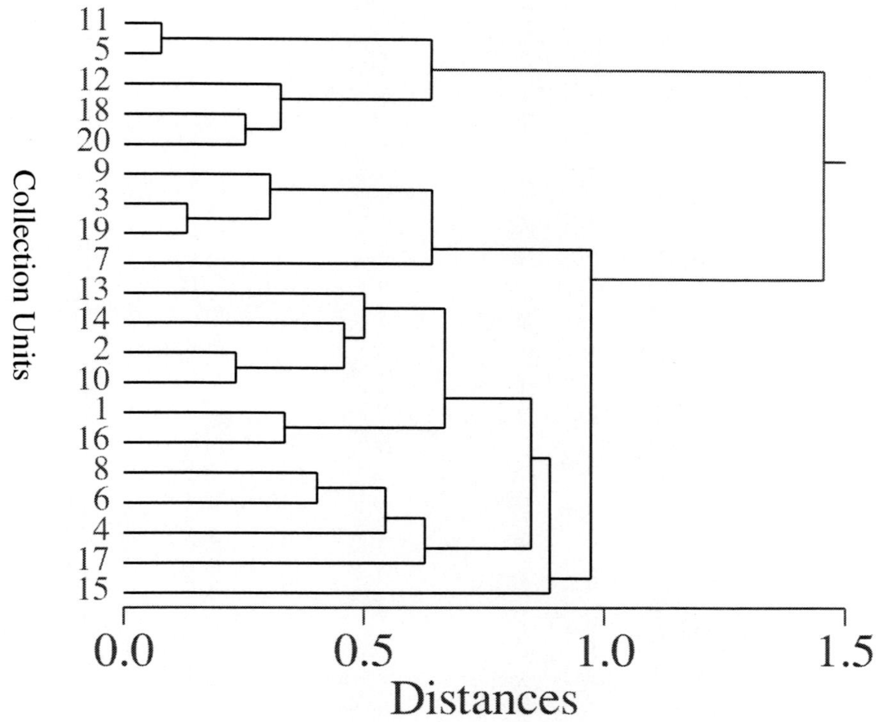

Figure 6.10. Hierarchical Cluster Analysis of Culture-Historical Type Frequencies of Holden Lake Collection Units.

Similarity of units was estimated from sherd counts and by calculating the distances as the chi-square measure of independence of rows and columns on 2-by-n frequency tables, formed by pairs of cases. Dendrogram was generated by Ward's method.

Table 6.14. Counts of Culture-Historical Types for Nickel Surface Collection Units.

Unit	Parkin Punctated	Barton Incised	Old Town Red	Fortune Noded	Ranch Incised	Kent Incised	Walls Engraved	Wallace Incised	Rhodes Incised	Mound Place Incised	Vernon Paul Appliqué	Carson Red On Buff	Hull Engraved	Hollywood White	Σ
1	7	1	0	0	1	0	0	0	0	1	0	0	0	0	10
2	2	1	0	1	0	0	0	1	0	0	0	0	0	0	5
4	2	1	0	0	0	0	0	0	0	0	0	0	0	0	3
5	11	3	0	0	0	0	0	0	0	0	0	0	0	0	14
6	4	1	1	0	0	0	0	0	0	0	0	0	0	0	6
7	8	2	1	0	0	0	0	1	0	0	0	0	0	0	12
8	4	1	1	1	0	0	0	0	0	0	0	0	0	0	7
9	0	0	0	0	0	0	0	0	0	0	0	0	0	0	0
10	13	10	5	0	0	0	0	1	0	0	1	0	2	0	32
14	2	2	1	0	0	0	0	0	0	0	0	0	0	0	5
15	13	6	0	0	0	0	0	0	0	1	0	0	0	0	20
16	14	11	0	0	0	0	1	1	0	0	0	0	0	0	27
17	10	7	4	0	0	0	1	1	0	0	0	0	0	0	23
18	7	3	2	0	0	0	0	1	0	0	1	0	0	0	14
19	13	4	1	0	0	0	0	0	0	0	0	0	0	0	18
20	4	1	1	0	0	1	0	0	0	0	0	4	0	0	11
21	3	1	1	0	0	0	0	0	0	0	1	0	0	0	6
22	17	0	5	1	0	1	0	1	0	0	0	0	0	0	25
24	34	23	11	0	1	1	1	0	1	0	0	0	0	0	72
25	56	21	10	2	1	0	0	1	0	0	0	0	0	0	91
26	17	2	3	0	0	1	0	0	0	1	1	0	0	0	25
27	29	13	8	0	0	1	0	0	0	0	0	0	0	0	51
28	4	1	1	0	0	0	0	0	0	0	0	0	0	0	6
29	5	2	1	0	0	0	0	0	0	0	0	3	0	0	11
30	4	0	0	0	0	0	0	0	0	0	0	1	0	0	5
31	7	4	1	0	0	0	0	0	0	0	0	0	0	0	12
32	2	2	1	0	0	0	0	0	0	0	0	0	0	0	5
33	7	7	2	0	1	0	0	0	0	0	0	0	0	1	18
34	13	8	7	0	0	0	0	1	0	0	0	0	0	0	29
35	22	9	4	0	0	0	1	0	0	0	0	0	0	0	36
36	6	3	3	0	0	0	0	0	0	0	0	0	0	0	12
37	22	9	6	0	0	2	0	1	0	0	0	0	0	0	40
38	8	2	5	0	0	0	1	1	0	0	0	0	0	0	17
39	33	10	2	0	0	1	0	0	0	0	0	0	0	0	46
40	19	12	3	1	0	0	1	0	0	1	1	0	0	0	38
44	3	6	0	0	0	0	0	1	0	0	2	0	0	0	12
45	34	9	3	0	0	2	0	0	0	0	0	0	1	0	49
46	44	11	2	0	0	0	0	1	0	0	0	0	0	1	59
47	21	12	15	1	0	1	1	0	0	1	0	0	0	0	52
48	37	10	11	0	0	0	1	0	0	0	0	0	1	0	60
49	27	6	5	0	1	0	0	0	0	1	0	0	0	0	40
50	6	5	1	0	0	1	0	0	0	0	0	1	0	0	14
51	12	6	2	1	0	0	0	0	0	0	0	7	0	0	28
52	6	2	3	0	0	1	0	0	0	0	0	1	0	0	13
53	10	6	3	0	0	1	0	1	0	0	0	0	0	0	21
Σ Units	622	256	136	8	5	14	8	14	1	6	7	17	4	2	1101

Table 6.15. Proportions of Culture-Historical Types for Nickel Surface Collection Units.[a]

Unit	Parkin Punctated	Barton Incised	Old Town Red	Fortune Noded	Ranch Incised	Kent Incised	Walls Engraved	Wallace Incised	Rhodes Incised	Mound Place Incised	Vernon Paul Appliqué	Carson Red On Buff	Hull Engraved	Hollywood White	Σ
1	0.700	0.100	0.000	0.000	0.100	0.000	0.000	0.000	0.000	0.100	0.000	0.000	0.000	0.000	1.000
2	0.400	0.200	0.000	0.200	0.000	0.000	0.000	0.200	0.000	0.000	0.000	0.000	0.000	0.000	1.000
4	0.667	0.333	0.000	0.000	0.000	0.000	0.000	0.000	0.000	0.000	0.000	0.000	0.000	0.000	1.000
5	0.786	0.214	0.000	0.000	0.000	0.000	0.000	0.000	0.000	0.000	0.000	0.000	0.000	0.000	1.000
6	0.667	0.167	0.167	0.000	0.000	0.000	0.000	0.000	0.000	0.000	0.000	0.000	0.000	0.000	1.000
7	0.667	0.167	0.083	0.000	0.000	0.000	0.000	0.083	0.000	0.000	0.000	0.000	0.000	0.000	1.000
8	0.571	0.143	0.143	0.143	0.000	0.000	0.000	0.000	0.000	0.000	0.000	0.000	0.000	0.000	1.000
9	0.000	0.000	0.000	0.000	0.000	0.000	0.000	0.000	0.000	0.000	0.000	0.000	0.000	0.000	1.000
10	0.406	0.313	0.156	0.000	0.000	0.000	0.000	0.031	0.000	0.000	0.031	0.000	0.063	0.000	1.000
14	0.400	0.400	0.200	0.000	0.000	0.000	0.000	0.000	0.000	0.000	0.000	0.000	0.000	0.000	1.000
15	0.650	0.300	0.000	0.000	0.000	0.000	0.000	0.000	0.000	0.050	0.000	0.000	0.000	0.000	1.000
16	0.519	0.407	0.000	0.000	0.000	0.000	0.037	0.037	0.000	0.000	0.000	0.000	0.000	0.000	1.000
17	0.435	0.304	0.174	0.000	0.000	0.000	0.044	0.044	0.000	0.000	0.000	0.000	0.000	0.000	1.000
18	0.500	0.214	0.143	0.000	0.000	0.000	0.000	0.071	0.000	0.000	0.071	0.000	0.000	0.000	1.000
19	0.722	0.222	0.056	0.000	0.000	0.000	0.000	0.000	0.000	0.000	0.000	0.000	0.000	0.000	1.000
20	0.364	0.091	0.091	0.000	0.000	0.091	0.000	0.000	0.000	0.000	0.000	0.364	0.000	0.000	1.000
21	0.500	0.167	0.167	0.000	0.000	0.000	0.000	0.000	0.000	0.000	0.167	0.000	0.000	0.000	1.000
22	0.654	0.000	0.192	0.039	0.000	0.039	0.000	0.039	0.000	0.000	0.000	0.000	0.000	0.000	1.000
24	0.472	0.319	0.153	0.000	0.014	0.014	0.014	0.000	0.014	0.000	0.000	0.000	0.000	0.000	1.000
25	0.615	0.231	0.110	0.022	0.011	0.000	0.000	0.011	0.000	0.000	0.000	0.000	0.000	0.000	1.000
26	0.680	0.080	0.120	0.000	0.000	0.040	0.000	0.000	0.000	0.040	0.040	0.000	0.000	0.000	1.000
27	0.569	0.255	0.157	0.000	0.000	0.020	0.000	0.000	0.000	0.000	0.000	0.000	0.000	0.000	1.000
28	0.667	0.167	0.167	0.000	0.000	0.000	0.000	0.000	0.000	0.000	0.000	0.000	0.000	0.000	1.000
29	0.455	0.182	0.091	0.000	0.000	0.000	0.000	0.000	0.000	0.000	0.000	0.273	0.000	0.000	1.000
30	0.800	0.000	0.000	0.000	0.000	0.000	0.000	0.000	0.000	0.000	0.000	0.200	0.000	0.000	1.000
31	0.583	0.333	0.083	0.000	0.000	0.000	0.000	0.000	0.000	0.000	0.000	0.000	0.000	0.000	1.000
32	0.400	0.400	0.200	0.000	0.000	0.000	0.000	0.000	0.000	0.000	0.000	0.000	0.000	0.000	1.000
33	0.389	0.389	0.111	0.000	0.056	0.000	0.000	0.000	0.000	0.000	0.000	0.000	0.000	0.056	1.000
34	0.448	0.276	0.241	0.000	0.000	0.000	0.000	0.035	0.000	0.000	0.000	0.000	0.000	0.000	1.000
35	0.611	0.250	0.111	0.000	0.000	0.000	0.028	0.000	0.000	0.000	0.000	0.000	0.000	0.000	1.000
36	0.500	0.250	0.250	0.000	0.000	0.000	0.000	0.000	0.000	0.000	0.000	0.000	0.000	0.000	1.000
37	0.550	0.225	0.150	0.000	0.000	0.050	0.000	0.025	0.000	0.000	0.000	0.000	0.000	0.000	1.000
38	0.500	0.316	0.079	0.026	0.000	0.000	0.026	0.000	0.000	0.026	0.026	0.000	0.000	0.000	1.000
39	0.250	0.500	0.000	0.000	0.000	0.000	0.000	0.083	0.000	0.000	0.167	0.000	0.000	0.000	1.000
40	0.694	0.184	0.061	0.000	0.000	0.041	0.000	0.000	0.000	0.000	0.000	0.000	0.020	0.000	1.000
44	0.746	0.186	0.034	0.000	0.000	0.000	0.000	0.017	0.000	0.000	0.000	0.000	0.000	0.017	1.000
45	0.404	0.231	0.289	0.019	0.000	0.019	0.019	0.000	0.000	0.019	0.000	0.000	0.000	0.000	1.000
46	0.617	0.167	0.183	0.000	0.000	0.000	0.017	0.000	0.000	0.000	0.000	0.000	0.017	0.000	1.000
47	0.675	0.150	0.125	0.000	0.025	0.000	0.000	0.000	0.000	0.025	0.000	0.000	0.000	0.000	1.000
48	0.429	0.357	0.071	0.000	0.000	0.071	0.000	0.000	0.000	0.000	0.000	0.071	0.000	0.000	1.000
49	0.429	0.214	0.071	0.036	0.000	0.000	0.000	0.000	0.000	0.000	0.000	0.250	0.000	0.000	1.000
50	0.462	0.154	0.231	0.000	0.000	0.077	0.000	0.000	0.000	0.000	0.000	0.077	0.000	0.000	1.000
51	0.476	0.286	0.143	0.000	0.000	0.048	0.000	0.048	0.000	0.000	0.000	0.000	0.000	0.000	1.000
52	0.565	0.233	0.124	0.007	0.005	0.013	0.007	0.013	0.001	0.005	0.006	0.015	0.004	0.002	1.000
53	0.577	0.233	0.125	0.007	0.005	0.010	0.008	0.013	0.001	0.006	0.007	0.001	0.004	0.002	1.000
$\overline{\Sigma}$ All Units	0.439	0.225	0.112	0.010	0.000	0.041	0.000	0.010	0.000	0.000	0.000	0.163	0.000	0.000	1.000

[a]No decorated ceramics were identified in Units 3,11,12,13.

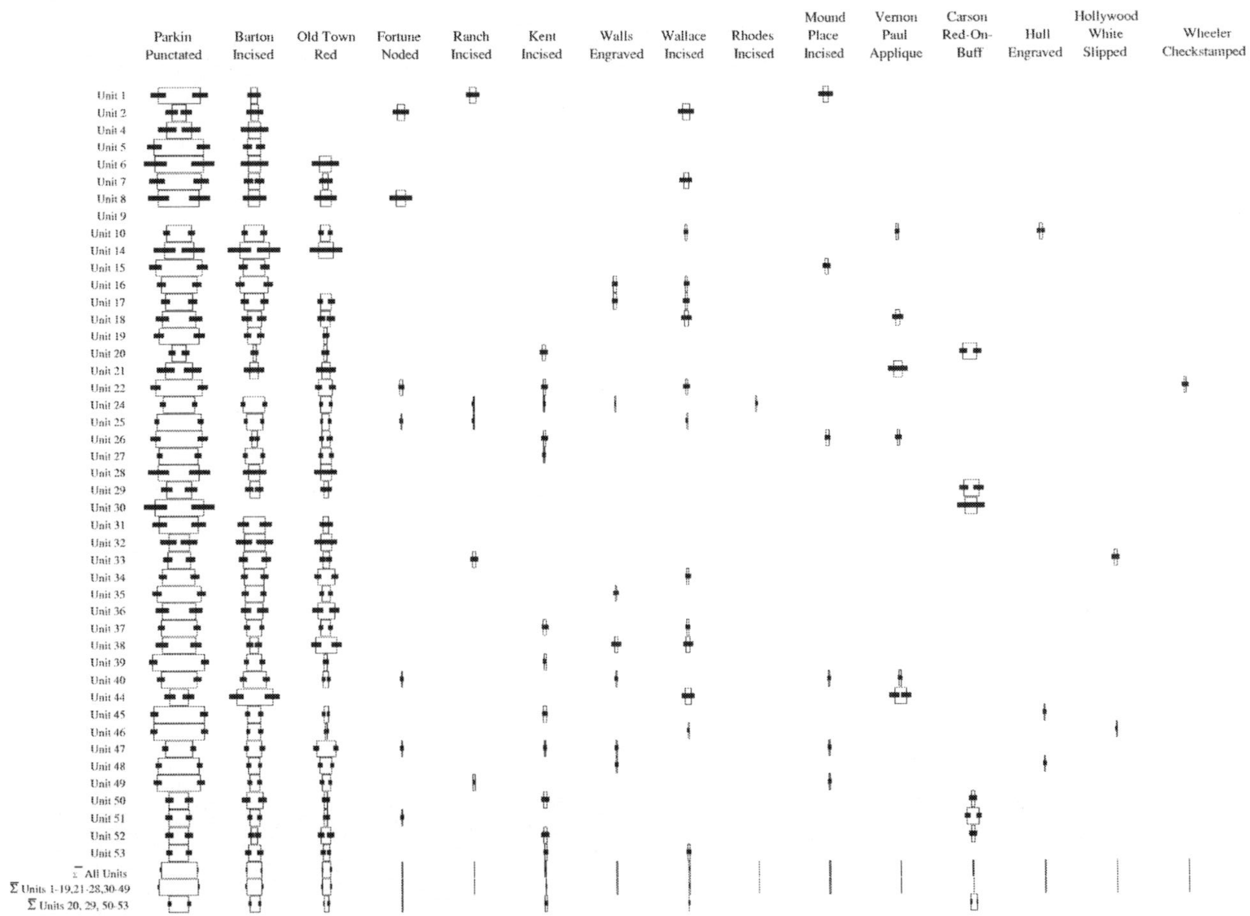

Figure 6.11. Comparison of the Proportions of Culture-Historical Type for the Nickel Surface Collection Units.

Confidence intervals of α=0.05 are shown for the relative proportion of each type.

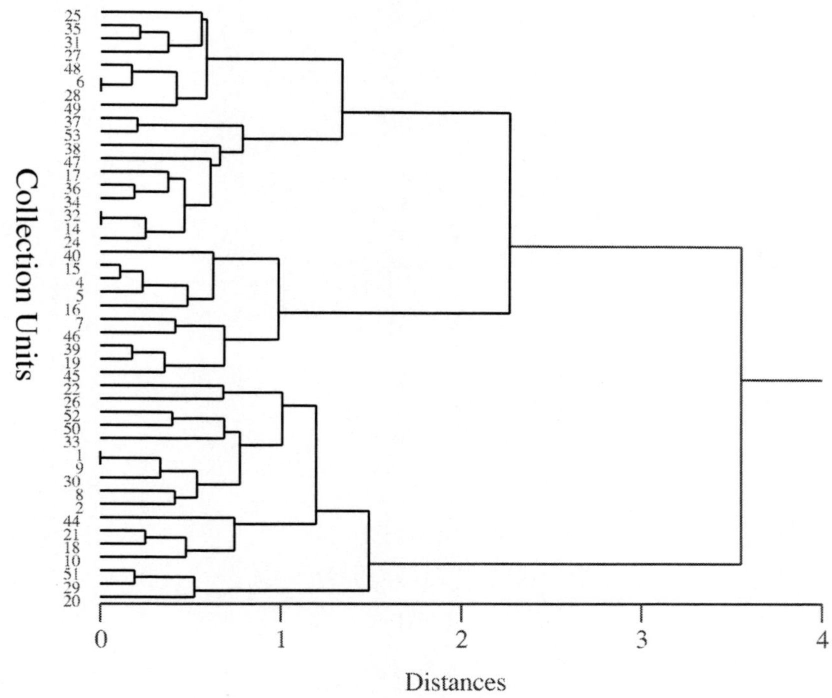

Figure 6.12. Hierarchical Cluster Analysis of Shell Tempered Culture-Historical Type Frequencies from Nickel Collection Units.

Clustering was conducted by the type counts and by calculating the distances as the chi-square measure of independence of rows and columns on 2-by-n frequency tables, formed by pairs of cases. Clusters were formed using Ward's method which averages all distances between pairs of objects in different clusters, with adjustments for covariances, to decide how far apart the clusters are.

Table 6.16. Counts of Culture-Historical Types for Rose Mound Surface Collection Units.

Unit	Parkin Punctated	Barton Incised	Old Town Red	Fortune Noded	Ranch Incised	Kent Incised	Wallace Incised	Rhodes Incised	Mound Place Incised	Vernon Paul Appliqué	Carson Red On Buff	Σ
1	7	9	1	0	0	0	0	0	0	0	0	17
2	5	8	0	1	0	1	0	0	0	0	0	15
3	16	6	0	0	1	0	2	0	0	0	0	25
4	32	9	2	0	0	0	0	0	0	0	0	43
5	11	7	3	0	0	0	0	0	1	0	0	22
6	17	14	0	2	0	0	0	0	0	0	0	33
7	5	4	0	2	0	0	0	0	0	0	0	11
8	4	2	0	0	0	0	0	0	0	0	0	6
9	5	7	2	0	0	0	0	0	0	0	0	14
10	1	1	2	0	0	0	0	0	0	0	0	4
11	7	7	1	2	0	0	0	0	0	0	0	17
12	13	16	4	1	0	0	0	0	0	0	0	34
13	29	27	1	0	0	0	0	0	0	0	0	57
14	10	12	1	3	0	0	0	0	0	0	0	26
15	19	3	2	1	0	0	1	0	0	0	0	26
16	31	20	3	1	0	0	0	0	0	0	0	55
17	32	23	6	2	0	1	0	0	0	0	1	66
18	18	13	1	0	0	0	0	0	2	0	0	34
19	16	10	3	1	0	0	0	0	0	0	0	30
20	9	14	3	0	0	0	0	0	0	0	0	26
21	25	13	6	1	0	0	0	0	0	0	0	45
22	49	16	8	0	1	0	0	0	0	0	0	74
23	19	10	3	1	1	0	0	1	0	1	0	36
24	15	8	3	0	0	0	0	0	0	0	2	31
Σ Units	395	259	55	18	3	2	3	1	3	1	3	743

Table 6.17. Proportions of Culture-Historical Types for Rose Mound Surface Collection Units.

Unit	Parkin Punctated	Barton Incised	Old Town Red	Fortune Noded	Ranch Incised	Kent Incised	Wallace Incised	Rhodes Incised	Mound Place Incised	Vernon Paul Appliqué	Carson Red-On-Buff	Σ
1	0.412	0.529	0.059	0.000	0.000	0.000	0.000	0.000	0.000	0.000	0.000	1.000
2	0.333	0.533	0.000	0.067	0.000	0.067	0.000	0.000	0.000	0.000	0.000	1.000
3	0.640	0.240	0.000	0.000	0.040	0.000	0.080	0.000	0.000	0.000	0.000	1.000
4	0.744	0.209	0.047	0.000	0.000	0.000	0.000	0.000	0.000	0.000	0.000	1.000
5	0.500	0.318	0.136	0.000	0.000	0.000	0.000	0.000	0.046	0.000	0.000	1.000
6	0.515	0.424	0.000	0.061	0.000	0.000	0.000	0.000	0.000	0.000	0.000	1.000
7	0.455	0.364	0.000	0.182	0.000	0.000	0.000	0.000	0.000	0.000	0.000	1.000
8	0.667	0.333	0.000	0.000	0.000	0.000	0.000	0.000	0.000	0.000	0.000	1.000
9	0.357	0.500	0.143	0.000	0.000	0.000	0.000	0.000	0.000	0.000	0.000	1.000
10	0.250	0.250	0.500	0.000	0.000	0.000	0.000	0.000	0.000	0.000	0.000	1.000
11	0.389	0.389	0.056	0.111	0.000	0.000	0.000	0.000	0.000	0.000	0.000	1.000
12	0.382	0.471	0.118	0.029	0.000	0.000	0.000	0.000	0.000	0.000	0.000	1.000
13	0.509	0.474	0.018	0.000	0.000	0.000	0.000	0.000	0.000	0.000	0.000	1.000
14	0.385	0.462	0.039	0.115	0.000	0.000	0.000	0.000	0.000	0.000	0.000	1.000
15	0.731	0.115	0.077	0.039	0.000	0.000	0.039	0.000	0.000	0.000	0.000	1.000
16	0.564	0.364	0.055	0.018	0.000	0.000	0.000	0.000	0.000	0.000	0.000	1.000
17	0.492	0.354	0.092	0.031	0.000	0.015	0.000	0.000	0.000	0.000	0.015	1.000
18	0.529	0.382	0.029	0.000	0.000	0.000	0.000	0.000	0.059	0.000	0.000	1.000
19	0.533	0.333	0.100	0.033	0.000	0.000	0.000	0.000	0.000	0.000	0.000	1.000
20	0.346	0.539	0.115	0.000	0.000	0.000	0.000	0.000	0.000	0.000	0.000	1.000
21	0.556	0.289	0.133	0.022	0.000	0.000	0.000	0.000	0.000	0.000	0.000	1.000
22	0.662	0.216	0.108	0.000	0.014	0.000	0.000	0.000	0.000	0.000	0.000	1.000
23	0.528	0.278	0.083	0.028	0.028	0.000	0.000	0.028	0.000	0.028	0.000	1.000
24	0.536	0.286	0.107	0.000	0.000	0.000	0.000	0.000	0.000	0.000	0.071	1.000
$\overline{\Sigma}$ All Units	0.531	0.348	0.074	0.024	0.004	0.003	0.004	0.001	0.004	0.001	0.004	1.000

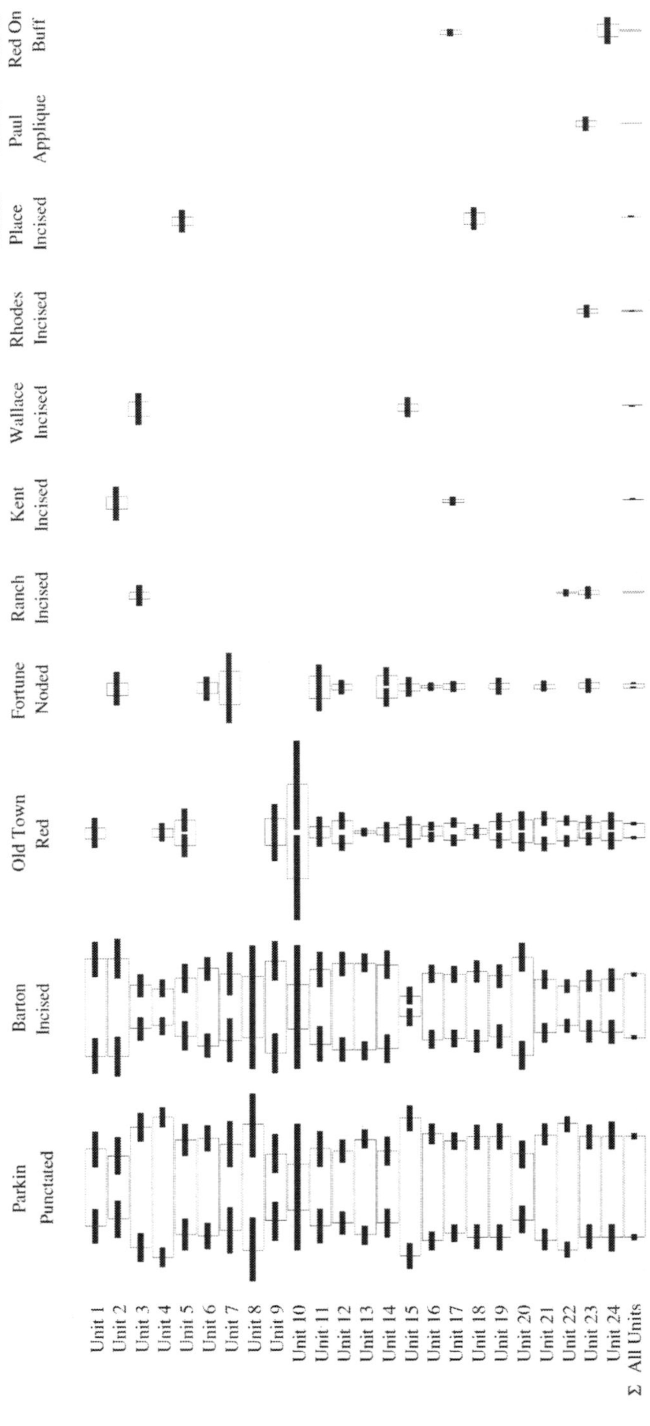

Figure 6.13. Comparison of the Proportions of Culture-Historical Types for the Rose Mound Surface Collection Units.

Confidence intervals of α=0.05 are shown for the relative proportion of each type.

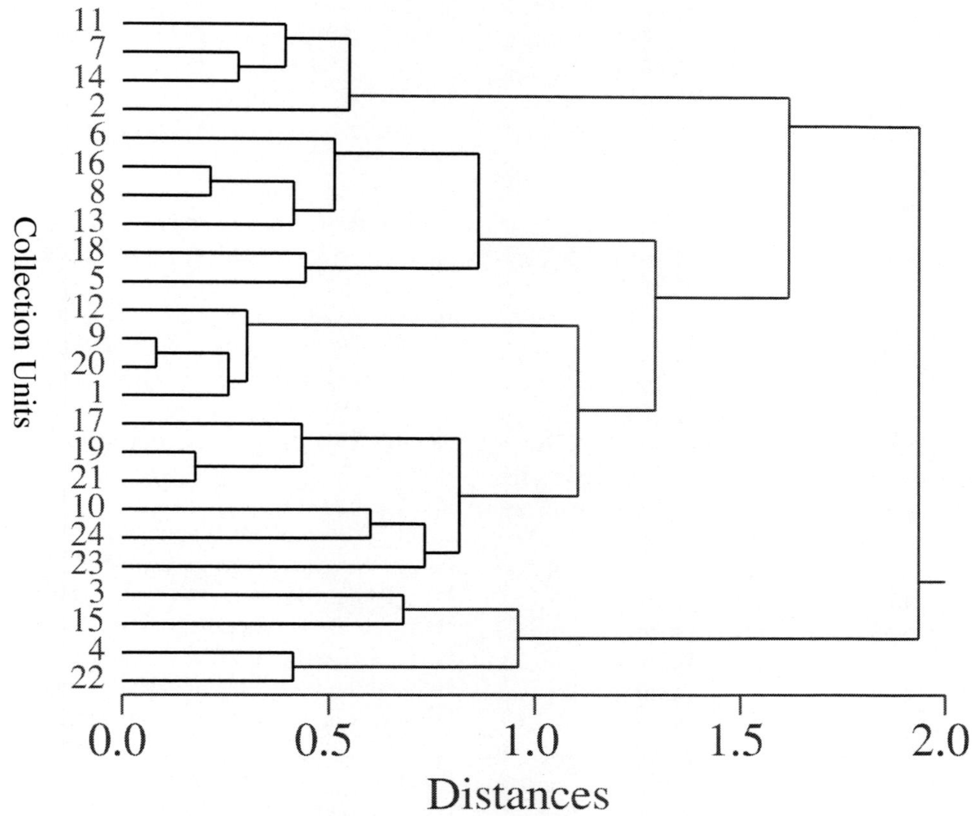

Figure 6.14. Hierarchical Cluster Analysis of Culture-Historical Type Frequencies from Rose Mound Collection Units.

Clustering was conducted by the type counts and by calculating the distances as the chi-square measure of independence of rows and columns on 2-by-n frequency tables, formed by pairs of cases. Clusters were formed using Ward's method which averages all distances between pairs of objects in different clusters, with adjustments for covariances, to decide how far apart the clusters are.

strong spatial patterning indicative of more than one underlying population.

KS tests for the distribution of culture-historical types of each unit compared with each other unit in Rose Mound produced only nine statistically distinct pairs of assemblages with p-values less than 0.05. Of the 300 unit comparisons, this number of significant p-values is less than one would expect by chance alone (3.0% of the comparisons). These values indicate that the distribution of PFG types for each unit is similar to each other unit statistically. The results are consistent with the hypothesis that differences in the frequencies of PFG types are not a function of differences (*e.g.*, functional or chronological) in the populations. Consequently, the units were aggregated to form a single larger sample for the remainder of these analyses.

CASTILE LANDING

The excavated collection from Castile Landing (3SF12) poses problems different from the surface collections. The collection strategy for all of the deposits other than Castile Landing assumes that the probability of collecting related sets of samples will be low since the collections are made from tilled surfaces and these samples have low levels of autocorrelation (Dunnell and Simek 1995). As discussed in Chapter 5, in the case of surface collection, it is assumed that if collections are carefully generated, descriptions of surface assemblages can be quantitatively compared between areas. Excavation samples, however, may suffer from problems of autocorrelation since samples collected from excavations have a high probability of being related. Multiple sherds in a sample are likely to be derived from a single larger piece of pottery. In addition, the amount of breakage across assemblages may not be consistent since the amount of breakage is a function of sherd form, temper, firing, and thickness (Orton and Tyers 1990, 1992) as well as tillage and other mechanical processes. Excavated samples, therefore, have extremely high variances rendering them idiosyncratic samples of the local deposit with characteristics derived from their own, largely unique, depositional history.

The compositional differences between the surface collections and the excavated material would ordinarily make comparisons difficult if not impossible since the frequencies of types would be driven by very different processes. Fortunately, in the case of St. Francis-type sites, the conditions of deposition may be such that comparison is possible. St. Francis-type sites consist of large piles of debris that were piled, moved, and shaped by the prehistoric occupants into thick, rectangular-shaped middens that served as an elevated foundation for occupation.

Deposits in these sites have been subject to mixing and, as a result, exhibit significant homogenization. In addition, the action of historic and modern tilling also functions to vertically homogenize samples by continually moving buried material to the surface. Thus, excavated plowzone samples collected at any location across these middens are potentially representative of the deposit as a whole and a comparable to other samples taken of the deposit. As reviewed in Chapter 5, Castile Landing is certainly a St. Francis-type site. The Castile Landing deposit was plowed before it was buried by the construction of a levee by the Army Corps of Engineers in the 1950s (Margaret Linn, personal communication; Figure 5.34). This information suggests that excavations made here could sample a midden deposit that was part of an active plowzone. Thus, collections made at Castile Landing have the potential for being comparable to collections made from tilled surfaces of midden material like the other locations in this study. However, since the construction activity has massively reshaped the deposit, there is no way to know *a priori* whether any particular excavation unit is located in a midden area, intact material exposed by construction, material redeposited by construction, or a mix of these. Consequently, it is necessary to evaluate the compositions and characteristics of the units to determine if there is reason to reject the (null) hypothesis that these units are derived from a homogenous deposit.

In order to test this hypothesis and to insure that the Castile Landing sample is comparable to the surface collection made from other locations, I performed a number of tests. The first step in the analysis is to refit all the decorated ceramics across all of the samples and to recalculate new decorated sherd totals for each unit/level. Refitting decorated ceramics greatly reduces or eliminates autocorrelation within the samples arising from breakage. In the Castile Landing collections, the presence of complicated design elements in late Mississippian ceramics made this task relatively easy. In addition, I calculated the proportion of refits to original sherds (r/n), the number of culture-historical types assigned (n), and the mean size of sherds in each level (s). These data are given in Table 6.18. I then totaled the number of refits for each set of levels. At the Castile Landing excavations, there are eight 0-20/25-cm units, four 20/25-40-cm units, and one 0-40-cm unit. For the set of 0-20/25-cm and 20/25-40-cm units, I calculated the mean proportion of refits as well as the density of the refits, the mean number culture-historical types, and mean sherd size. Table 6.19 displays these data.

In case of *de facto* house deposits that are only disturbed by the actions of plowing, one would expect that the number and proportion of refits would be approximately constant or would increase with depth as the probability of including intact deposits increases (Dunnell and Simek 1995). In addition, one expects that sherd size would increase with depth, as tilling processes would be more likely to impact sherds closer to the surface. As the standard deviations show, the sample sizes are far too small to place any confidence in the apparent differences. The results, however, are consistent with the thesis that mixing or related processes dominate the deposit. Importantly, there is no change in number or frequencies (at α=0.05 level of significance) of culture-historical types between the upper and lower levels suggesting that the contents of each level were derived from the same mixed deposit.

While the deposit is not intact, the characteristics of the deposit do not support the notion that the material was part of an active plowzone. In an active plowzone, one would expect extremely few refits (*i.e.*, none), mainly small sherds

Table 6.18. Sherd Characteristics for Each Unit and Level in Castile Landing Excavations.

Unit	Level	Number of Refits (*r*)	Original Number of Decorated Sherds (*n*)	Recalculated Number of Decorated Sherds	Proportion of Refits(*r/n*)	Excavated Volume (cubic *m*)	Density of Refits (refits/ cubic meter)	Number of Culture-Historical Types	Mean Size Class (*s* - phi size)
1	0 - 20 cm	1	20	19	0.050	0.2	5.000	2	-4.25
2	0 - 20 cm	1	19	18	0.053	0.2	5.000	5	-3.81
3	0 - 20 cm	1	12	11	0.083	0.2	5.000	4	-3.76
4	0 - 20 cm	1	13	12	0.077	0.2	5.000	3	-3.94
4	20 - 40 cm	1	27	26	0.037	0.2	5.000	3	-4.19
6	0 - 25 cm	4	7	3	0.571	0.25	16.000	2	-4.35
6	25 - 40 cm	1	14	13	0.071	0.15	6.667	4	-4.33
7	0 - 20 cm	1	44	43	0.023	0.2	5.000	4	-3.88
7	20 - 40 cm	3	40	37	0.075	0.2	15.000	4	-4.35
8	0 - 20 cm	0	23	23	0.000	0.2	0.000	5	-3.79
8	20 - 27 cm	1	18	17	0.056	0.07	14.286	4	-3.95
9	0 - 20 cm	0	18	18	0.000	0.2	0.000	4	-4.41
10	0 - 40 cm	2	10	8	0.200	0.4	5.000	4	-3.98

Table 6.19. Summary of Sherd Characteristics for Excavation Levels of Castile Landing. [a]

Level (cm)	Number of Levels	Mean Proportion of Refits	Std. Dev. Of Mean Proportion of Refits	Mean Density of Refits	Std. Dev. Of Density of Refits	Mean Number of Culture-Historical Types	Std. Dev. Of Number of C.H. Types	Mean Sherd Size (phi)	Std. Dev. Of Sherd Size (phi)
0–20	8	0.107	0.190	5.125	4.941	3.625	1.188	-4.024	0.268
20–40	4	0.060	0.017	10.238	5.140	3.750	0.500	-4.205	0.184

[a]The 0-40-cm level in Unit 10 is not included here.

and sherds with very rounded edges. In the Castile Landing deposits, found low numbers of refits, many large sherd sizes (mean phi sizes of –4.03 [16mm]) and fresh, sharp edges on the sherds. Thus, it is likely that the excavations encountered St. Francis-type deposits but that these were not part of the earlier plowzone. This difference indicates that it will be important to use only the tabulations of the refitted ceramics in further analyses so that the counts will be comparable to the collections from tilled surfaces. In addition, it is also necessary to evaluate the impact that differences in mean sherd size might have on the differences in proportion of types assigned in Castile Landing and the other surface assemblages. I turn to this task in Chapter 8.

In addition to examining characteristics of sherds in the deposit, the excavated material still must be evaluated to see if each unit can be combined together to form a single assemblage. Combining the units requires that (1) the collections were derived from the same kind of deposit and (2) that the within-group frequency of types is not greater than the between group type frequencies. To meet the first requirement, I had planned to place the excavations units in areas that would be compositionally homogeneous. PFG (Phillips 1940) described Castile Landing as a St. Francis-type site that was composed of mixed, dump deposits with large piles of house debris surrounding an open "plaza" area. Due to the levee construction, it was unclear which portion of the deposit my excavations intersected. Units may have been dug into a St. Francis-type midden, *de facto* house debris, structural features or a combination of the these kinds of deposits. During my excavations, I was careful not to include deposits that were obviously part of intact features. All excavations ceased when we encountered pit features, hearths, or anything that appeared not to be a mixed deposit. A number of characteristics can be used to distinguish the mixed, St. Francis-type deposits from *de facto* house debris. Unlike dump deposits, in *de facto* house debris one would expect large numbers of relatively large-sized sherds that are derived from the same vessels. In this kind of deposit, one would also expect a relatively large number of refits and that these refits would be fresh. Frequencies of types would be heavily dependent on post-depositional processes and the amount of breakage. Mixed deposits, on the other hand, are expected to have smaller sherd sizes and few, ancient refits.

Overall, the characteristics of the excavated deposits at Castile Landing are similar to one another and, for the most part, resemble mixed St. Francis-type deposits. These data (Table 6.19) are consistent with the hypothesis that the Castile Landing sample as a unit is derived from an early stage (or truncated min-max zone) tillage regime (Dunnell and Simek 1995) not yet approaching equilibrium. Since the principal source of non-comparability of Castile Landing to other collections lies in breakage, refitting makes it comparable.

With the exception of Unit 6/0-20cm, the proportion of refits is low ($\mu = 0.060$, $\sigma = 0.052$). The breaks on the sherds, while not ancient, appear to be fresh and due to recent activity. The distinctive composition of Unit 6/0-20cm is explicable as the unit included a relatively large piece of a red-slipped bowl. Several small fragments from this bowl were found in nearby units and were apparently broken by recent subsurface disturbance. Since the level in Unit 6 below 0-20cm (*i.e.,* 20-40cm) has characteristics of a St. Francis deposit and historic material is found below 40 cm, it is likely that the presence of these bowl fragments is anomalous and caused by recent disturbance. Once the bowl fragments are refit, the characteristics of the level are like that of other units (*i.e.,* one refit, low density of refits) and it can be reasonably included with the remainder of the units and treated as part of a St. Francis-type deposit.

Since we can account for variability in the distribution of attributes as differences in sample size or chance, we cannot reject the hypothesis that the material from each of these units was derived from a single mixed deposit. In addition, it appears that all of the units were subjected to prehistoric mixing processes consistent with St. Francis-type deposits and some degree of post-depositional mixing resulting from minimal plowing and in some cases (*e.g.,* Unit 6) from recent construction activity. This conclusion indicates that the units are derived from a relatively homogenous deposit of uniform composition and that we can potentially use the material from the units as samples of a single deposit.

Combining Units

Next, we must determine if the collections from each unit can be pooled together to form a single sample. Lumping of the material from excavation units will produce samples representative of the entire deposit if the variance in frequencies of material between excavation units is the same as variance within units. If significant differences in the variance of materials are detected then the entire set of assemblages cannot be used as a single unit in these analyses. It is possible in this scenario, however, that a subset of the units can be aggregated.

Counts and proportions of culture-historical types for the excavated units at Castile Landing are shown in Tables 6.20 and 6.21. Figure 6.15 graphically shows the distribution of the proportions for each unit and the averaged set of frequencies across all units. A distinctive pattern emerges. Driven by disproportionally large numbers of Old Town Red ceramics, Units 2 and 3 have distinctive type distributions. The results of hierarchical clustering shown in Figure 6.16 using a chi-square metric and Ward's method to aggregate clusters neatly separates units into two groups. The clusters show that the within-group variance of Units 2, 3, 6, and 8 is less than the between-group variance of that set of units and the remainder of the excavated units. This result suggests that there are at least two distinct compositional types within these collections.

There are several potential factors that can account for the differences in the composition of Units 2, 3, 6, and 8 that are detected by the clustering algorithm. First, differences in the size of samples may mean that some of the differences detected in the distribution of the types are not statistically significant. With small sample sizes, it is very possible that the samples are simply outliers of a multinomial distribution of frequencies from a single sample and that sample size is driving the differences. Indeed, all of the samples are relatively small (mean sample size=27.7). We can evaluate the sample size hypothesis by conducting a two-sample KS test for the contents of each unit with each other unit. Table 6.22 displays probability values from the KS test for each unit combination and shows that Units 2 and 3 are statistically distinct from all of the other units but not from each other ($p < 0.05$ values are shown in bold). Units 6 and 8, however, are similar to the other units. Thus, it is likely that the differences seen in Units 6 and 8 are a function of sample size alone. Unit 6, in particular, has only 15 sherds.

Second, temporal or functional differences in the deposit could also produce variability in the distribution of types. Importantly, Units 2 and 3 (along with Units 1-5) are located on the eastern side of the levee road while Units 6 through 10 are located on the western side. The location of Units 2 and 3 next to each other and close to the slope of the levee fill suggests that the units are located on a functionally or temporally distinct area of the deposit. Additionally, unlike the dark, midden soils of the other units, the hard, compact sandy composition of the sediment in Units 2 and 3 is indicative of different formation processes.

Table 6.23 shows the counts of the PFG types for the two groups of units focusing in particular on the amounts of Parkin Punctated, Barton Incised, and Old Town Red. The differences in the distribution of types are highly significant ($p<0.0001$). Thus, in the case of Castile Landing excavation units, the requirement of homogeneity can only be met if Units 2 and 3 are excluded. Subsequent analyses, therefore, were conducted using only the material from Units 1 and 4 through 10. The sample size of Units 2 and 3 is too small ($n=33$) to serve as an independent analytic unit in the seriation studies.

Table 6.20. Counts of Culture-Historical Types for Castile Landing Excavated Units.[ab]

Unit	Parkin Punctated	Barton Incised	Old Town Red	Fortune Noded	Kent Incised	Wallace Incised	Mound Place Incised	Σ
1	2	0	0	0	0	0	0	2
2	3	2	10	0	1	0	0	16
3	2	1	14	0	0	0	0	17
4	31	1	4	0	0	0	0	36
6	6	2	7	0	0	0	0	15
7	67	6	10	0	0	0	0	83
8	22	5	10	0	0	1	0	38
9	11	1	2	0	0	0	0	14
10	21	3	2	1	0	0	2	29
Σ All Units	165	21	59	1	1	1	2	250
Σ Units 2,3	5	3	24	0	1	0	0	33
Σ Units 1,4-10	160	18	35	1	0	1	2	217

[a] Unit 5 produced no decorated ceramics.
[b] Counts are of units after refits.

Table 6.21. Proportions of Culture-Historical Types for Castile Landing Excavated Units.

Unit	Parkin Punctated	Barton Incised	Old Town Red	Fortune Noded	Kent Incised	Wallace Incised	Mound Place Incised	Σ
1	1.000	0.000	0.000	0.000	0.000	0.000	0.000	1.000
2	0.188	0.125	0.625	0.000	0.063	0.000	0.000	1.000
3	0.118	0.059	0.824	0.000	0.000	0.000	0.000	1.000
4	0.861	0.028	0.111	0.000	0.000	0.000	0.000	1.000
6	0.400	0.133	0.467	0.000	0.000	0.000	0.000	1.000
7	0.807	0.072	0.121	0.000	0.000	0.000	0.000	1.000
8	0.579	0.132	0.263	0.000	0.000	0.026	0.000	1.000
9	0.786	0.071	0.143	0.000	0.000	0.000	0.000	1.000
10	0.724	0.103	0.069	0.035	0.000	0.000	0.069	1.000
$\overline{\Sigma}$ All Units	0.660	0.084	0.236	0.004	0.004	0.004	0.008	1.000
$\overline{\Sigma}$ Units 2, 3	0.152	0.091	0.727	0.000	0.030	0.000	0.000	1.000
$\overline{\Sigma}$ Units 1,4-10	0.737	0.083	0.161	0.005	0.000	0.005	0.009	1.000

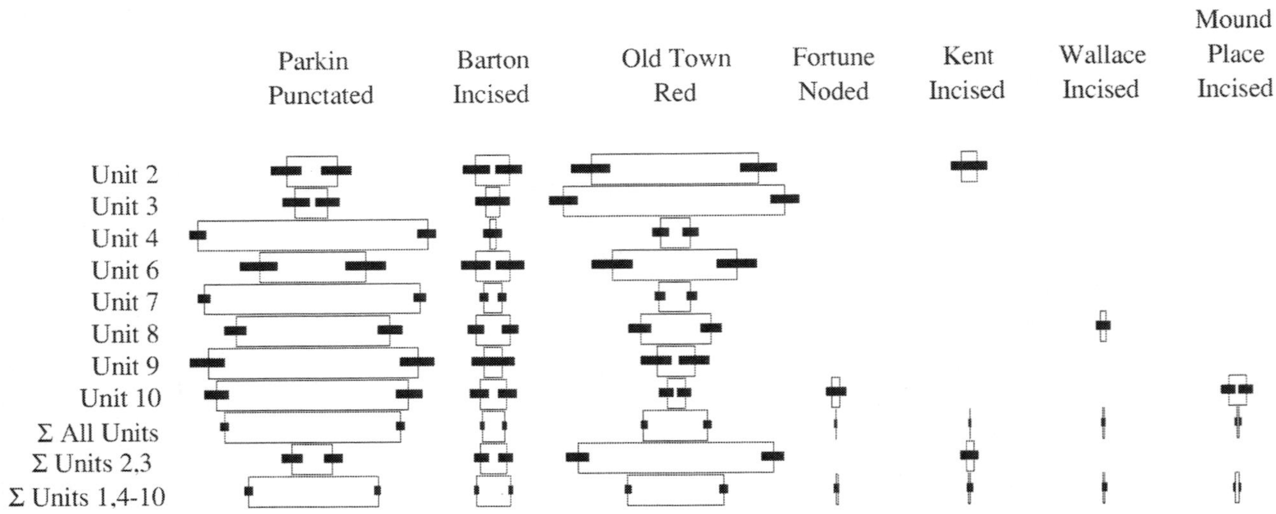

Figure 6.15. Comparison of the Proportions of Culture-Historical Types for the Castile Landing Excavation Units.

Confidence intervals of α=0.05 are shown for the relative proportion of each type. Unit 1 is not included as has only one type (Parkin Punctate). Unit 5, which did not produce any decorated ceramics is also excluded.

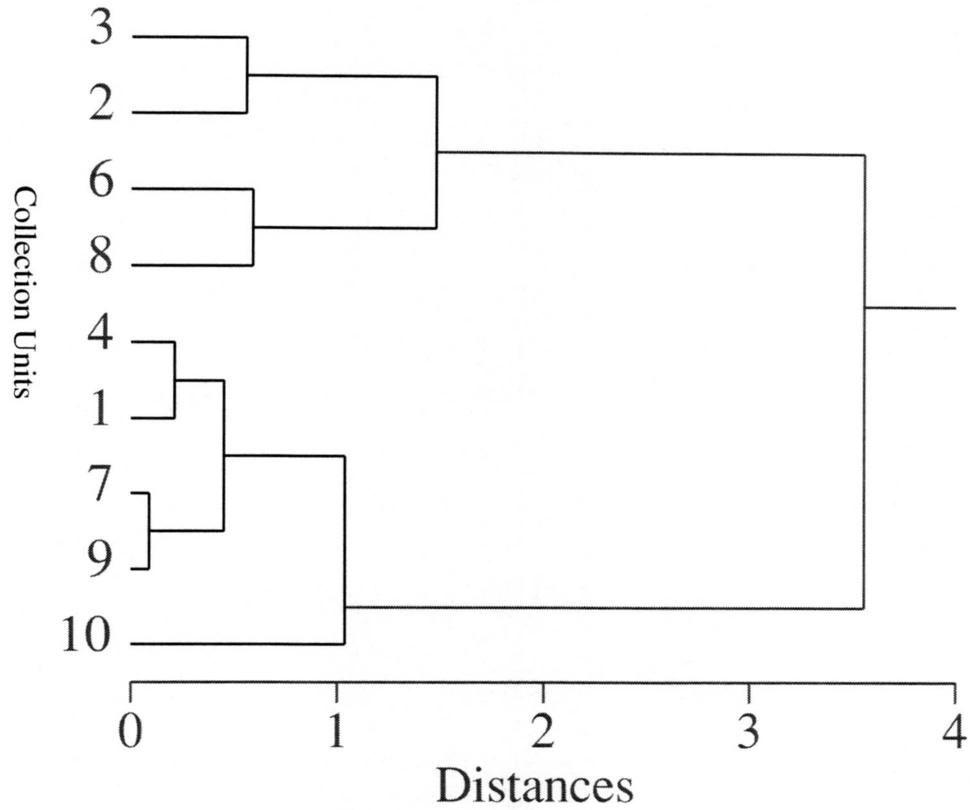

Figure 6.16. Hierarchical Cluster-Analysis of Culture-Historical Type Frequencies from Castile Landing Excavation Units.

Similarity of units was estimated from sherd counts and by calculating the distances as the chi-square measure of independence of rows and columns on 2-by-n frequency tables, formed by pairs of cases. Dendrogram was generated by Ward's method.

Table 6.22. Two-sided Probabilities that Pairs of Assemblages are Derived from the Same Distribution: Results of Two-Sample Kolmogorov-Smirnov Test for Castile Landing Excavation Units.[a]

Unit	1	2	3	4	6	7	8	9	10
1	.								
2	0.375	.							
3	0.235	0.854	.						
4	1.000	**0.000**	**0.000**	.					
6	0.800	0.818	0.131	0.018	.				
7	1.000	**0.000**	**0.000**	1.000	**0.026**	.			
8	1.000	**0.044**	**0.000**	0.092	0.852	0.133	.		
9	1.000	**0.006**	**0.000**	1.000	0.190	1.000	0.713	.	
10	1.000	**0.003**	**0.000**	0.886	0.242	0.996	0.842	1.000	.

[a]Pairs of units with distributions that differ significantly (p-value < 0.05) are shown in bold.

Table 6.23. Counts of Parkin Punctated, Barton Incised, Old Town Red, and other PFG Types for Two Groups of Castile Landing Excavation Units.[a]

Units	Parkin Punctated	Barton Incised	Old Town Red	Other Types	Σ
Units 2,3	5	3	24	1	33
Units 1,4-10	160	18	35	4	217
Σ	165	21	59	5	250

[a] The distribution of the cultural types for the two sets of units is significant at $p<0.0001$ ($X^2=53.97$.)

CONCLUSION

Archaeological deposits are dynamic bodies whose characteristics change as the result of biological and mechanical processes that operate over time (Schiffer 1987). When analyses consist of comparing counts made from samples drawn from independently formed deposits, it is necessary to assess if differences in formation processes influence deposit composition. Consequently, being able to evaluate and augment the PFG collections in order to use them in the study of transmission requires attention to the physical process of data collection (*i.e.,* field methods) as well as the process of description and counting.

In Chapter 5, I presented my field strategies for building comparable samples and the results of controlled surface collection and excavation from seven locations in the St. Francis basin area. In this chapter, I detailed my efforts at describing the sherds collected from this work using systematic criteria equivalent to the PFG culture-historical types. The criteria I used were designed to be comparable, reproducible, and usable solely from information gained from sherds. Since comparisons are the goal of this research, it was essential to avoid descriptive criteria that could not be uniformly applied to all of the samples. The assignment key used here is a starting point in the analyses. Subsequent studies of the types and comparisons with PFG may result in aggregation of types when consistent and comparable assignments are not possible. These tasks will be considered in Chapters 7 and 8.

One primary assumption made in the collection of the material from each deposit is that the material is spatially homogenous and that the composition of the material does not vary chronologically or functionally across space. Variation in measurements caused by either of these two factors contributes error to the measurements of interest: the variation resulting from cultural transmission. It is for this reason I collected material from the surface in 10x10-meter units. Using these units, I was able to evaluate this assumption by statistically assessing whether variance between spatially contiguous units was relatively the same as the variance within units. A summary of this effort is shown in Table 6.24.

In the cases of assemblages gathered from Belle Meade, Cranor, Nickel, Holden Lake, and Rose Mound, I could not define statistically valid and spatially contiguous groups of units. These results are consistent with our general assumptions that underlay the choice of St. Francis-type deposits. The analysis supports a hypothesis that these deposits have been constantly shuffled, redeposited, and managed by their inhabitants producing a homogenous mix of material.

Table 6.24. Summary of Collection Descriptions and 10x10-m Unit Aggregation for Lipo Assemblages.

Assemblage	10x10-m Surface Collection Units	1x1-m Excavation Units	Sherds Assigned to Cultural-Historical Types	Analysis Results
Beck	28	--	1075	Two clusters identified (Beck-A and Beck-B).
Belle Meade	12	--	1338	Compositionally homogenous samples.
Castile Landing	--	10	216	Two groups of units identified.
Cranor	13	--	291	Compositionally homogenous samples.
Holden Lake	20	--	306	Compositionally homogenous samples.
Nickel	53	--	1084	Compositionally homogenous samples.
Rose Mound	24	--	743	Compositionally homogenous samples.
Σ	160	10	4840	

Table 6.25. Counts of Sherds Assigned to PFG Cultural Historical Types for Lipo Assemblages from Aggregated Collection Units.

Assemblages	Parkin Punctated	Barton Incised	Old Town Red	Fortune Noded	Ranch Incised	Kent Incised	Walls Engraved	Wallace Incised	Rhodes Incised	Mound Place Incised	Vernon Paul Appliqué	Hull Engraved	Wheeler Check-stamped	Hollywood White	Carson-Red-On-Buff	Σ
Beck-A	305	138	36	7	14	14	12	5	5	1	4	0	0	2	1	544
Beck-B	253	208	25	3	6	13	6	3	7	2	3	1	0	0	1	531
Belle Meade	795	395	29	12	30	13	25	15	11	3	5	2	0	0	3	1338
Castile Landing	158	18	36	1	0	1	0	1	0	2	0	0	0	0	0	217
Cranor Place	121	106	35	6	7	4	4	2	1	4	0	0	0	1	0	291
Holden Lake	27	233	3	24	2	3	0	2	1	4	3	0	1	1	0	304
Nickel	622	256	136	8	5	14	8	14	1	6	7	4	1	2	17	1101
Rose Mound	395	259	55	18	3	2	0	3	1	3	1	0	0	0	3	743
Σ	2676	1613	355	79	67	64	55	45	27	25	23	7	2	6	25	5069

In the cases of Beck and Castile Landing, however, it was not possible to aggregate the collection units into single samples.

The Beck units separate into two compositionally distinct groups based on the relative abundance of Parkin Punctated, Barton Incised, and Old Town Red. The differences between the compositions of the two groups can be potentially explained by differences in the mean age of deposition. This hypothesis might indicate that not all St. Francis-type deposits are comprised of material that has been horizontally mixed across the midden and that exposed areas vary chronologically. Since time is a component of stylistic variability, I chose to treat each group of units as independent samples until additional information can falsify this hypothesis. It is very possible that additional analyses (in Chapter 7) will provide the warrant to aggregate these groups, but until such evidence is obtained, a conservative approach is warranted.

Unlike the other locations, the deposit at Castile Landing could not be systematically surface collected since it was not under cultivation but partially covered by a levee. Consequently, it was necessary to do data collection by excavation. Ordinarily, however, excavated samples are incomparable to surface collections since samples drawn from buried deposits suffer disproportionally from the effects of autocorrelation. Fortunately, the characteristics of at least some portion of the deposit were consistent with the identification of Castile Landing as mixed midden or a St. Francis-type site. The mixed characteristic of the deposit has one important effect. It means samples drawn from this deposit are consistent with samples drawn from surface contexts. My analyses of the number and density of refits within unit materials support the hypothesis that the Castile Landing samples were derived from a mixed deposit.

While at least part of the Castile Landing deposit appears to be mixed, I was not able to aggregate all of the excavated samples at Castile Landing into a single set. The analysis of the cultural historical type assignments between Castile Landing units showed that two spatially contiguous units (2 and 3) could be statistically distinguished from the remainder of the units. Like the case of the Beck units, it was not possible to eliminate the hypothesis that differences in the composition were generated by sampling two deposits with differences in mean age. Since the sample size for the two units is very small (n=33), I took a conservative approach and eliminated these samples from the subsequent analyses. This strategy produced a slightly smaller but statistically defensible assemblage.

Table 6.25 summarizes the pooled collection units for each of the locations in the project. In general, the evidence presented here support PFG's intuitively defined homogenous collection units. With the exception of Beck, it is not possible to divide the 10x10 meter collection units into statistically defensible sets. While the analysis of the composition of culture-historical types for collection units revealed substantial variability across the deposits, the variability measured for between-unit composition differences was explicable as the variance produced by the sampling of a single population. In these cases, individual units could be pooled to form larger assemblages. The results at Beck point out, however, that aggregation of surface material must be carefully considered. Field evaluation may not be sufficient to indicate differences in content that may be present within a collection area.

The results also demonstrate the considerations needed to construct assemblages from excavated contexts that are comparable to surface collected material. The situations in which this is possible are few, requiring mixed deposits in which autocorrelation within samples can be minimized through refitting. Castile Landing, fortunately, provides such a situation. Having controlled each assemblage for biases due to collection, observation, and description, these pooled descriptions form the basis for making comparisons between my collection and the original PFG set. These data are useful as a means for evaluating the degree to which the PFG data reflect the idiosyncrasies of their sampling and description procedures and the degree to which they can be regarded as reliable and representative stylistic descriptions of these archaeological deposits useful for the study of cultural transmission. These comparisons are the topic of Chapter 7.

In Chapter 4, I presented the results of seriation analyses based on a cultural transmission model and using ceramic data collected by PFG in the Mississippi river valley. The seriations detected spatially coherent clusters of assemblages that suggest structure to past interaction in the region. Explanation of these clusters, however, requires data from sources not available from PFG. Although PFG was concerned with the way in which their data were generated, they did not conduct systematic evaluation and *any* factor that creates differential representation of ceramic types across several assemblages potentially structures a seriation. My initial analyses made it clear that additional fieldwork was necessary to evaluate these factors. Thus, to test the hypothesis that the seriation groups are a product of cultural transmission and, therefore, represent lineages, I generated new data from seven deposits in the St. Francis basin. These collections were constructed with a goal of minimizing variability arising from the characteristics of the deposits, methods of collection and type assignment. Details of the collections process were given in Chapters 5 and 6. The first objective of this research, evaluation of the PFG results, is now possible.

COLLECTION COMPARISONS

The process of aggregating collection units formed eight stylistically homogenous assemblages. A total of 5,068 sherds were unambiguously assignable to decorated, shell-tempered PFG types. Counts and proportions using PFG types for each new collection and the original PFG collections are listed in Tables 7.1 and 7.2.

For these tabulations, I summed all painted types (Nodena Red-and-White, Carson Red-on-Buff, Hollywood White, Avenue Polychrome, and Old Town Red). The occurrence of painted sherds was very low across all assemblages. In

Table 7.1. Counts of Sherds Assigned to Late Prehistoric, Decorated, Shell-Tempered, PFG Types in PFG and Lipo Assemblages.

Assemblages	Parkin Punctated	Barton Incised	Painted[a]	Fortune Noded	Ranch Incised	Kent Incised	Walls Engraved	Wallace Incised	Rhodes Incised	Mound Place Incised	Vernon Paul Appliqué	Hull Engraved	Σ
Beck-A	305	138	39	7	14	14	12	5	5	1	4	0	544
Beck-B	253	208	26	3	6	13	6	3	7	2	3	1	531
Beck-PFG	32	23	2	0	1	20	1	0	0	0	0	0	79
Belle Meade-Lipo	795	395	32	12	30	13	25	15	11	3	5	2	1338
Belle Meade-PFG	128	77	10	0	3	32	2	0	2	0	0	0	254
Castile Landing-Lipo	158	18	36	1	0	1	0	1	0	2	0	0	217
Castile Landing-PFG	268	36	70	3	4	8	0	0	4	1	1	0	395
Cranor Place-Lipo	121	106	36	6	7	4	4	2	1	4	0	0	291
Cranor Place-PFG	91	27	7	3	1	9	0	0	0	0	0	0	138
Holden Lake-Lipo[b]	27	233	4	24	2	3	0	2	1	4	3	0	303
Nickel-Lipo	622	256	156	8	5	14	8	14	1	6	7	2	1099
Nickel-PFG	106	31	19	1	0	3	0	0	2	0	0	0	162
Rose Mound-Lipo	395	259	58	18	3	2	0	3	1	3	1	0	743
Rose Mound-PFG	154	62	22	1	1	1	0	0	0	0	1	0	242

[a] Painted category includes Nodena Red-and-White, Carson Red-and-Buff, Old Town Red, Hollywood White, and Avenue Polychrome.

[b] Holden Lake was not collected by PFG.

Table 7.2. Proportions of Sherds Assigned to Late Prehistoric Decorated PFG Types in PFG and Lipo Assemblages After Combining Only Statistically Homogenous Collection Units.

Assemblages	Parkin Punctated	Barton Incised	Painted[a]	Fortune Noded	Ranch Incised	Kent Incised	Walls Engraved	Wallace Incised	Rhodes Incised	Mound Place Incised	Vernon Paul Appliqué	Hull Engraved	Σ
Beck-A	0.561	0.254	0.072	0.013	0.026	0.026	0.022	0.009	0.009	0.002	0.007	0.000	1.000
Beck-B	0.476	0.392	0.049	0.006	0.011	0.024	0.011	0.006	0.013	0.004	0.006	0.002	1.000
Beck-PFG	0.405	0.291	0.025	0.000	0.013	0.253	0.013	0.000	0.000	0.000	0.000	0.000	1.000
Belle Meade-Lipo	0.594	0.295	0.024	0.009	0.022	0.010	0.019	0.011	0.008	0.002	0.004	0.002	1.000
Belle Meade-PFG	0.503	0.303	0.039	0.000	0.012	0.126	0.008	0.008	0.000	0.000	0.000	0.000	1.000
Castile Landing-Lipo	0.728	0.083	0.166	0.005	0.000	0.005	0.000	0.005	0.000	0.009	0.000	0.000	1.000
Castile Landing-PFG	0.679	0.091	0.177	0.007	0.010	0.020	0.000	0.010	0.000	0.003	0.003	0.000	1.000
Cranor Place-Lipo	0.416	0.364	0.124	0.021	0.024	0.014	0.014	0.007	0.003	0.014	0.000	0.000	1.000
Cranor Place-PFG	0.659	0.196	0.051	0.022	0.007	0.065	0.000	0.000	0.000	0.000	0.000	0.000	1.000
Holden Lake-Lipo[b]	0.089	0.769	0.013	0.079	0.007	0.010	0.000	0.007	0.003	0.013	0.010	0.000	1.000
Nickel-Lipo	0.566	0.233	0.142	0.007	0.005	0.013	0.007	0.013	0.001	0.005	0.006	0.002	1.000
Nickel-PFG	0.654	0.191	0.117	0.006	0.000	0.019	0.000	0.012	0.000	0.000	0.000	0.000	1.000
Rose Mound-Lipo	0.532	0.349	0.078	0.024	0.004	0.003	0.000	0.004	0.001	0.004	0.001	0.000	1.000
Rose Mound-PFG	0.636	0.256	0.091	0.004	0.004	0.004	0.000	0.000	0.000	0.000	0.004	0.000	1.000

[a] Painted category includes Nodena Red-and-White, Carson Red-and-Buff, Old Town Red, Hollywood White, and Avenue Polychrome.
[b] Holden Lake was not collected by PFG.

addition, it was clear that the recognition of painted types was problematic. As PFG (1951:134) point out, it is impossible to determine whether a sherd with white slip should be described as Hollywood White-Slipped or as a fragment of Nodena Red-and-White. This conundrum is also true for red slips since a red slipped sherd may be part of an Old Town Red, Carson Red-on-Buff, Nodena Red-and-White, or Avenue Polychrome. Combining painted types to form a single group represents the most conservative approach to type assignment and thus the strategy taken here.

If different painted styles reached popularity at different times, this aggregation would not behave as a style. However, the small numbers of sherds would likely render even the combined classes irrelevant The proportion of decorated types described by PFG in their collections maybe an artifact of chance, idiosyncratic sampling, or systematic bias. Resolving this issue is critical to our understanding of the seriation analyses: have we learned something about the archaeological record or are the seriation groups a reflection of the way in which the PFG samples were collected? If comparison of PFG type proportions in each of the assemblages shows that the distribution of PFG types are distinguishable statistically in each pair of assemblages, the hypothesis that the PFG seriation results measure an aspect of the archaeological record would be falsified. Instead, it would be necessary to attribute the seriation groups to differences arising from counting, collection strategies, and sampling are being measured.

Comparison of the type proportions in both the new and PFG collections can easily be made by calculating a *z*-score for the comparison of two population proportions using independent samples. This calculation is given in Equation 7.1:

Equation 7.1.
$$z = \frac{\hat{p}_1 - \hat{p}_2}{\sqrt{\hat{p}_p(1-\hat{p}_p)}\sqrt{(1/n_1)+(1/n_2)}}$$

where n_1 and n_2 are the total number of decorated sherds in the two assemblages, $\hat{p}_1$ and $\hat{p}_2$ are the proportions of a type to be compared ($\hat{p}_1 = x_1/n_1$, $\hat{p}_2 = x_2/n_2$), x_1 and x_2 are the number of sherds for the type in each sample and $\hat{p}_p = \frac{x_1 + x_2}{n_1 + n_2}$. If it is statistically impossible to distinguish proportions of types in the PFG data and my new assemblages at $\alpha = 0.05$, we can consider that these assemblages were derived from the same population or parent deposit.

As an example of this statistical evaluation, consider two assemblages (1 and 2) with sample sizes n_1=200 and n_2=500, and proportion of sherds of a single type, x, where, x_1 =50 and x_2=165. In order to determine if the proportion of this type is statistically distinct in the two assemblages at α=0.10, the null and alternative hypotheses in this test are:

H$_o$: $p_1 = p_2$ (The percentages of the type in the assemblages are equivalent.)
H$_a$: $p_1 \neq p_2$ (The percentages of the type in the assemblages are different.)

To test these hypotheses, we first calculate the proportions of the type in the two assemblages: $\hat{p}_1 = x_1/n_1 = 50/200 = 0.35$ and $\hat{p}_2 = x_2/n_2 = 165/500 = 0.33$. We then calculate the value of $\hat{p}_p = \frac{x_1 + x_2}{n_1 + n_2} = \frac{50+165}{200+500} = 0.3071$. Thus by Equation 7.1, $z = \frac{0.35 - 0.33}{\sqrt{0.3071(1-0.3071)}\sqrt{(1/200)+(1/500)}} = 0.51821$. The critical value for a 90% confidence level in a two-tailed test is 1.645. Since $z < 1.645$, we cannot reject the null hypothesis and thus cannot argue that the proportion of the type is different in the two assemblages.

Table 7.3 presents the *z*-scores calculated for each type in the seven new collections compared with the PFG data. Values that are significant at a 90% confidence level (critical value at 90% significance = $z_{a/10} = z_{0.05} = \pm 1.645$) are shown in bold. The two new Beck assemblages (Beck A and Beck B) are both compared with the single PFG assemblage. Remarkably, only a few type comparisons are statistically significant. For the most part, the PFG and new assemblages are quite similar in their distributions. Except for the Beck A, Beck B, and Belle Meade proportions of Kent Incised, the frequencies of types in the two collections cannot be statistically distinguished at a 90% confidence level. Given that only a single type is involved, the collections appear to be comparable with the exception of Kent Incised. The most likely cause for a discrepancy in only one type lies in how PFG and I assigned sherds to Kent Incised.

As Lumb and McNutt (1988:26) have noted, distinguishing between Barton Incised and Kent Incised is difficult in sherd populations. Kent Incised is defined by vertical incised lines while Barton Incised has by oblique incised lines. Thus, these assignments require knowledge of sherd orientation. For PFG, however, Kent Incised was not considered problematic because their definition, in this case, took advantage of inferred vessel form to make type assignments. PFG (1951:141) state that the Kent Incised type consists of vertical incised lines extending from "lower rim area to base." This information is rarely available for sherds. In the case of the ceramics of the Mississippi river valley, this problem is compounded as most sherds originate from roughly spherically vessels and are, consequently, difficult to orient. During my sherd descriptions, I was aware that Kent Incised would likely be under represented in my collections relative to the PFG type assignments since vertically oriented lines are particularly difficult to identify without subjective inference to the entire vessel.

Other workers have noted the problem of distinguishing Kent Incised from Barton Incised. Phillips (1970:44-46), for example, subsequently demoted Kent Incised into a "variety" of Barton Incised since Kent sherds are recognizable only when information other than that necessary to recognize it as Barton is present.

Table 7.3. *Z*-Score Values for the Comparison of Type Proportion between Lipo and PFG Collections.[a]

Assemblages	Parkin Punctated	Barton Incised	Painted	Fortune Noded	Ranch Incised	Kent Incised	Walls Engraved	Rhodes Incised	Wallace Incised	Mound Place Incised	Vernon Paul Appliqué	Hull Engraved
Beck PFG-Beck-A	0.90	-0.25	0.54	0.35	0.25	**-2.89**	0.19	0.30	0.30	0.13	0.27	--
Beck PFG-Beck-B	0.41	0.60	0.33	0.23	-0.04	**-2.91**	-0.04	0.23	0.36	0.19	0.23	0.13
Belle Meade	0.70	-0.07	-0.37	0.40	0.28	**-2.68**	0.32	0.12	0.38	0.20	0.26	0.16
Castile Landing	0.40	-0.09	-0.14	-0.13	-0.43	-0.45	--	-0.21	--	0.33	-0.22	--
Cranor Place	-1.52	1.13	0.76	-0.02	0.39	-0.93	0.44	0.31	0.22	0.44	--	--
Nickel	-0.62	0.34	0.25	0.05	0.25	-0.17	0.32	0.01	0.11	0.27	0.30	0.16
Rose Mound	-0.78	0.72	-0.17	0.54	-0.01	-0.10	--	0.27	0.16	0.27	-0.23	--

[a] Significant values that are greater than the critical value, $z_{0.05} = \pm 1.645$, are in bold. Significant values indicate that the proportions can be distinguished at a 90% significance level.

In order to make assemblage descriptions comparable to PFG, the Barton and Kent Incised counts must be combined into a single incised category. This insures that differences in type assignments due to sherd orientation do not impact type frequencies. This strategy has been taken by others trying to build recent collections comparable to the PFG assemblages. Pugh (1990), for example, aggregates counts from Kent and Barton Incised to form a single incised type in his surface collection studies at Belle Meade.

Mound Place Incised, defined by the presence of parallel, horizontally incised lines on shell-tempered ceramics, also requires sherd orientation relative to its vessel in order to make assignments. Without knowledge of the location of the rim on convex sherds, it is often not possible to determine whether a set of lines should be assigned to Barton, Kent, or Mound Place Incised. Thus, I aggregated the counts for these three incised types to form a single incised type. It is important to note that this decision is necessitated by the conditions of these collections and not by the nature of variability.

Table 7.4 presents the count of PFG types in the new assemblages after combining the counts of Barton, Kent, and Mound Place Incised. This new category also includes Lipo collection sherds that were previously unassignable due to ambiguous orientation. Table 7.5 gives the proportion of these new type categories within each collection.

Table 7.4. Counts of Decorated, Shell-Tempered PFG Types in New and PFG Assemblages after Combining Barton, Kent, and Mound Place Incised Counts.

Assemblages	Parkin Punctated	Barton / Kent/ MPI[c]	Painted[a]	Fortune Noded	Ranch Incised	Walls Engraved	Wallace Incised	Rhodes Incised	Vernon Paul Appliqué	Hull Engraved	Σ
Beck-A Lipo	305	189	39	7	14	12	5	5	4	0	580
Beck-B Lipo	253	266	26	3	6	6	3	7	3	1	574
Beck-PFG	32	43	2	0	1	1	0	0	0	0	79
Belle Meade-Lipo	795	528	32	12	30	25	15	11	5	2	1455
Belle Meade-PFG	128	109	10	0	3	2	2	0	0	0	254
Castile Landing-Lipo	158	24	36	1	0	0	1	0	0	0	220
Castile Landing-PFG	268	44	70	3	4	0	4	0	1	0	394
Cranor Place-Lipo	121	120	36	6	7	4	2	1	0	0	297
Cranor Place-PFG	91	36	7	3	1	0	0	0	0	0	138
Holden Lake-Lipo[b]	27	294	4	24	2	0	2	1	3	0	357
Nickel-Lipo	622	330	156	8	5	8	14	1	7	2	1153
Nickel-PFG	106	34	19	1	0	0	2	0	0	0	162
Rose Mound-Lipo	395	265	58	18	3	0	3	1	1	0	744
Rose Mound-PFG	154	63	22	1	1	0	0	0	1	0	242

[a] Painted category includes Nodena Red-and-White, Carson Red-and-Buff, Old Town Red, Hollywood White, and Avenue Polychrome.
[b] Holden Lake was not collected by PFG.
[c] MPI = Mound Place Incised.

Table 7.6 presents the *z*-scores for the comparison of the frequencies between the two collections. At a 90% level of confidence (critical value of $z = \pm1.645$), the effect of combining Barton, Kent Incised, and MPI into a single type, in the original PFG collections and my collections is to make them statistically indistinguishable. By combining the types together, none of the *z*-scores exceed or is less than the critical value of ±1.645 for a 90% level of confidence.

CONCLUSIONS

That carefully generated assemblages produce frequencies of culture-historical types that are statistically indistinguishable from the PFG results has two important implications. First, it is clear that with careful considerations made towards counting criteria, repetitive sampling of the surfaces of St. Francis-type deposits produces representative and increasingly larger samples of the deposits. Recollections by Pugh (1990) at Belle Meade provide further evidence in this regard. Pugh made surface collections from the same area of Belle Meade from which the PFG assemblages and my assemblages were generated. Pugh found that his frequencies of PFG types from this area were indistinguishable from those reported by PFG. An analysis of the Pugh results compared with the results obtained from my assemblages (Table 7.7) shows that they cannot be distinguished from one another at a 90% confidence level (as none of the z-scores exceed or are less than a critical value of ±1.645). Careful recollection can provide, therefore, comparable samples of deposits to form larger assemblages. These larger samples, in turn, are useful for making distinctions that are more precise in the seriation analyses.

In addition, the analyses presented here demonstrate the importance of counting considerations when comparisons are being made. Differences in the criteria used for type assignments contribute to variability in the composition of types. Assignment criteria, therefore, must be explicit and unambiguous. In both these analyses and in the work of Pugh (1990), it was only when a simplified classification was used that combined varieties of incised decoration into a single type that comparable descriptions to PFG could be made. Once the assemblages were made with a simplified classification, however, the new descriptions, those made by Pugh (1990) and PFG became isomorphic and statistically indistinguishable from one another.

Table 7.5. Proportions of Decorated Shell-Tempered PFG Types in Lipo and PFG Assemblages after Combining Barton, Kent and Mound Place Incised Counts.

Assemblages	Parkin Punctated	Barton/ Kent/ MPI[c]	Painted[a]	Fortune Noded	Ranch Incised	Walls Engraved	Wallace Incised	Rhodes Incised	Vernon Paul Appliqué	Hull Engraved	Σ
Beck-A Lipo	0.526	0.326	0.067	0.012	0.024	0.021	0.009	0.009	0.007	0.000	1.000
Beck-B Lipo	0.441	0.463	0.045	0.005	0.011	0.010	0.005	0.012	0.005	0.002	1.000
Beck-PFG	0.405	0.544	0.025	0.000	0.013	0.013	0.000	0.000	0.000	0.000	1.000
Belle Meade-Lipo	0.546	0.363	0.022	0.008	0.021	0.017	0.010	0.008	0.003	0.001	1.000
Belle Meade-PFG	0.504	0.429	0.039	0.000	0.012	0.008	0.008	0.000	0.000	0.000	1.000
Castile Landing-Lipo	0.718	0.109	0.164	0.005	0.000	0.000	0.005	0.000	0.000	0.000	1.000
Castile Landing-PFG	0.680	0.112	0.178	0.008	0.010	0.000	0.010	0.000	0.003	0.000	1.000
Cranor Place-Lipo	0.407	0.404	0.121	0.020	0.024	0.013	0.007	0.004	0.003	0.000	1.000
Cranor Place-PFG	0.659	0.261	0.051	0.022	0.007	0.000	0.000	0.000	0.000	0.000	1.000
Holden Lake-Lipo[b]	0.076	0.824	0.011	0.067	0.006	0.000	0.006	0.003	0.008	0.000	1.000
Nickel-Lipo	0.539	0.286	0.135	0.007	0.004	0.007	0.012	0.001	0.006	0.002	1.000
Nickel-PFG	0.654	0.210	0.117	0.006	0.000	0.000	0.012	0.000	0.000	0.000	1.000
Rose Mound-Lipo	0.531	0.356	0.078	0.024	0.004	0.000	0.004	0.001	0.001	0.000	1.000
Rose Mound-PFG	0.636	0.260	0.091	0.004	0.004	0.000	0.000	0.000	0.004	0.000	1.000

[a] Painted category includes Nodena Red-and-White, Carson Red-and-Buff, Old Town Red, Hollywood White, and Avenue Polychrome.

[b] Holden Lake was not collected by PFG.

[c] MPI = Mound Place Incised.

Table 7.6. *Z*-Score Values for the Comparison of Proportions between the Lipo and PFG Collections with Barton, Kent and Mound Place Incised Counts Combined.[a]

Assemblages	Parkin Punctated	Barton/Kent/ MPI	Painted[b]	Fortune Noded	Ranch Incised	Walls Engraved	Rhodes Incised	Wallace Incised	Vernon Paul Appliqué	Hull Engraved
Beck-A	0.70	-1.32	0.50	0.34	0.22	0.17	0.29	0.29	0.26	--
Beck-B	0.21	-0.47	0.28	0.22	-0.06	-0.06	0.22	0.34	0.22	0.13
Belle Meade	0.33	-0.53	-0.43	0.38	0.25	0.29	0.09	0.36	0.24	0.15
Castile Landing	0.28	-0.03	-0.13	-0.13	-0.43	--	-0.22	--	-0.22	--
Cranor Place	-1.57	0.93	0.74	-0.03	0.38	0.44	0.31	0.22	--	--
Nickel	-0.80	0.59	0.18	0.03	0.24	0.31	-0.01	0.11	0.29	0.15
Rose Mound	-0.78	0.75	-0.17	0.54	-0.01	--	0.27	0.16	-0.23	--

[a] Missing values (as indicated by '--') result when counts of both collections are zero for a particular type.
[b]Painted category includes Nodena Red-and-White, Carson Red-and-Buff, Old Town Red, Hollywood White, and Avenue Polychrome.
[c] MPI = Mound Place Incised.

Table 7.7. Comparison of Pugh Belle Meade Collections with Lipo Collections.

Collection	Parkin Punctated	Barton/Kent/ MPI[b]	Painted[a]	Fortune Noded	Ranch Incised	Walls Engraved	Rhodes Incised	Wallace Incised	Vernon Paul Appliqué	Hull Engraved	Σ
Lipo - *N*	795	528	32	12	30	25	15	11	5	2	1455
(Proportion)	(0.546)	(0.363)	(0.022)	(0.008)	(0.021)	(0.017)	(0.010)	(0.008)	(0.003)	(0.001)	(1.00)
Pugh - *N*	141	108	7	1	0	4	6	0	1	1	269
(Proportion)	(0.524)	(0.401)	(0.026)	(0.004)	(0.000)	(0.015)	(0.022)	(0.000)	(0.004)	(0.004)	(1.00)
z-score	0.17	-0.31	-0.11	0.20	0.61	0.07	-0.42	0.37	-0.02	-0.22	--

[a]Painted category includes Nodena Red-and-White, Carson Red-and-Buff, Old Town Red, Hollywood White, and Avenue Polychrome.
[b] MPI = Mound Place Incised.

Further analyses of the new descriptions, however, show that the PFG and the new Beck samples, however, cannot be simply combined since the new collections have been split into two spatially discrete samples. The PFG sample pooled the variability at the collection stage so is indistinguishable from both Beck-A and Beck-B. Based on the *z*-scores (Table 7.6) the PFG sample is slightly more similar to Beck A than Beck B but the difference is slight. The inability to resolve the PFG frequencies from either Beck-A or Beck-B is consistent with the hypothesis, presented in Chapter 6, that PFG sample is a mix gathered from across the entire deposit. Thus, is likely that using only intuition to guide them and given the smaller samples PFG were unable to detect small compositional the differences in the field. Consequently, to make the new collections comparable to the PFG collections, both Beck A and Beck B will be combined to form a single, larger sample.[1]

[1] Further resolution about the relationship between Beck-A and Beck-B will require absolute dating information. Chapter 9 discusses a program to collect this information using luminescence dating.

A second conclusion is that the PFG samples are reliable, accurate, and precise estimates of the composition of the archaeological record at this scale. It is not possible to use Ford's seriation analyses without speculating about collection methods, sampling, and counting procedures. The results demonstrate that, with modifications introduced here, PFG samples and those that I generated can be combined and the remaining St. Francis and Memphis PFG samples can be treated as representative samples of decorated pottery.

SERIATIONS

We can now use the new collections to examine the seriation results presented in Chapter 4 in a new light. In these analyses, I examine seriations created using only the original PFG samples. Second, I build seriations using only data from the Lipo collections. Third, I compare these results with seriations composed of combined Lipo and PFG assemblages. Finally, I evaluate the effect of adding the new, larger samples to the combined St. Francis and Memphis assemblages collected by PFG.

PFG SOLUTIONS

Figure 7.1 presents the seriation solution of the original PFG assemblages. At a confidence level of 99.5%, two groups are formed. The first group is comprised of Beck, Belle Meade, and Rose Mound while Castile Landing, Cranor Place, and Nickel form the second group. Spatially, the distribution of the two groups is coherent, though the inclusion of Rose Mound with Group 1 potentially falsifies a tightly bound spatial solution. The spatial distribution of the assemblages and the seriation groups is shown in Figure 7.2. The inclusion of Rose Mound with Group 1 is potentially caused by low number of samples (n=6) and relatively small sample sizes. Additional collections (such as those afforded by the Lipo collections described here) should result in our ability to compositionally distinguish Rose Mound from the Beck/Belle Meade group.

LIPO SOLUTIONS

The larger assemblages generated during the recollection of the St. Francis Basin area deposits provide an opportunity to increase the precision of the seriation solutions and the spatial distributions of assemblages. Figure 7.3 shows a seriation using the new assemblages described with PFG types with solved with a significance of α=0.005. Three distinct groups are formed. The first group consists of Cranor, Nickel, and Castile Landing, the second group is composed of Beck and Belle Meade, and the third group is composed of only Rose Mound, which does not fit into the other two groups. Holden Lake appears distinct from all of the other assemblages and fits at the ends of all groups. This evidence suggests that Holden Lake is earlier and potentially "ancestral" to the other assemblages.[2] Spatially, these seriation groups form two cohesive sets: Group 1 and 2 (Figure 7.4).

SOLUTIONS WITH COMBINED PFG AND LIPO SAMPLES

In the first part of this chapter, I determined that with a simplified classification scheme, the descriptions of the new collections were statistically isomorphic with the PFG assemblages. This result supports a hypothesis that the PFG assemblages were generated in a comparable fashion as the new assemblages and thus can be treated as samples from the same populations. Consequently, the PFG and new samples can be pooled together to form larger assemblages. In this way, we can further examine the effect of increasing sample size on our ability to resolve the seriation solutions.

Tables 7.8 and 7.9 present the counts and relative frequencies of the combined Lipo and PFG datasets. Figure 7.5 presents the results from the iteratively created seriations of the seven assemblages ordered with a $\alpha = 0.005$ significance level. As in the seriation of the Lipo collections, three groups are distinguishable when the sample sizes are combined to form a single set of larger assemblages. In addition, I found that as in the previous seriations, Holden Lake could be added to the end of any of the groups. Consequently, the results support the hypothesis that Holden Lake is earlier than and ancestral to the other assemblages in the analyses. The composition of Groups 1 and 2 are the same as the seriation results when the PFG and Lipo assemblages are treated separately with a single exception: Rose Mound no longer forms a part of Group 2 and cannot be combined with either set of assemblages. Thus, the sample size hypothesis can be falsified. Spatial cohesiveness among deposits is maintained with larger samples.

We can also calculate the likelihood that these spatially cohesive groups were formed by chance alone. Using Equation 4.1, we find that the probability that these three groups were formed by chance is 0.043, a low value. Although larger sample sizes minimize the role of chance in the relative order of the stylistic frequencies of decorated types, before we can conclude that the structure of the groupings is not due to chance we must increase the number of assemblages. We can pose a null hypothesis that states that the spatial grouping of the assemblages occurred by chance alone. An alternative hypothesis is that chance did not play a role in forming the groups of assemblages but that other processes (*i.e.,* transmission) were involved. We can assess the first hypothesis by increasing the number of assemblages in the analyses. In doing so, the probability of creating groups by chance alone, as calculated by Equation 4.1, decreases drastically. If we find that the seriation groups are still coherent when we increase the number of

[2] More information is needed to evaluate the ancestry relations of Holden Lake. With denser data, it is possible that Holden Lake will seriate along only one lineage.

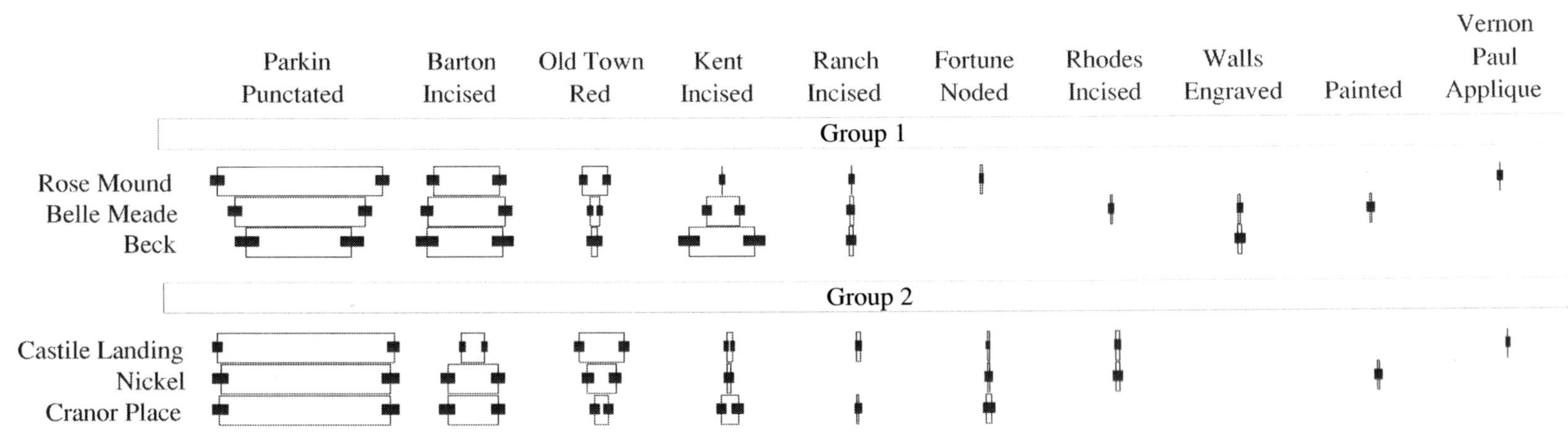

Figure 7.1. Seriations Created with the PFG Assemblages Described with Shell-Tempered, Decorated PFG Types for the Six Recollected St. Francis Basin Area Deposits.

Using the PFG collections from the six St. Francis Basin area deposits recollected in this study, I built two seriation groups were created with significance of α=0.005. These were the largest groups that could be ordered while meeting the expectations of the seriation model. As seen in the following figure (Figure 7.2), the groups are remarkably spatially coherent. Note that no Hull Engraved sherds were assigned by PFG in these six assemblages.

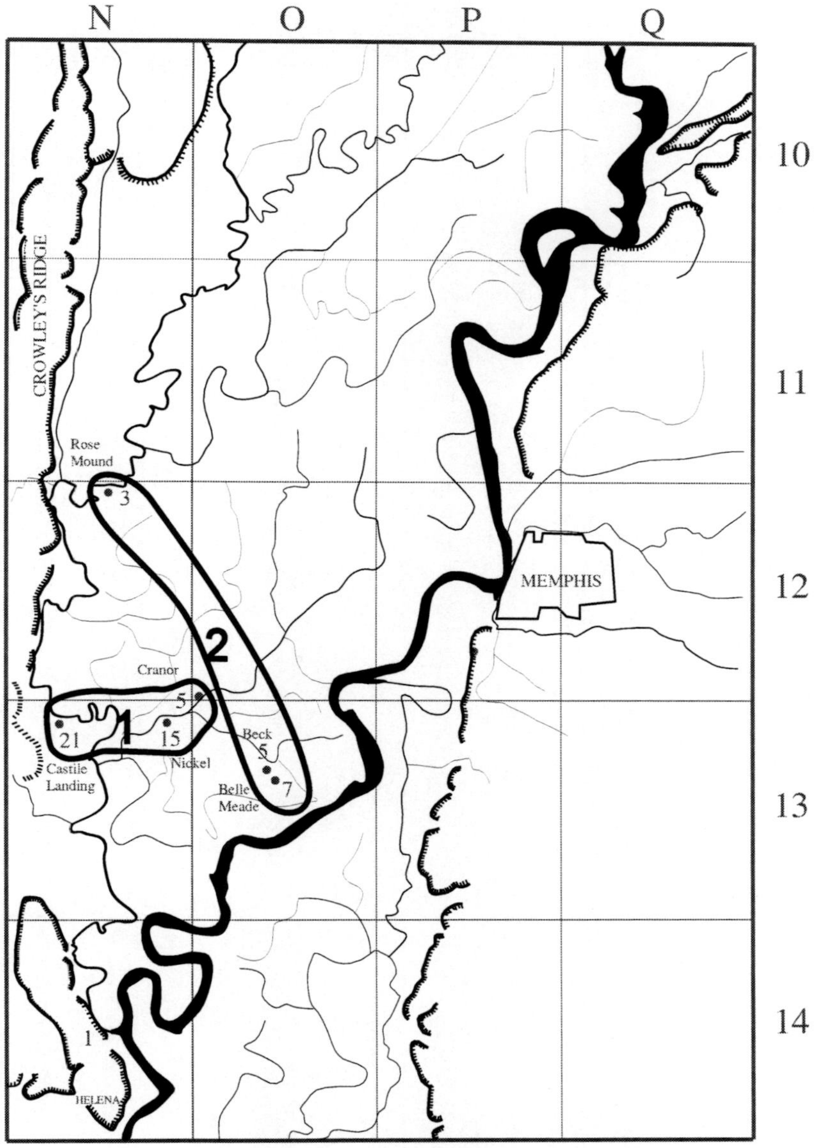

Figure 7.2. Spatial Groupings of Seriation Solutions Created Using the Shell-Tempered, Decorated PFG Types from the PFG Assemblages of the Six Recollected St. Francis Basin Area Deposits.

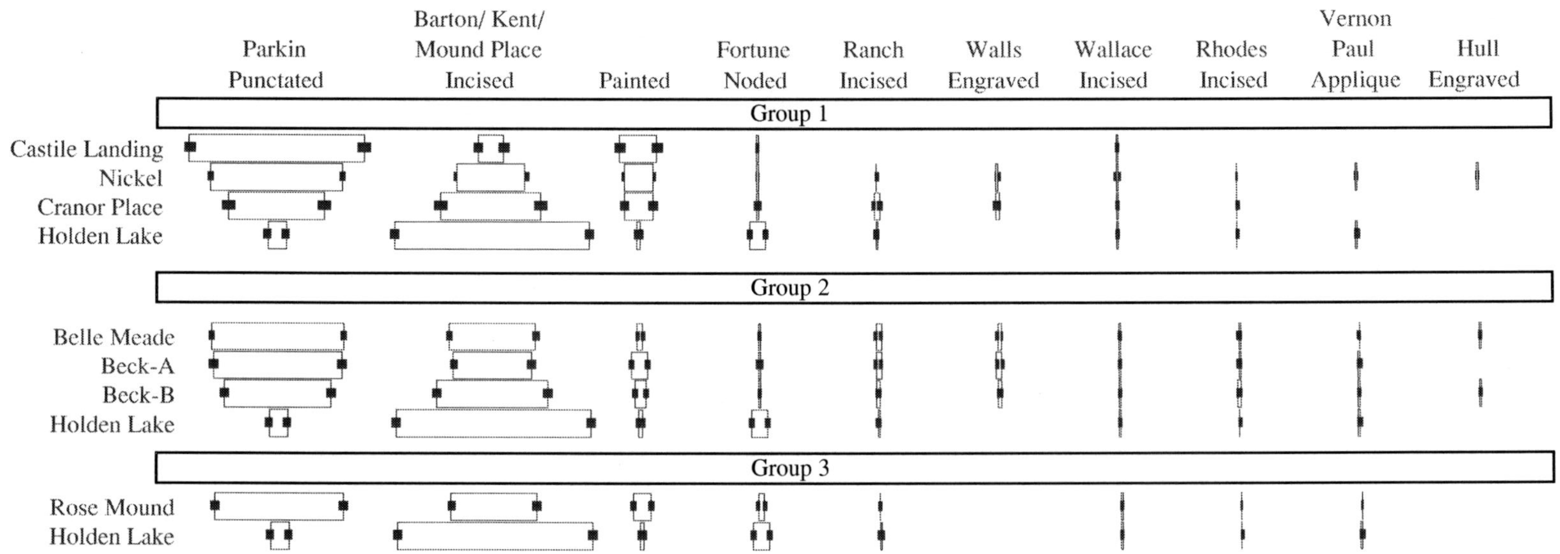

Figure 7.3. Seriation Groups Built of Lipo Collections Described with Shell-Tempered, Decorated PFG Types.

Using the new collections made from the St. Francis Basin area deposits and described by PFG shell-tempered, decorated types, I built three seriation orders at a significance of α=0.005. Compare these results with those obtained using the original PFG assemblages: the groups are remarkably similar. Holden Lake fits in all three groups. This suggests that Holden Lake is earlier and "ancestral" to the other assemblages.

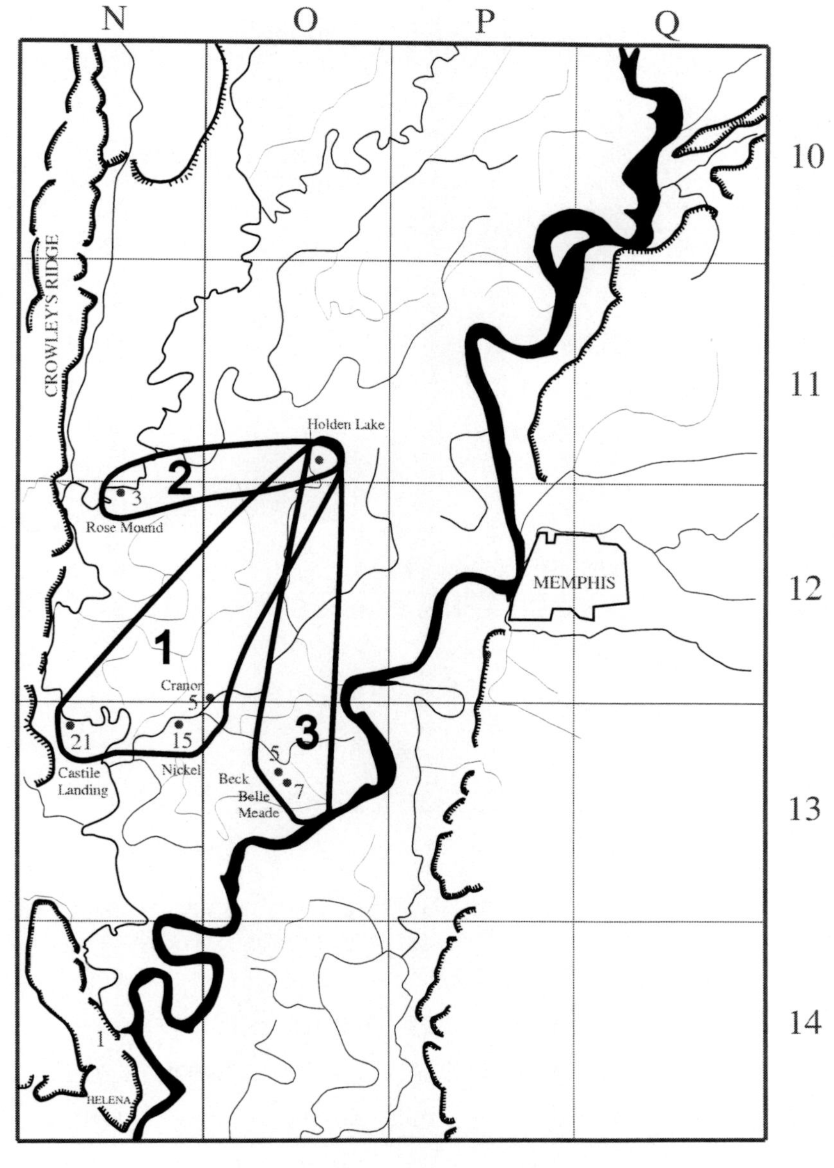

Figure 7.4. Spatial Groupings of Seriation Solutions for Lipo Collections Described with Shell-Tempered, Decorated PFG Types.

Holden Lake fits in all three groups. The relationship of Holden Lake to the other assemblages in the seriations (Figure 7.3) suggests that this assemblage is earlier than those collected at the other locations.

Table 7.8. Combined Totals of Decorated, Shell-Tempered PFG Types for Combined Lipo and PFG Assemblages.

Assemblage	Parkin Punctated	Barton/ Kent/ MPI[b]	Painted	Fortune Noded	Ranch Incised	Walls Engrav ed	Wallace Incised	Rhodes Incised	Vernon Paul Appliqué	Hull Engraved	Σ
Beck	590	498	67	10	21	19	8	12	7	1	1233
Belle Meade	923	637	42	12	33	27	15	13	5	2	1709
Castile Landing	426	69	105	4	4	0	1	4	1	0	614
Cranor Place	204	156	42	7	8	4	2	1	0	0	424
Holden Lake	27	294	7	24	2	0	2	1	3	0	360
Nickel	728	364	160	9	5	8	14	3	7	2	1300
Rose Mound	549	328	77	19	4	0	3	1	2	0	983
Σ	3415	2303	498	85	76	57	45	35	25	5	6544

[a]Painted category includes Nodena Red-and-White, Carson Red-and-Buff, Old Town Red, Hollywood White, and Avenue Polychrome.
[b] MPI = Mound Place Incised.

Table 7.9. Proportions of Decorated, Shell-Tempered PFG Types for Combined Lipo and PFG Assemblages.

Assemblage	Parkin Punctated	Barton/ Kent/MPI[b]	Painted[a]	Fortune Noded	Ranch Incised	Walls Engraved	Wallace Incised	Rhodes Incised	Vernon Paul Appliqué	Hull Engraved
Beck	0.479	0.404	0.054	0.008	0.017	0.015	0.007	0.010	0.006	0.001
Belle Meade	0.540	0.373	0.025	0.007	0.019	0.016	0.009	0.008	0.003	0.001
Castile Landing	0.694	0.112	0.171	0.007	0.007	0.000	0.002	0.007	0.002	0.000
Cranor Place	0.481	0.368	0.099	0.017	0.019	0.009	0.005	0.002	0.000	0.000
Holden Lake	0.075	0.817	0.019	0.067	0.006	0.000	0.006	0.003	0.008	0.000
Nickel	0.560	0.280	0.123	0.007	0.004	0.006	0.011	0.002	0.005	0.002
Rose Mound	0.559	0.334	0.078	0.019	0.004	0.000	0.003	0.001	0.002	0.000

[a]Painted category includes Nodena Red-and-White, Carson Red-and-Buff, Old Town Red, Hollywood White, and Avenue Polychrome.
[b] MPI = Mound Place Incised.

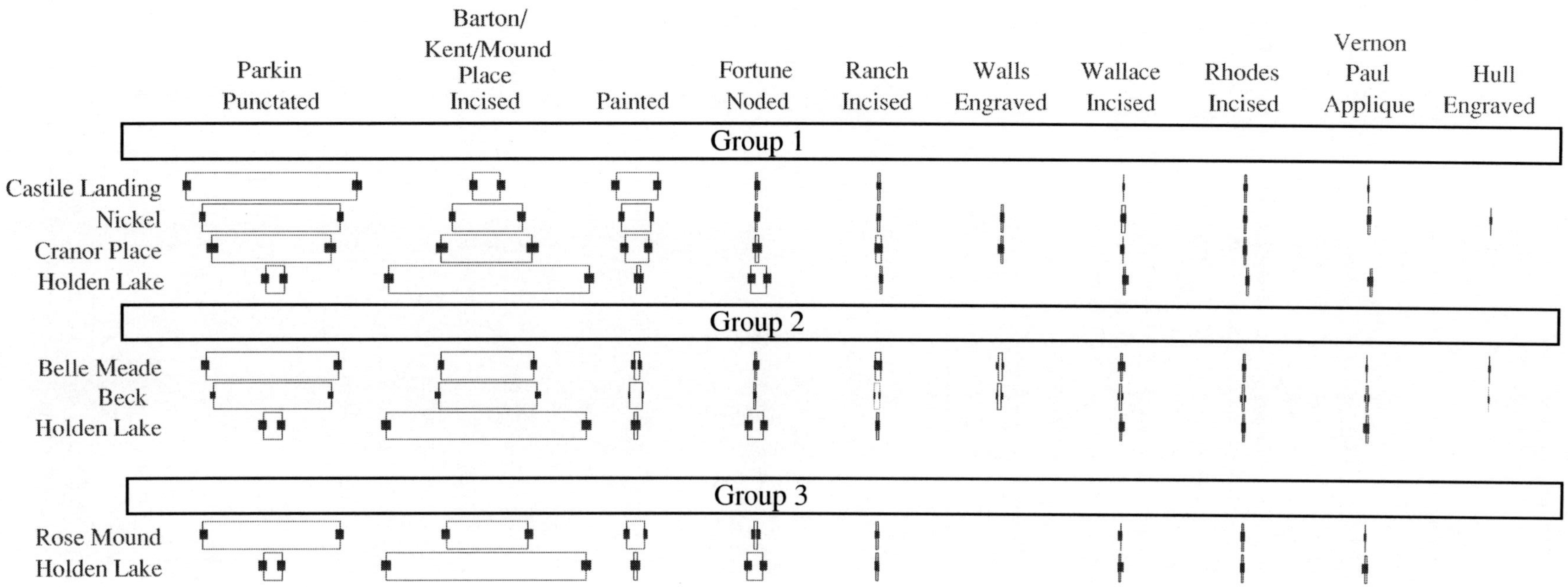

Figure 7.5. Seriation Solution for Combined PFG and Lipo Assemblages Described with Shell-Tempered, Decorated PFG Types.

These three groups were formed with a significance of α=0.005. The orders and groups presented here use larger sample sizes but are equivalent to those shown Figure 7.3. This support the hypothesis that these are not the result of chance.

assemblages, then we would have reason to reject the null hypothesis that the groups were formed by chance. We would then be in a position to argue that the best model for explaining the structure of stylistic similarities among the deposits is transmission.

SOLUTIONS WITH ST. FRANCIS AND MEMPHIS PFG ASSEMBLAGES INCLUDING LIPO COLLECTIONS

We can evaluate these hypotheses by including the set of Memphis and St. Francis assemblages collected by PFG that meet our sample size criteria ('A' Assemblages) in the seriation analyses and examining how the increased numbers of samples impact the results obtained in these seriations. This analysis also provides an opportunity to evaluate the results of the seriation studies conducted in Chapter 4 with the original PFG samples: how do larger samples gained in my collections affect the seriation groups obtained? Tables 7.10 and 7.11 present the total counts and proportions of decorated types for all the Memphis and St. Francis assemblages. The seven new aggregated PFG/Lipo assemblages are included.

The results of the seriation using the PFG data from the St. Francis and Memphis combined with the Lipo assemblage data are shown in Figure 7.6. At a 99.95% confidence level, eight groups are distinguishable that take much of the same form as in the previous analyses in terms of composition and spatial configuration. However, with the addition of the full set of assemblages, the probability of obtaining the set of groups with a spatial contiguity is tremendously low. Using Equation 4.1, we can calculate the probability value to be less than 0.00001. In addition, unlike the seriation solution obtained in Chapter 4 (Figures 4.10 and 4.11) additional detail can now be resolved in the spatial groups. Castile Landing, Nickel, and Cranor Place are distinguished as a seriation group (Group 5). This conclusion resolves the hazy picture we had obtained with the smaller assemblages in Chapter 4. Larger samples provide a solid theoretical and

Table 7.10. Combined Counts of Decorated, Shell-Tempered PFG Types for Lipo and PFG Collections in St. Francis and Memphis Area Collections.[a]

Assemblages	Parkin Punctated	Barton/Kent/MPI	Painted[b]	Fortune Noded	Ranch Incised	Walls Engraved	Wallace Incised	Rhodes Incised	Vernon Paul Appliqué	Hull Engraved	Σ
10-P-1	39	62	46	0	0	0	0	0	0	6	153
11-N-9	528	198	13	0	19	0	0	0	0	0	758
11-N-1	865	323	59	17	35	0	0	0	4	0	1303
11-O-10	404	208	6	16	4	0	0	0	0	0	638
11-N-4	764	470	18	5	9	0	0	0	0	0	1266
13-N-5	35	11	33	0	0	0	0	0	0	0	79
13-N-4/B,C,D,E	71	67	96	0	3	4	0	0	0	0	241
13-N-16	42	56	69	0	1	3	0	0	0	0	171
13-O-11	35	65	24	0	0	2	0	1	0	1	128
13-O-10	61	74	79	0	2	8	0	2	0	0	226
13-P-1	244	40	18	1	16	21	0	14	0	6	360
13-P-8	83	25	43	0	18	17	0	3	0	3	192
13-P-10	30	15	12	0	12	12	0	7	2	1	91
13-O-7 (Belle Meade)	590	498	67	10	21	19	12	8	7	1	1233
13-O-5 (Beck)	923	637	42	12	33	27	15	13	5	2	1709
13-N-21 (Castile Landing)	426	69	105	4	4	0	1	4	1	0	614
12-O-5 (Cranor)	204	156	42	7	8	4	2	1	0	0	424
Holden Lake	27	294	7	24	2	0	2	1	3	0	360
13-N-15 (Nickel)	728	364	160	9	5	8	14	3	7	2	1300
12-N-3 (Rose Mound)	549	328	77	19	4	0	3	1	2	0	983

[a] In order to compare the new assemblages with the existing PFG samples from the St. Francis and Memphis areas that were not recollected, I modified the decorated types appropriately; all painted types were included in the Painted class and Barton/Kent/Mound Place Incised were aggregated.

[b]Painted category includes Nodena Red-and-White, Carson Red-and-Buff, Old Town Red, Hollywood White, and Avenue Polychrome.

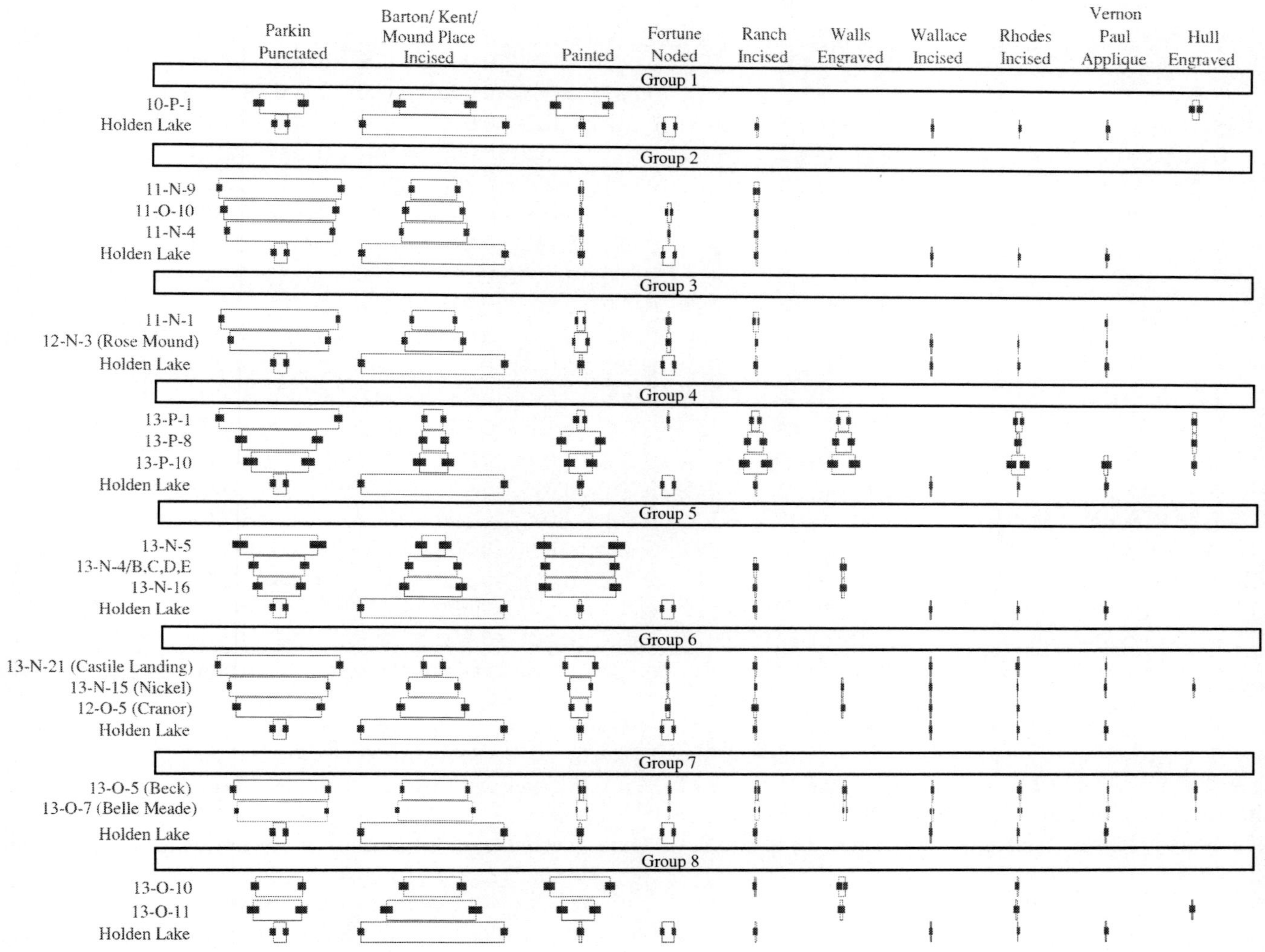

Figure 7.6. Seriation Groups Built from the Combined Lipo and PFG St. Francis and Memphis Assemblages ('A' Assemblages) and Described with Shell-Tempered, Decorated PFG Types.

Table 7.11. Proportions of Shell-Tempered, Decorated PFG Types for Lipo and PFG Collections in St. Francis and Memphis Area Collections.

Assemblages	Parkin Punctated	Barton/Kent/MPI	Painted	Fortune Noded	Ranch Incised	Walls Engraved	Wallace Incised	Rhodes Incised	Vernon Paul Appliqué	Hull Engraved	Σ
10-P-1	0.255	0.405	0.301	0.000	0.000	0.000	0.000	0.000	0.000	0.039	1.000
11-N-9	0.697	0.261	0.017	0.000	0.025	0.000	0.000	0.000	0.000	0.000	1.000
11-N-1	0.664	0.248	0.045	0.013	0.027	0.000	0.000	0.000	0.003	0.000	1.000
11-O-10	0.633	0.326	0.009	0.025	0.006	0.000	0.000	0.000	0.000	0.000	1.000
11-N-4	0.604	0.371	0.014	0.004	0.007	0.000	0.000	0.000	0.000	0.000	1.000
13-N-5	0.443	0.139	0.418	0.000	0.000	0.000	0.000	0.000	0.000	0.000	1.000
13-N-4/B,C,D,E	0.295	0.278	0.398	0.000	0.012	0.017	0.000	0.000	0.000	0.000	1.000
13-N-16	0.246	0.328	0.404	0.000	0.006	0.018	0.000	0.000	0.000	0.000	1.000
13-O-11	0.273	0.508	0.188	0.000	0.000	0.016	0.000	0.008	0.000	0.008	1.000
13-O-10	0.270	0.327	0.350	0.000	0.009	0.035	0.000	0.009	0.000	0.000	1.000
13-P-1	0.678	0.111	0.050	0.003	0.044	0.058	0.000	0.039	0.000	0.017	1.000
13-P-8	0.432	0.130	0.224	0.000	0.094	0.089	0.000	0.016	0.000	0.016	1.000
13-P-10	0.330	0.165	0.132	0.000	0.132	0.132	0.000	0.077	0.022	0.011	1.000
13-O-7 (Belle Meade)	0.479	0.404	0.054	0.008	0.017	0.015	0.010	0.007	0.006	0.001	1.000
13-O-5 (Beck)	0.540	0.373	0.025	0.007	0.019	0.016	0.009	0.008	0.003	0.001	1.000
13-N-21 (Castile Landing)	0.694	0.112	0.171	0.007	0.007	0.000	0.002	0.007	0.002	0.000	1.000
12-O-5 (Cranor)	0.481	0.368	0.099	0.017	0.019	0.009	0.005	0.002	0.000	0.000	1.000
Holden Lake	0.075	0.817	0.019	0.067	0.006	0.000	0.006	0.003	0.008	0.000	1.000
13-N-15 (Nickel)	0.560	0.280	0.123	0.007	0.004	0.006	0.011	0.002	0.005	0.002	1.000
12-N-3 (Rose Mound)	0.559	0.334	0.078	0.019	0.004	0.000	0.003	0.001	0.002	0.000	1.000

[a]Painted category includes Nodena Red-and-White, Carson Red-and-Buff, Old Town Red, Hollywood White, and Avenue Polychrome.

empirical basis to argue that the deposits in the area of the Memphis and St. Francis areas form a distinct group of interacting populations (Figure 7.7). In addition, like the solution in Figure 7.3, Holden Lake fits at one end of all of the groups.

PARKIN AREA SERIATION SOLUTIONS

Interestingly, what was formerly a single group of assemblages in the quadrants 11-O and 11-N now breaks down into two distinct sets, Group 2 consisting of 11-O-10, 11-N-4, and 11-N-9 and Group 3 consisting of 12-N-3 (Rose Mound) and 11-N-1. This result is surprising since it suggests that what is commonly referred to as a single cultural unit, the Parkin phase (*e.g.,* P. Morse 1981, 1990; Phillips 1970), may involve more complicated patterns of transmission. Significantly, 11-N-1 (Parkin) fits into both groups yet the groups themselves cannot be merged. On one hand, 11-N-1 fits with Rose Mound on one hand, and in the middle of Group 2. However, on the other hand, Rose Mound cannot be added to Group 2. Figure 7.8 shows the two possible solutions for the assemblages in Groups 2 and 3 and how 11-N-1 can be placed in both seriation solutions. The central location of 11-N-1 is significant as it reflects the same result that was obtained in the seriation solutions with poorer sample sizes: two lineages that share the Parkin site assemblage.

Four possible hypotheses shown in Figure 7.9 can account for the patterns of stylistic similarity between Rose Mound (12-N-3), the southern component, Parkin (11-N-1) itself and a group of Parkin "phase" components (11-O-10, 11-N-4, 11-N-9) (Dunnell and Lipo 1999). First, a linear, purely chronological arrangement with one component preceding the other, could account for this pattern of similarity in either of the two orders. This was the PFG view. Second, the pattern is potentially explained by divergent relations between assemblages in which the Parkin site is ancestral to both Rose and the locations north of Parkin. This hypothesis would be supported if the mean age of the Parkin deposit was earlier than the other settlements. The pattern of similarity could indicate convergence of transmission between previously disparate communities. This scenario is consistent with a model involving population increase and community fission. Third, the pattern could be attributed to a convergent relationship in which Parkin is a descendant amalgamation of the two previously separate lineages. This hypothesis would be supported if Parkin were found to be later than other settlements in the groups.

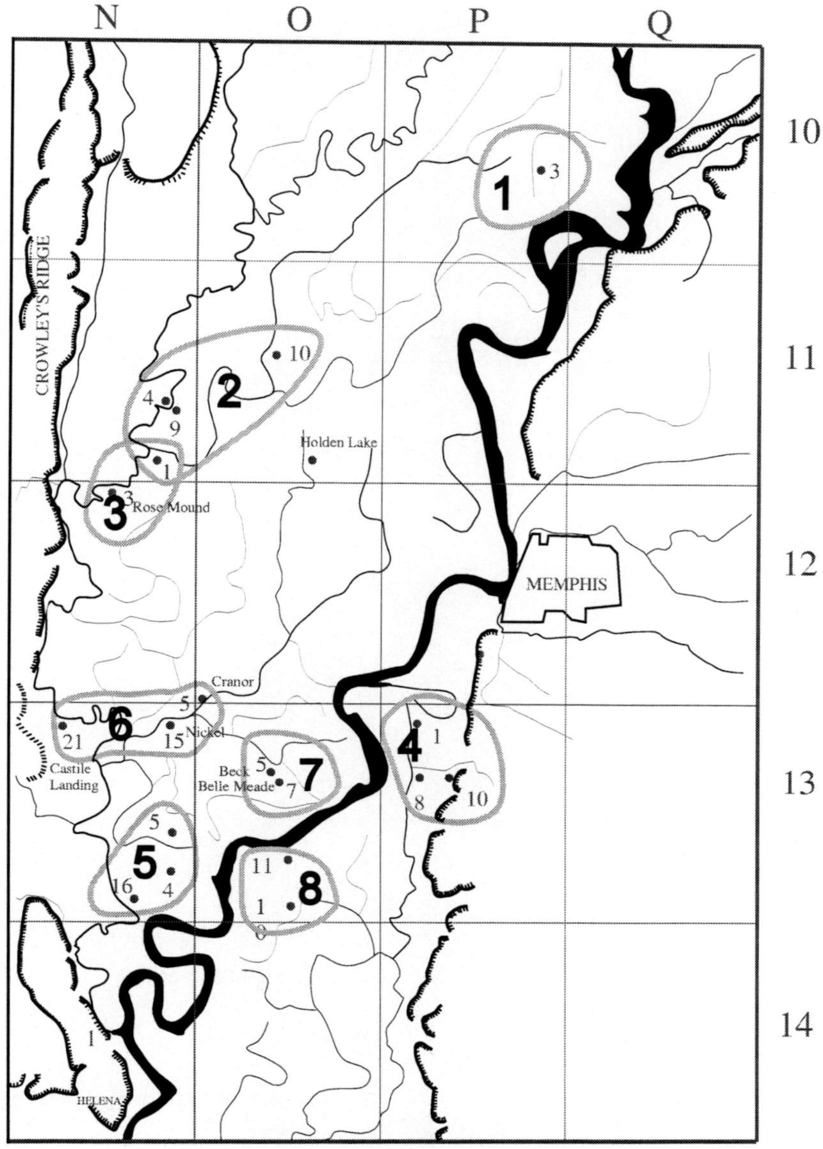

Figure 7.7. Spatial Distribution of Seriation Groups with St. Francis and Memphis Assemblages ('A' Assemblages) with Combined Lipo and PFG Samples.

Labels for groups refer to seriation solution in Figure 7.6. As shown in the seriation solutions, each group should also include Holden Lake. The group boundaries do not include Holden Lake here for visual clarity.

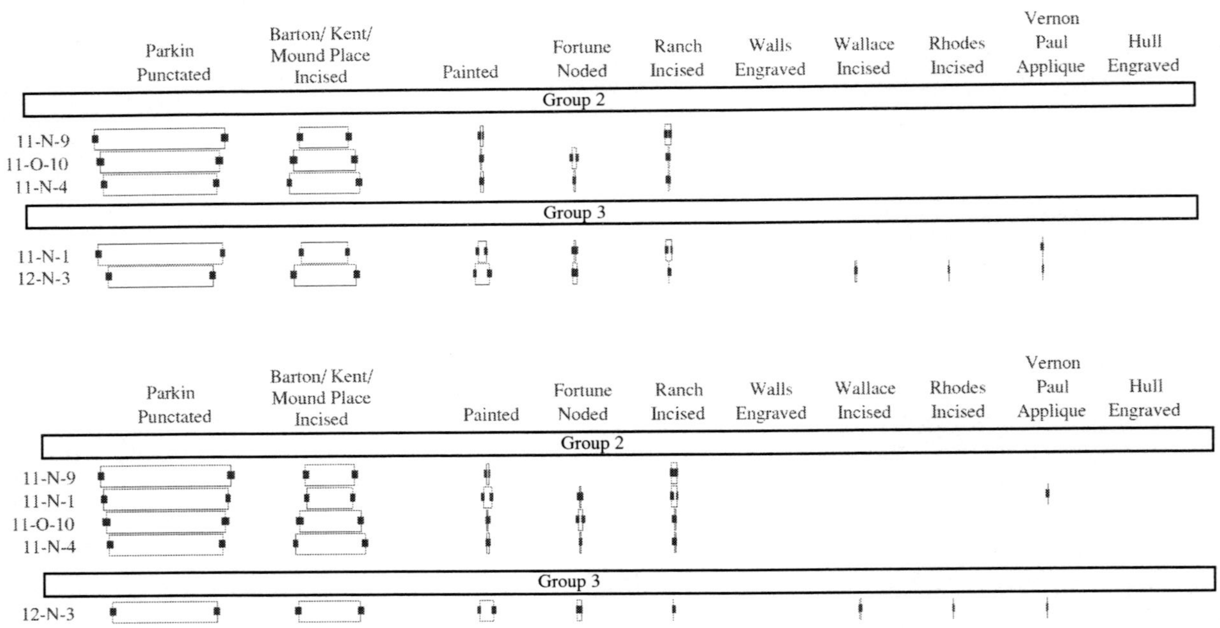

Figure 7.8. Detail of the Two Seriation Solutions for Assemblages in Groups 2 and 3.

Assemblages are comprised of PFG and Lipo collections and are described by Shell-Tempered, Decorated PFG Types. Here, two seriation solutions are possible. In the first, 11-N-1 (Parkin) is part of Group 2. In the second, 11-N-1 is part of Group 3. This results supports a hypothesis that two lineages share the Parkin site assemblage.

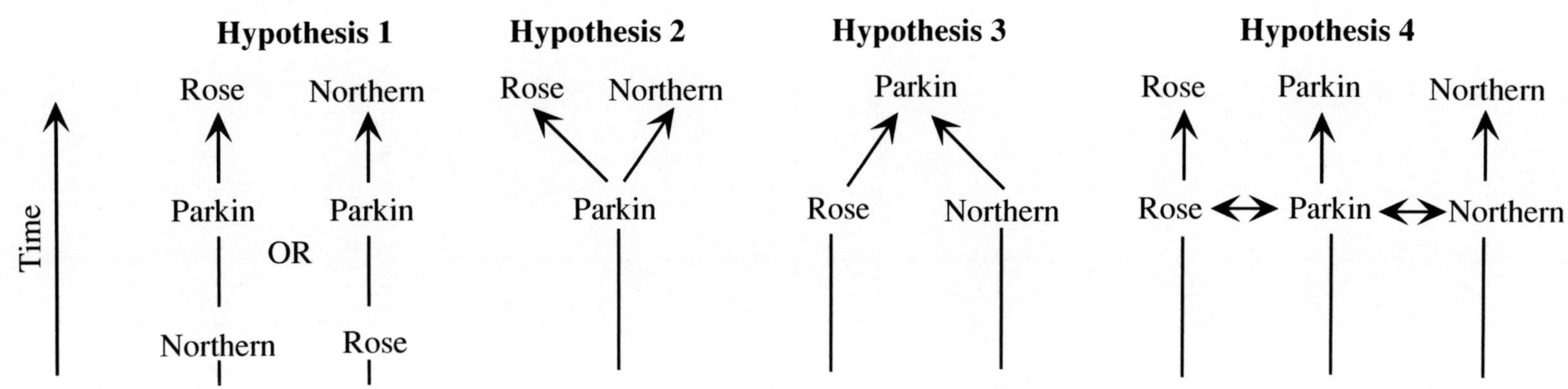

Figure 7.9. Four Hypotheses That Can Account for the Stylistic Similarity Among the Parkin Area Assemblages.

"Northern" is composed of assemblages 11-O-10, 11-N-4, 11-N-9. Parkin is 11-N-1. Rose is comprised of location 12-N-3. Descriptions of hypotheses are given in the text.

Lastly, the pattern among the assemblages is consistent with a model of hierarchic relations between locations in which both Rose Mound and the components to the north interact with one another only or largely through the Parkin node. That is, if the deposits are contemporaneous with one another, the pattern of stylistic similarity could represent model in which a single node is more connected to other locations than all the locations are with each other.

The first hypothesis is readily falsified. The two groups of assemblages do not form a single, statistically valid seriation. Hypothesis 3 also seems unlikely. Only in the case that the derivative assemblages have no pre-Contact (*i.e.,* prior to AD 1542) elements could Hypothesis 3 be correct. However, Parkin (11-N-1) is certainly Contact age and both Rose Mound (12-N-3) and the northern components have contact materials (Moore 1910, 1911; P. Morse 1981, 1990; Morse and Morse 1990). Hypothesis 2 is a mirror image of Hypothesis 3. Only if Parkin had no pre-Contact occupation could Hypothesis 2 be likely. If Parkin can be identified as a settlement visited by DeSoto (Morse and Morse 1990), this possibility can be ruled out. Consequently, Hypothesis 4 seems the most likely account without a better chronology. In this model, Parkin is a central node held in common by distinct lineages. The Parkin ceramic data, therefore, potentially represent evidence of the emergence of social complexity among late Mississippian populations (Dunnell and Lipo 1999; Dunnell and Wenke 1980; Lipo and Madsen 1999; Wenke 1981). Significantly, no other set of seriation groups displayed this pattern in the Memphis and St. Francis areas. In each of the other groups, only single solutions were possible; no overlapping sets were noted.

The case for regarding the Parkin assemblages and *only* the Parkin assemblages in the Memphis and St. Francis areas of Mississippi river valley as having characteristics of complex societies is by no means settled. Much larger and representative sample sizes that render small frequency differences statistically significant are essential. In addition, more assemblages that are better distributed in space are critical. Furthermore, accurate, independent dating is required to go beyond presence/absence of European artifacts used here. Distinguishing these alternatives requires fine-scaled, absolute chronological resolution between settlements and detailed information about functional relationships, information that is not available from the current dataset.

CONCLUSIONS

A substantial portion of our archaeological knowledge of the southeastern United States is built upon the work of PFG (Dunnell 1985a). In addition to providing the fundamental chronological framework for Mississippi valley prehistory, the data from the PFG survey provide the basis upon which cultural associations across space have been mapped. Yet, despite the centrality of this body of work, virtually no systematic analyses have been done to determine the degree to which the PFG data are accurate and reliable samples of the archaeological record. With the exception of House (1991, 1993, 1995), for example, the majority of phase interpretations rely on the conclusions drawn from the PFG samples. This lack of critical evaluation is particular problematic when we seek to utilize the PFG data in new analyses. As a large set of stylistic measurements made from several hundred locations, the PFG data are potentially an excellent source of information for studying cultural transmission in the archaeological record. The recollections and descriptions of six of the PFG locations presented here shed substantial light on the degree to which the PFG sample can be relied upon. The results of the analyses presented here suggest that the PFG assemblages are reliable samples of the archaeological record and that in at least a gross way using simplified classes the original PFG 'A' assemblages can be used in transmission studies.

This conclusion has a number of implications. First, it highlights the technical excellence with which the original PFG samples were collected. Although guided mostly by intuition, PFG field methods were sufficient to build samples that could be used even 50 years later. Second, the degree to which the composition of the PFG samples are substantiated by the new collections lends weight to the seriation groups built in Chapter 4; the seriation solutions are supported with independently generated data.

Finally, is clear that one limiting factor in the study of transmission – and one that can be addressed – is sample size. Increasing the size of samples produces new information about relationships between assemblages. In analyses presented here, increasing sample size resulted in an ability to statistically distinguish several groups of assemblages in the St. Francis Basin that were not previously identified by culture historians in their "phase" definitions. In addition, I was able to resolve new details about the interaction among groups not previously visible. These results suggest that at least the Parkin area populations may have been hierarchically organized and that this kind of organizational structure is not evident for the other sets of assemblages. Consequently, using the detailed information formed from these seriation analyses it is possible to pose hypotheses about the evolution of interacting populations.

With the information gathered here, we can now turn to evaluating whether the patterns noted in the Parkin group of assemblages represent the emergence of complex organization and, if so, what conditions and processes are involved in this change. Does the archaeological record support a hypothesis of sympatric evolution in which larger-scaled, functionally specialized phenotypic variants differentially succeeded within the populations of the valley? Alternatively, is the prehistoric record consistent with an allopatric model where functionally integrated populations out-competed less integrated groups: a model suggestive of group selection? Additionally, we can begin to examine the historical constraints that are driving the selection of phenotypic attributes within these populations. Finally, we can examine whether the other seriation groups detected are representative of spatially bounded, interacting populations or are more likely due to the effects of distance and geography.

Most archaeological analyses require comparisons between sets of independently generated data. Consequently, one never knows *a priori* the degree to which observed frequencies are a function of the target phenomena and the degree to which they reflect characteristics of the classification used to tabulate phenomena. In the present analyses, my ability to make statements beyond my own assemblages and to a wider area in the central Mississippi river valley depends on combining my samples with those made by PFG into a single coherent and comparable set. However, simply confirming that the results that I generated in Chapter 7 with my collections are compatible with those generated with PFG's data does not guarantee that there are not biases common to both. Sampling issues and variation at the level of classification higher than PFG types have already been examined. Here, I take up the issue of classification at levels *smaller* than types and the interaction between measurement, sherd size, and type assignments.

It is imperative from the outset that one recognizes that there is no "correct" abundance measure as often presumed implicitly (*e.g.,* Orton 1993; Phillips 1970) or explicitly (*e.g.,* McNutt 1973; Newell and Krieger 1949). There is, for example, no reason to prefer vessels as a measure of abundance over any other (beyond the tyranny of common sense) and many reasons not to (*e.g.,* the physical impossibility of such reconstructions [see Orton 1993] except under extreme conditions [Newell and Krieger 1949]). What is required is that samples included in a given analysis be *comparable.*

The evaluation of comparability is assessed by evaluating the factors that contribute to the measurement of type proportions for each sample in the analyses. These factors can be specified as:

Equation 8.1. $T_p = f(S_a, E_a, E_r, D_a, V_a, V_n)$

where T_p = the proportion of a type measured in a sample, S = sherd, E = decorative element, D = design (sets of elements), a = area, r = robusticity (factors comprised of a combination of a preservation constant and a type assignment constant). In addition, T_p is a product of the number of vessels bearing the set of traits that define the type (V_n) and the area of vessel that bears the decoration (V_a).

The virtue of such an approach is immediately apparent in that it forces systematic consideration of all the variables that might influence the generation of a number including the relationship between such factors and how they interact with the sample properties. For example, it is clear from Equation 8.1 that the minimum size of countable sherds is a function of decoration. Rules-of-thumb like collecting sherds that are at least "2 square centimeters" or "quarter-size" have no rationale basis when quantification of abundance is at issue. It is also apparent that at least two approaches exist for "correcting" non-comparable samples: mathematical normalization and altering classification. Both approaches are applicable to these data, but in the end, the creation of comparable assemblages, not "right" numbers, is the target. The tactic is familiar. When one cannot use absolute values, it is necessary to rely upon the analysis of differences (as in *e.g.,* atomic weight).

CLASSIFICATION LEVEL AND SCALE

Equation 8.1 specifies that type proportions (T_p) are calculated based on the recognition of sets of classificatory attributes that define types on sherd instances. These type proportions are strongly influenced by the set of decorative elements (E) and designs (D) used as definitive features through the properties of *level* and *scale* (Dunnell 1971). Though conceptually related, the concepts of level and scale are ontologically distinct. Level is an ideational property (*sensu* Dunnell 1971) that refers to the inclusiveness, rank, or "size" of measurement classes. In a single classificatory scheme, for example, the classes with the fewest dimensions are the highest level while those classes with greater number of definitive features of a lower level. A scale, on the other hand, is defined as the set of *objects* that display the same degree of inclusiveness. Scale is not, however, inherent in phenomena but is *stipulated* for classifications by the definition of the classification root (sherds vs. vessels vs. assemblages of vessels or design element vs. design vs. design set). Scale stipulates the counting units. Level, on the other hand, is determined by the number of criteria for class membership. The larger the number of criteria, the lower the level, and the smaller the number of cases that meet the criteria. The two are independent, one can use any number of membership criteria (one to very many) for any scale. The criteria chosen, though, are influenced by the scale selected. Since scale dictates the counting units, it impacts the kind of phenomena that can be counted and thus places empirical constraints on sampling. In the case of level, the lower the level, the finer resolution possible but the larger the sample needed.

After PFG, the type-variety system (*e.g.,* Gifford 1960; Sabloff and Smith 1969; Smith 1960; Wheat *et al.* 1958) came into wide use to control problems of classification *level.* Phillips (1958) played an important role in its formative stage. Because level was uncontrolled in types like those used by PFG, it was possible to build some "types" that would cover large amounts of time and/or space while others had members only in much smaller time/space areas. Other workers, specifically Phillips, used varieties in an attempt to account for empirical variation not treated by the dimensions defining types. Indeed, in the absence of theory to justify the selection of criteria, this was commonly perceived as a problem. Workers like Krieger (1944), in the nearby Caddoan area, lamented such variation even though he expressly ruled out "subtypes" to do this work. Subtypes for Krieger were what most would call "provisional" types:

attribute combinations, the historicity (neutrality) of which had yet to be demonstrated.

The "unaccounted variation" issue, however, is a result of treating types as empirical entities rather than archaeological constructs (*a la* Chang [1967] where he acknowledges that types are constructed but asserts, *sans* any possibility of testing, that the constructions match those of the subject populations, consciously or unconsciously; *cf.* Rouse [1939] vs. Rouse [1964]). No one expects a ruler to weigh even though things that have length also have weight. However, in the case of historical classes, researchers made just this logical error by attributing empirical properties to arbitrary measurement units. This failure to understand the relationship between the empirical properties associated with phenomena identified by measurement units from the measurement units themselves created significant confusion. Consequently, culture historians began to chase "right" types rather than construct types to do some specific work. This confusion greatly retarded theory development and resulted in the mixture of style and function that continues to plague many areas in archaeology (*e.g.,* Dunnell 1971, 1986; Jelinek 1976).

Despite the clarity of Ford on the role of types as measurement tools (*e.g.*, Ford 1954a, 1961), the confusion between the ideational status of types and the empirical properties of objects identified with types has roots in the archaeology of the Mississippi river valley. PFG (1951:68) and Phillips (1970) realized that they did not exploit all of the variability that might potentially be historical and thus useful for building chronologies. In this effort, PFG (1951:68) created several tentative or "provisional types" that encompassed modes of stylistic variability in the "considerable residue of unclassified material" left over from their classification. In some cases, PFG (1951:69) saw the problem as "insufficient material for satisfactory type definition." PFG (1951:68) mention "it was felt that these [types] would no doubt attain the status of types with more material, so they are useful in pointing to areas for further investigation." In others, however, PFG point to empirical properties of types as the reason behind creating the provisional types. They state (PFG 1951:69, emphasis mine) that in some cases the provisional types "failed to show significant differences in time space distribution from closely related types in full status and, therefore, could be thought as *variants* of those types." In this way, instead of viewing these types as units at different levels and scales of measurement, PFG began to treat types as if they were effectively empirical entities and used empirical properties of these entities to define variants.

Continuing in this tradition and based on his empirical work in the valley (*e.g.,* Phillips 1939) and his research in *Archaeological Survey in the Lower Yazoo Basin*, Phillips (1970) expanded the original twenty PFG types into a system of over forty *types* and eighty-eight *varieties.* Phillips introduced the type-variety system into the Mississippi river valley as an attempt to refine the existing ceramic categorization system and integrate the variability observed but not used in PFG. Phillips (1970:25; see also Gifford 1960; Phillips 1958; Smith 1960; Wheat *et al.* 1958) envisioned the type-variety system as a nested set of units in which types were primarily defined based on paste, decoration techniques. and design while varieties consisted of "areal and temporal variants in the norms of the types" seen in modes of form and decoration. Phillips argued that although varieties are transitional forms and thus not "as readily sortable as types" (Phillips 1970:2), the recognition of varieties is significant because intergrading units represent narrow intervals on a sliding scale of time and area (Phillips 1970:26). He also claimed many examples of ceramics exhibit variability that made identification arbitrary. In order to address this additional variability not addressed by the original PFG types, Phillips proposed the use of *varieties* in the Mississippi valley ceramics. It is significant to note that varieties were not introduced to specifically help sort issues of scale and measurement level embedded in PFG types. Rather, Phillips' introduction of varieties into the PFG classification scheme was prompted by a recognition that measurements made with PFG types treated empirical variability at different levels and scales *and* it was clear that more classes *could* be created. Phillips provides no evidence that his varieties are better measures of stylistic variability than their predecessors. That decorative and formal variability *could* be resolved with finer measurement levels and with measurement units of varying scales served the primary justification for the introduction of varieties.

In Phillips' system, types are predominately based on the attributes used by PFG: temper and decoration technique. To these two dimensions, Phillips added definitional attributes consisting of decoration design and vessel form. This system produced units at different levels and scales than the PFG type. For example, Phillips created type-varieties such as "Parkin Punctated, variety Castile" to describe shell-tempered ceramics with punctates arranged in a linear pattern and "Barton Incised, variety Togo" to account for shell-tempered ceramics with incising on the neck and punctates on the body.

The type-variety system is widely used in the Mississippi valley and across the Southeast (Lyman *et al.* 1997:112). Despite the ubiquity of type-varieties in contemporary archaeological discussion (*e.g.,* House 1991; O'Brien 1994; Smith 1990; Steponaitis 1983), no chronologies have been constructed using these units. In the creation of varieties out of the PFG types, Phillips (1970; see also Davis 1981; Gifford 1960) implicitly assumes that the added attributes that define varieties continue to measure homologous similarity. However, since his interest is in treating what he regarded as *empirical* variability observed in the Mississippian ceramic decoration, Phillips was not compelled to provide a theoretical basis for making this claim. In addition, the lack of chronologies built with varieties suggests that assumptions of historical significance may not be warranted. The attributes used in building Phillips' varieties may not contribute the ability of the PFG types to measure stylistic variability.

While the rationale for building any particular set of varieties has been unwarranted, the concept behind varieties raises the possibility of decreasing the *level* of types and the *scale* of measurement units in order to increase sensitivity to change for descriptions. In this way, varieties might provide added detail about chronology and spatial interaction. As discussed in Chapter 3, changing the number of dimensions affects the degree to which differences in measurements can be resolved. In addition, adding dimensions such as those implemented by Phillips (1970), decoration design and vessel form, to the type definitions to change the measurement units used in these analyses should produce units that describe the empirical world over a smaller area and through a more narrow span of time. Both level and scale potentially provide a means of getting higher resolution results for the seriation analyses.

We can evaluate whether units using additional criteria beyond the PFG type definitions are useful for measuring stylistic (*i.e.*, neutral) distance by examining three alternative hypotheses. First, if varieties contribute to our ability to measure variability in homologous similarity, then the inclusion of additional definitional attributes to types should add meaningful detail to chronological orders. In this case, varieties will produce the same seriation order but could resolve previously indeterminate relationships between assemblages. A second hypothesis poses that the change in level could produce a valid order but the increased sample size requirements could mean that additional resolution between assemblages could remain statistically ambiguous. In this case, the overall structure of the seriation order(s) will not be falsified but no additional detail will be resolvable. Third, it is possible that a change in scale in units and/or level with the attributes used by Phillips (1970) simply creates noise and increases the error terms of the seriation solutions. In this case, no statistically significant orders can be constructed when the assemblages are described with the lower level/scale types.

Testing these hypotheses consists of assessing the original seriation orders obtained in Chapter 7 with new descriptions of the assemblages obtained by lower-level classification attributes or by changing the root of the classification. If adding criteria to definitions or changing the classification root provides useful modifications to types and sample size is adequate, it will be possible to distinguish previously indistinct assemblages; relationships between assemblage pairs will be more clearly resolved. It is in this way, that it is possible to evaluate that the hypothesis (implied by Phillips 1970) that the decrease in classification level and scale implied in varieties provides new detail about the homologous similarity between the assemblages.

Since PFG only recorded data as sherd counts by type, it is not possible to evaluate the effect of altering the combination of attributes below the level of the culture historical type or at different scales. Changing classification level or scale has a significant cost in terms of analytic precision: increasing the requirements for class membership mechanically results in smaller samples assigned for any class. Changing scale potentially changes these physical requirements for counting as one moves from describing elements to sets of elements to designs. This can also have an impact on the sample size. Any reduction in sample size has a corresponding impact on one's ability to make statistical distinctions between assemblages (Ford 1936a). Since large sample sizes are required to make these evaluations, I restrict my analyses to sherds assigned to the two most abundant types, Parkin Punctated and Barton Incised.

My analyses included making detailed studies of decorative attributes for each sherd. My design analysis, detailed in Appendix E and shown in Figure 8.1, starts by describing *elements*, the minimum repeated modification to the surface of a pottery vessel that comprises decoration. Elements are identified as any repeated deviation to the planar surface of a sherd that is visible to the naked eye. Elements are analyzed along dimensions described below. At the next scale of analyses, the relationship and arrangement of elements to each other are described as *element sets*. Finally, combinations of element sets and their relationships form *designs*. With these data, it is possible to test the value of more divisive stylistic descriptions.

PARKIN PUNCTATED VARIABILITY

Stylistic ceramic types are constructed of attributes that are insignificant to vessel function but affect appearance. Thus, the challenge for archaeologists is to build classification dimensions that parse appearance differences. It is essential to note, however, that such a classification does not necessarily reflect the decoration of *vessels* (*contra* McNutt 1973). What pots looked like is not the target of classification, but how style varies. Our focus, therefore, is on building types that describe comparable physical samples of style. In the case of the Mississippian deposits, these samples come in the form of decorated sherds. In addition, as long as they are stylistic, the physical forms of descriptions are free to vary. There are no constraints on the possible dimensions that can be used to describe samples. For example, all of the PFG authors (*e.g.*, Ford 1935, 1936a; Griffin 1938, 1939; Phillips 1939) examined a wide variety of descriptive dimensions for building types including decoration patterns, paste hardness, and sherd thickness. For defining types, however, they used three primary dimensions: temper, decoration technology and decoration design.

In late Mississippian ceramics, differences in decoration technology and decorative design comprise the major components the PFG classification. For shell-tempered ceramics, one of the decoration technology modes defined by PFG is "punctation," non-linear indentations into the exterior surface of the ceramics. PFG created three types to describe variability among shell-tempered, punctated sherds: Parkin Punctated, Owens Punctated, and Manly Punctated. While all three types are defined on the presence of punctated decoration, they differ based on decoration design. Manly Punctated is defined as sherds with punctates arranged in an "incised meander, usually a straight line on upper shoulder area making an arcaded effect around the vessel" (PFG 1951:147). Owens Punctate, on the other hand, is composed of "fine incised lines both curved and straight, enclosing bands of fine, scattered punctations" (PFG 1951:136). Parkin Punctated is, in contrast, defined by the *lack* of either of these designs (PFG 1951:110). Parkin is defined as shell-tempered sherds with decoration consisting of "punctation

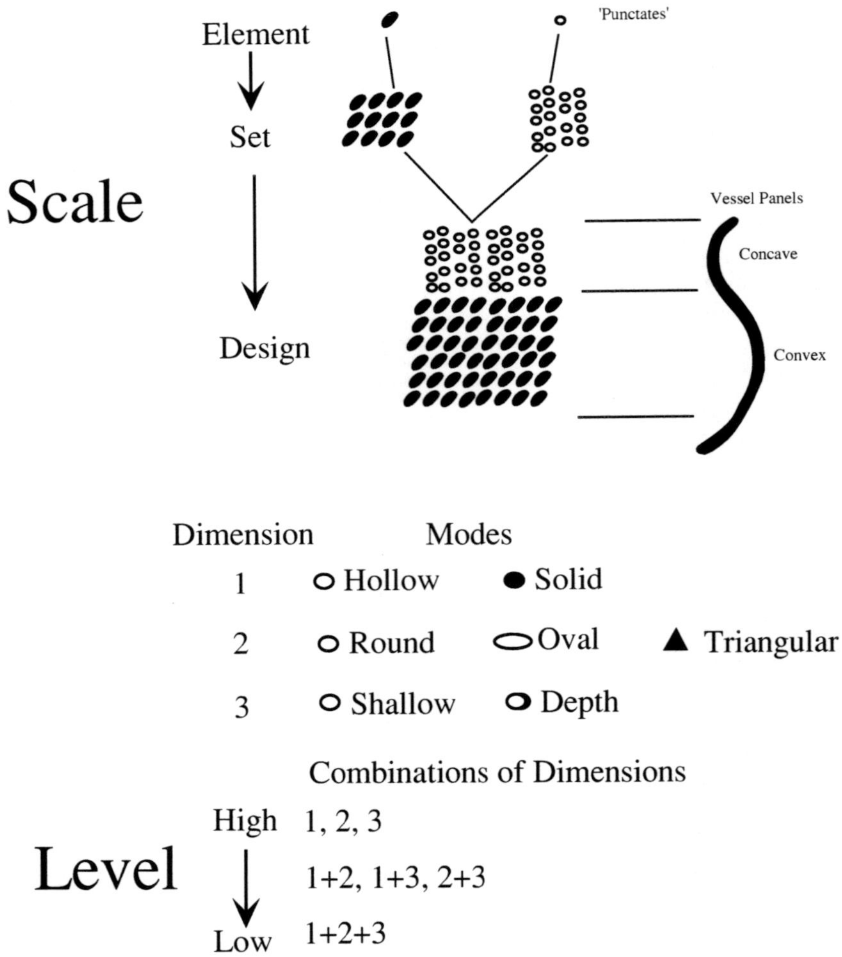

Figure 8.1. Decorative Analyses: Scale and Level.

Scale is adjusted in the design analysis starts by changing the counting units for measurement. The lowest highest scale is the *element*, the minimum identifiable modification to the surface of a pottery vessel. Elements consist of any identified modification to the surface of a sherd such as punctates or incisions. Elements are analyzed along a number of dimensions. The relations and arrangements of elements to each other, element *sets*, are at a lower scale. At the lowest scale are designs that consist of combinations of element sets and their relationships. Level in measurement is changed by changing the number of dimensions used to describe decorations. High-level measurements are made with a low number of dimensions while low-level measurements make use of a large number of dimensions.

produced by a variously shaped instruments, resulting in a wide variety of size, shape, and arrangement" (PFG 1951:110). Since Owens and Manly Punctate form lower-scale *and* lower-level classes than types simply composed of "the presence of punctates," they have greater requirements for type assignment than Parkin Punctate. In particular, the assignment of Owens and Manly Punctated requires sherd sizes large enough to encompass punctate *designs* and bear information on the origin of sherds from vessels forms (*i.e.,* shoulders).

It should not be surprising, therefore, that Owens and Manly Punctated were of limited use in the PFG seriations when used in conjunction with other higher-level types. Even PFG (1951:136) noted that in the case of Owens Punctated "sherds are small and fragmentary, and it is to be doubted whether we are actually dealing with a 'type' or merely a collection of sherds which share similar observable characteristics." Chapter 4 analyses confirm this impression, finding that the frequencies of Owens Punctate did not conform to the neutral model. Likewise, due to an unclear distribution in assemblages, Manly Punctated remained "provisional" or a type not considered by PFG (1951:69) to "show significant differences in time-space." PFG (1951:147) point out "because of the small number of examples and limited distribution, this type [*i.e.*, Manly] was not placed on the seriation graphs." The ambiguity of Owens and Manly Punctated relative to Parkin Punctated is due to differences in classification level and empirical requirements for type assignment. Since these differences are enhanced by factors related to post-depositional processes and vessel form, the abundances of these types would not be useful for building chronologies since seriations frequencies need to be predominately tied to cultural transmission.

Parkin Punctated, however, is central to the seriation analyses. As seen in the PFG analyses, (Chapters 4 and 7) assemblages described at this classification level and scale form reliable frequencies that meet the requirements of the neutral model. PFG (1951:110) define Parkin Punctated on the presence of punctation but also note that sherds described as Parkin Punctated encompass variability in punctate form and design. For example, they comment that Parkin punctations are arranged without pattern, are positioned in horizontal or vertical rows, outline bands, create a corrugated effect, and are aligned into psuedo-incised patterns. In addition, punctates are formed by pinching, and are made with fingernails, sticks, and hollow reeds. PFG (1951:110) note that these patterns and forms have some variability but not "enough" to warrant distinguishing these differences as separate types in their samples. In the case of hollow tools for making punctates, for example, PFG (1951:112) mention that the type is "not particularly common anywhere" and that "it seems advisable to treat this as a variant included in Parkin Punctated, at least until more material is obtained." In another case, linear alignment of punctates is "an extreme variant of the more fundamental punctated type and is therefore included in Parkin Punctated" (PFG 1951:110). A footnote included in the PFG text (PFG 1951:110) about this variant of Parkin Punctate is especially informative:

> In the early stages of classification, the type, "Castile Linear Punctated," was set up to take care of this variant. Sorting produced only a handful of sherds, however, and these occurred invariably in association with Parkin Punctated with no apparent significance from the standpoint of chronology or distribution, so the type was abandoned.

Thus, although PFG recognize variability in sets of punctates that is not described by Parkin Punctate they find they do not have warrant to append definitional criteria to their types. The definition of Parkin Punctated, therefore, remains at the scale of the decorative element, the punctate.

Phillips (1970), on the other hand, treats variability in punctate patterning as *real* and, therefore, uncritically includes this dimension as a basis for one of his varieties, "Parkin Punctate, variety Castile." Phillips (1970:151) simply asserts that "its abandonment as a type may or may not have been justified, but there is sufficient reason for keeping it as a variety" without providing any additional rationale. Phillips also defined other type-varieties based on his observations of punctate arrangement variability. In the creation of type-varieties, grid arrangements of punctates on jars became "Parkin Punctated, variety Harris," sherds with curved lines of punctates as "Parkin Punctated, variety Transylvania" and simple linear patterns as "Parkin Punctate, variety Parkin." Although he recognizes that Owens Punctated is poorly defined, Phillips likewise remains convinced that this type is a good empirical unit as well. Consequently, based on a drawing of a whole vessel from Beck, Phillips defines Owens Punctated as punctates filling bands and triangle zones in rectilinear patterns (Phillips 1970:149). Phillips redefines Manly Punctated as a variety of Matthews Incised (Phillips 1970:128). Mathews Incised, variety Manly describes sherds that have incised shoulders with accompanying punctate in an arcaded pattern. However, other than observations of descriptive differences between specific examples of PFG types, the reasoning behind these type definitions remains cryptic. Beyond the observation that variability exists in punctate design (an *ad hoc* observation), Phillips offers no warrant for accepting varieties as historically useful.

Does the classification level and scale that encompasses descriptive variability in sets of decorative elements as suggested by Phillips' varieties and noted in types such as "Castile Linear Punctated" provide additional, useful detail about homologous similarity than is available through the use of types? New collections provide an opportunity to evaluate this question. My analysis of these assemblages describes all sherds with punctate decorations at the scale of the punctate "set": combinations of individual punctate elements. Descriptions were composed of three features, one drawn of each of three dimensions. The first dimension consists of the arrangement of punctates in relationship to one another. Punctates were oriented into a Cartesian grid using the direction of application for orientation. As shown in Figure 8.2, the direction of application is easily identified by the alignment of linear striations on the interior surface of punctate and by the presence of a characteristic 'lip' at the terminus of the tool mark. The y-axis is defined as the

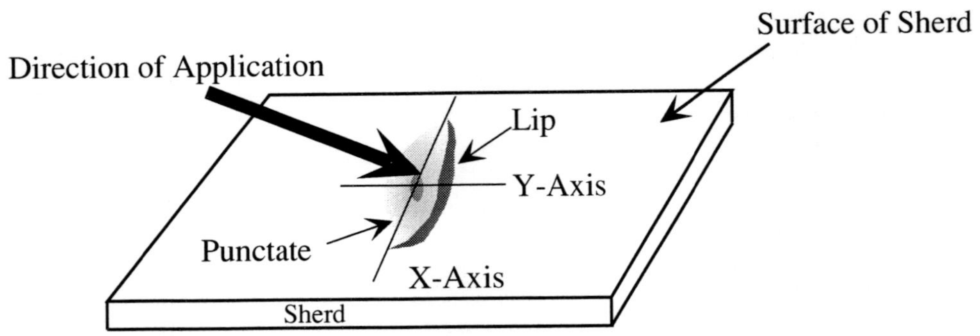

Figure 8.2. Punctate Orientation.

Punctates are described using the direction of application for orientation. The direction of application is easily identified by the alignment of linear striations on the surface of the sherd and by the characteristic presence of a 'lip' at the terminus of the tool mark. The X-axis is perpendicular to the direction striations, while the Y-axis is defined as the application direction.

Table 8.1. Combinations of Punctated Attribute Sets.[ab]

Type	Example	Description	Phillips (1970)
4-B-III	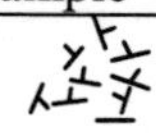	4 – No consistent alignment along the X- and Y-axes. B – No overlapping punctates. III – Neither rectilinear nor curvilinear alignment.	
3-B-I		3 – Punctates aligned along both the X- and Y-axes. Creates "grid-like" pattern. B – No overlapping punctates. I – Rectilinear alignment.	Parkin Punctated, variety Harris.
2-B-I		2 – Punctates aligned along the Y-axis. B – No overlapping punctates. I – Rectilinear alignment.	Parkin Punctated, variety Parkin.
1-B-I		1 – Punctates aligned along the X-axis. B – No overlapping punctates. I – Rectilinear alignment.	
1-A-I		1 – Punctates aligned along the X-axis. A – The margins of the punctate overlap one another. I – Rectilinear alignment.	Parkin Punctated, variety Castile Landing. PFG (1951:110) previously described this provisional type as "Castile Linear Punctated."
2-B-II		2 – Punctates aligned in Y-axis. B – No overlapping punctates. II – Curvilinear alignment.	
2-A-I		2 – Punctates aligned in Y-axis. A – The margins of the punctate overlap one another. I – Rectilinear alignment.	
2-B-III		2 – Punctates aligned in Y-axis. B – The margins of the punctate overlap one another. III – Neither rectilinear nor curvilinear alignment.	
1-B-III		1 – Punctates aligned along the X-axis. B – The margins of the punctate overlap one another. III – Neither rectilinear nor curvilinear alignment.	

[a] Punctates are represented by dashes.
[b] The alignment of punctate was determined by the direction of punctate application. The direction of application was defined to be as the Y-axis. The X-axis was defined as perpendicular to the direction of application.

direction of application and the x-axis is defined as perpendicular to this direction. Using this grid to orient punctates, four features can be defined: (1) alignment along the x-axis only, (2) y-axis only, (3) x- and y-axes and (4) neither axis. The second dimension describes whether the punctates in the set are (A) overlapped or are (B) distinct (B). Finally, I defined the third dimension to indicate whether punctate sets are aligned with one another in (I) rectilinear or (II) curvilinear patterns. Class definitions were built by combining features from each dimension. Class 1AI, for example, is defined by punctate sets that are aligned along the (1) x-axis only, are (A) overlapping and (I) aligned in rectilinear patterns. Table 8.1 provides several examples of combinations of punctate set dimensions.

Although twenty-four combinations are possible using these three dimensions, only nine were observed in the assemblages (Table 8.2). Figure 8.3 displays the frequencies of the punctate set alignment modes. Holden Lake is not included as the number of punctated sherds is very low for this assemblage (n=13). This order, while not ideal in terms of mean values, does not falsify the seriation order (α = 0.005) obtained earlier in Chapter 7 (Figure 7.5). However, no new information about the order can be discerned using

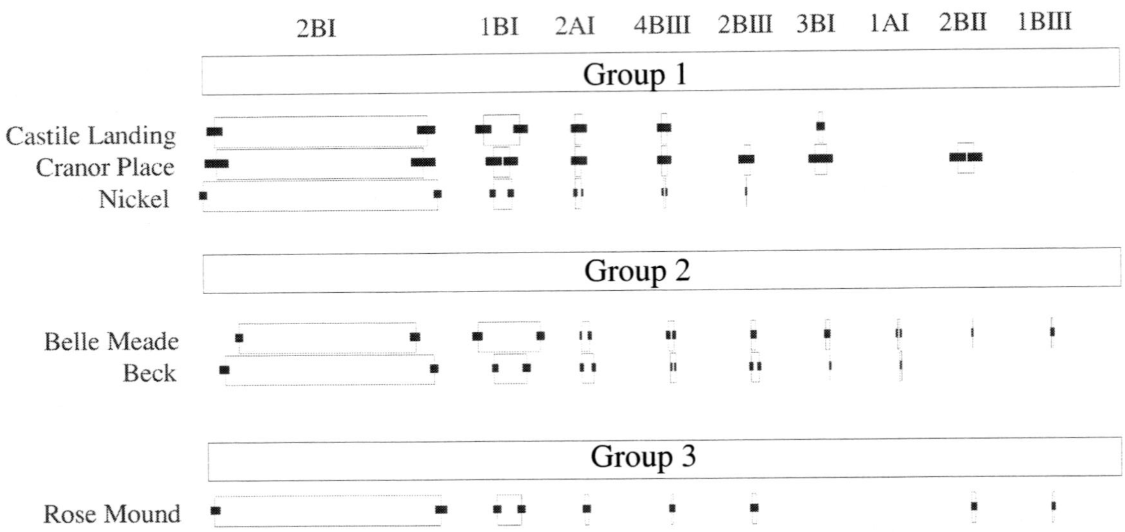

Figure 8.3 Seriation Solution for Lipo Assemblages Described by Punctate Set Types.

Solutions given at α=0.005 level of significance. Holden Lake is not included here as the number of punctate decorated sherds is small (n=13) and results in error terms that place it anywhere.

Table 8.2. Counts of Punctate Sets in Lipo Assemblages.[a]

Assemblage	2BI	1BI	2AI	4BIII	2BIII	3BI	1AI	2BII	1BIII	Σ
Beck	237	35	13	5	10	1	2	0	0	303
Belle Meade	253	90	13	9	5	5	5	1	2	383
Castile Landing	69	12	3	2	0	1	0	0	0	87
Cranor Place	37	3	1	1	1	2	0	3	0	48
Holden Lake	8	1	1	0	1	2	0	0	0	13
Nickel	256	21	7	4	1	0	0	0	0	289
Rose Mound	147	16	3	2	3	0	0	2	1	174
Σ	1007	178	41	23	21	11	7	6	3	1297

[a]Type designations are described in the text. In order to provide consistent measurements across the assemblages, the tabulations include only sherds in which all three dimensions could be identified. Instances in which more than one punctate set on a single sherd were identified were counted multiple times.

these lower scale measures than the PFG types. If anything, resolution is less at this scale and level of measurement due to the smaller sample sizes. It is important to note, though, that the degree to which the assemblages can be seriated measure of the descriptive types being built from *stylistic* (*i.e.*, neutral) attributes. Consequently, this result does not mean that types built with punctate alignment information are not *potentially* useful for measuring homologous similarity, only that they do not contribute new information given these assemblages and sample sizes.

Given their smaller samples, the pattern of frequencies supports PFG's decision to abandon "Castile Linear Punctated" as a chronological type: frequency differences among assemblages with this type are too small to be useful in sorting. As shown in Table 8.3, the frequencies of Type 2AI (equivalent to their "Castile Landing Punctated") are statistically indistinguishable from one another. Therefore, at smaller sample sizes PFG (1951:110) were correct to assert that Castile Linear Punctate is so rare (n=47) that it is not useful for measuring homology.

Table 8.3. Frequencies of Castile Linear Punctate in Lipo Assemblages.[a]

Assemblage	Frequency
Beck	0.043 ±0.034
Belle Meade	0.034 ±0.027
Castile Landing	0.034 ±0.061
Cranor Place	0.021 ±0.069
Nickel	0.024 ±0.027
Rose Mound	0.017 ±0.031

[a]Frequencies calculated with a 99.5% (α=0.005) confidence level. Holden is not included here as it has a very small sample size of punctate decorated sherds (n=13).

Analyses at the scale of punctate sets and the work in Chapter 4 show that variability in sherd decorations described at higher classification levels does not provide detail on chronological relationships between the St. Francis and Memphis assemblages beyond what was found using the PFG types. It does not follow that descriptions using additional measurement criteria and producing lower level and scale types cannot provide information on homologous similarity. For example, it is reasonable to suggest that variability in the relationship of punctate sets to each other (*i.e.*, *designs*) could be useful for examining stylistic variability. However, neither PFG nor Phillips (1970) systematically explored the use of descriptions using classes at this scale. Given the lack of interest in this measurement scale by these researchers and relatively small sample sizes of the original collections, we expect PFG found that variability in descriptions built from types of this scale did not provide additional information than descriptions made with lower scales. Again, my large assemblages allow this expectation to be evaluated.

Critical to evaluating this expectation is building classes that have multiple dimensions to minimize the effect of chance in the descriptions. In addition, the attributes must be common enough in the assemblages such that relatively large numbers of sherds can be assigned to the classes. This helps to insure the greatest sample size possible for the analyses. In my study of the assemblages, I described each of the punctate decorated sherds along four dimensions related to the punctate designs. The first dimension describes the relationship of punctates sets to one another. Five features were identified along this dimension: multiple sets of punctates with applications (as determined in Figure 8.1) that are oriented (1) ninety degrees to one another, (2) multiple sets of punctates with applications acute (less than ninety degrees ±5 degrees) to one another, (3) multiple sets of punctates with applications at obtuse (greater than 90 degrees) angles to one another multiple and (4) sets of punctates with applications one hundred and eighty degrees to each other and (5) multiple sets of punctates that were applied in the same direction. The second dimension describes whether the sets of punctates are (A) adjacent to one another or (B) overlap. The third dimension, similar to a defining criterion of Owen's Punctated, indicates whether the punctate set(s) are bordered by an incised line or not (I = no incised line, II = incised line). Finally, the fourth dimension describes whether the designs (a) form a band or (b) cover the entire sherd area. Following the strategy of building types constructed of punctate set dimensions, punctate design

Table 8.4. Attribute Combinations for Punctate Designs.[ab]

Type	Example	Definition	Notes
5AIIa	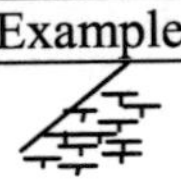	1 – One punctate set in a single band/shape. A – Punctate set adjacent to empty space. II – Punctate set bounded by incised line. a – Punctate set forms a band.	Since PFG (1951:110) generally ignored variability at the level of design, sherds belonging to these design types were lumped together with "Barton Incised."
3AIb	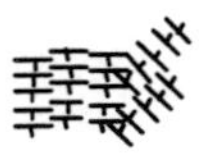	3 – Two punctate sets with applications at obtuse angles to one another. A – Punctate sets adjacent to one another. I – Punctate sets not bounded by incised lines. b – Punctate set covers entire sherd.	
2AIb		2 – Two punctate sets with applications that are at acute angles to one another. A – Punctate sets that are adjacent to one another. I – Punctate sets that are not bounded by incised lines. b – Punctate set covers entire sherd.	
4BIb		4 – Two punctate sets with applications 180 degrees to one another. B – Punctate sets overlap each other. I – Punctate sets not bounded by incised lines. b – Punctate set covers entire sherd.	
1BIb		1 – Two punctate sets are at an acute angle to one another. B – Punctate sets overlap each other. I – Punctate sets not bounded by incised lines. b – Punctate set covers entire sherd	
1AIb		1 – Two punctate sets with applications 90 degrees to one another. A – Punctate sets are adjacent to each other. I – Punctate sets not bounded by incised lines. b – Punctate set covers entire sherd	

[a] Punctates are represented by dash in which the direction of the application of the punctate is presented as perpendicular to the direction of the dash.
[b] The alignment of punctate was determined by the direction of application. The direction of application was defined as the Y-axis. The X-axis was defined as perpendicular to the direction of application.

types are constructed by combining features from each dimension. These sets of descriptions Type 2AI, for example, describes two adjacent punctate sets that consist of punctates that are applied at acute angles to one another and are not bordered by an incised line. Table 8.4 presents observed examples of types built from these dimensions.

Table 8.5 provides the counts of these types in my assemblages. The results support the hypothesis that PFG were unable to pursue the use of classifications that incorporated attributes at the scale of designs because of insufficient sample size. While punctate set design types are potentially useful for examining stylistic variability, the small numbers of sherds that can be described with these types, even with these large collections, make them unusable here. In the case of the PFG collections, the relatively small samples would only have exacerbated any attempts to make use of this scale of classification for studying homologous similarity.

In addition to exploring the use of descriptive attributes related to punctate sets and designs for building measures with additional criteria (*i.e.*, building lower level and scale classes), PFG also examined features related to the manufacture of punctates. PFG (1951:110) described punctates, for example, as being:

> produced by variously shaped instruments, resulting in a wide variety of size, shape, and arrangement. The commonest characteristic, to which there are

few exceptions, is that the instrument is jabbed obliquely into the clay, producing a ridge or "burr" alongside the punctations of corresponding size and shape. Most frequently, the shape is oval or semi-lunar, often suggesting "nail-marking," by which term this type of treatment is often described. However, there is a wide variety of shapes including round, square, triangular, U-shaped, etc., most of which could not have been produced by the unaided fingers.

Table 8.5. Punctate Set Designs in Lipo Assemblages.[a]

Assemblage	2AIb	5AIIa	4BIb	1BIb	1AIb	Σ
Beck	14	4	3	1	0	22
Belle Meade	10	2	3	1	0	16
Castile Landing	7	1	2	0	0	10
Cranor Place	2	3	0	0	0	5
Nickel	12	1	4	2	1	20
Rose Mound	3	0	0	0	0	3
Σ	48	11	12	4	1	77

[a]No punctate set design types were identified in the Holden Lake sample.

Based on these features, PFG inferred that punctates were made by a variety of instruments including 'fingernails,' 'sticks,' and 'hollow canes or reeds.' However, despite this variability, PFG found that types made with these traits did not appear to usefully resolve similarity between assemblages in time and space with the sample sizes they collected. They note (1951:112), for example, that "the use of a hollow cane or reed as a punctating instrument produces a variant type, not particularly common anywhere, but more apt to occur in the southern portion of the area." Rather than use this variability as the basis of a separate type, PFG chose to include the material in Parkin Punctated. PFG's recognition of punctate tool type in their type construction, though, presents the possibility that lower-level types than Parkin Punctated made with features related to the manufacturing of punctates *could* be useful as historical types but at sample sizes larger than those that were available to PFG.

To evaluate this hypothesis, I described punctate decorative elements using three potentially stylistic dimensions that captured the kinds of features used by PFG when they discussed punctate forms in their collections. The first dimension defines the shape of the punctate outline. Punctates were described as (A) circular, (B) crescent-shaped, (C) lenticular, (D) oval, (E) semi-circular, (F) rectangular, or (G) triangular. The second dimension describes whether the form of the punctate is (1) "solid" (an indentation formed by pressing a solid object into clay) or (2) "hollow" (an indentation with an interior raised area that would be formed by impressing a hollow object such as a bone or a reed into clay). Finally, I described punctates in terms of the size of the 'burr' formed by the insertion of the manufacturing tool into the wet clay. 'Burrs' were described as none (I, no burr, only indentation in surface), small (II, minimal presence of burr as simple ridge of clay less than depth of punctate extending out from surface), medium (III, ridge of clay as large as depth of punctate, no overlapping or curling), or large (IV, large ridge of clay, curled-over). Combining these three dimensions, I formed fifty-six types. Thirty-six instances of these types were identified in my new assemblages. The counts for the punctate form types are presented in Table 8.6.

Figure 8.4 presents the seriations using the punctate form types. The order shown in Figure 8.4 does not falsify the seriation order ($\alpha = 0.005$) obtained earlier in Chapter 7 (Figure 7.5). However, no new information about the order can be discerned using this level of classification than can be determined using the PFG types. Again, this conclusion does not indicate that types built with descriptions of punctate forms are not potentially useful for measuring homologous similarity, only that they do not contribute new details in this instance.

Although the potential role of chance is increased, removing one dimension from the punctate form classification provides a means of testing the seriation solution at a slightly higher classification level. Table 8.7 presents the tabulation of punctate forms in my assemblages where types consist of only the dimension of outline shape (A-E) and the dimension defining whether the punctate has a hollow or solid form (1-2). As shown in Figure 8.5, the use of these types for assemblage descriptions produces a clear solution for the seriation groups; the composition of three assemblage groups are distinctly separated using this classification. One difference between the solution presented with the punctate form types and the PFG types, however, is the relationship between the Group 2 assemblages. Due to the error terms on the samples, the order between Castile Landing and Nickel is unclear. However, the orders do not falsify the order produced when the assemblages were measured using the PFG types with the full samples.

BARTON INCISED VARIABILITY

PFG's types that encompass incised decorations also make use of multiple classification levels and scales. As shown in Figure 6.1, incised PFG types incorporate five different classification levels: paste hardness, rectilinear/curvilinear form, width of line, angle of incision relative to the vessel, and location of the decorations on vessel interior/exterior surfaces. As is clear from the results of the original PFG seriations and the seriations conducted in Chapter 7, types made at a number of these classification levels resolved patterns of homologous similarity between the late prehistoric assemblages. However, in my analyses (Chapter 6; see also Pugh 1990), I found that including incision orientation resulted in ambiguous and inconsistent type assignments between multiple researchers. In specific, when I compared my descriptions of collections that used Kent Incised (vertical incisions), Barton Incised (angular incisions), and Mound Place Incised (horizontal incisions) with the same assemblages described by PFG, I found that I could not evaluate my type assignments with respect to those

Table 8.6. Punctate Forms in New Assemblages.

Assemblage	E1III	E1II	C1II	D1II	E1IV	C1III	D1I	C1I	D1III	E1I	A1II	B1II	D1IV	C1IV	A1I	B1I	F1II	E2IV	B1III	B2III
Beck	60	35	84	61	33	37	30	27	18	15	13	7	13	12	9	9	8	6	2	1
Belle Meade	132	102	78	81	51	45	38	21	37	20	9	6	2	3	2	5	10	14	11	16
Castile Landing	20	16	13	12	14	7	8	13	7	8	1	6	4	5	4	2	3	1	5	0
Cranor Place	10	21	10	3	12	6	10	6	5	6	1	2	1	0	2	3	0	0	2	0
Holden Lake	2	1	1	3	2	2	1	5	1	0	1	0	1	1	4	0	0	0	0	0
Nickel	56	66	80	54	54	40	44	34	32	30	9	13	12	9	3	8	2	5	5	2
Rose Mound	99	71	40	13	17	23	14	18	14	13	4	4	3	5	8	2	5	1	2	1
Σ	379	312	306	227	183	160	145	124	114	92	38	38	36	35	32	29	28	27	27	20

Assemblage	A1III	B1IV	G1II	B2IV	G1I	F1III	G1III	A1IV	G1IV	F1I	F1IV	B2II	E2II	G2IV	F2III	C2III	A2II	A2I	C2II	B2I	D2III	E2III
Beck	6	4	6	4	6	5	2	3	1	2	1	0	1	1	0	1	0	1	1	0	0	4
Belle Meade	5	5	5	5	4	2	2	1	0	3	0	1	2	1	2	0	0	0	0	0	0	10
Castile Landing	0	4	2	0	0	0	0	1	0	0	0	1	0	0	0	0	0	0	0	0	0	2
Cranor Place	1	2	1	0	1	0	1	1	3	0	0	0	0	0	0	0	0	0	0	0	0	0
Holden Lake	0	0	0	0	0	0	0	0	1	1	0	0	0	0	0	0	0	0	0	0	0	0
Nickel	1	4	1	5	2	4	3	0	1	1	3	0	0	0	0	0	0	0	0	1	1	0
Rose Mound	3	0	1	1	2	2	1	2	2	0	1	2	0	0	0	0	1	0	0	0	0	0
Σ	16	19	16	15	15	13	9	8	8	7	5	4	3	2	2	1	1	1	1	1	1	16

Assemblage	Σ
Beck	529
Belle Meade	731
Castile Landing	159
Cranor Place	110
Holden Lake	27
Nickel	585
Rose Mound	375
Σ	2516

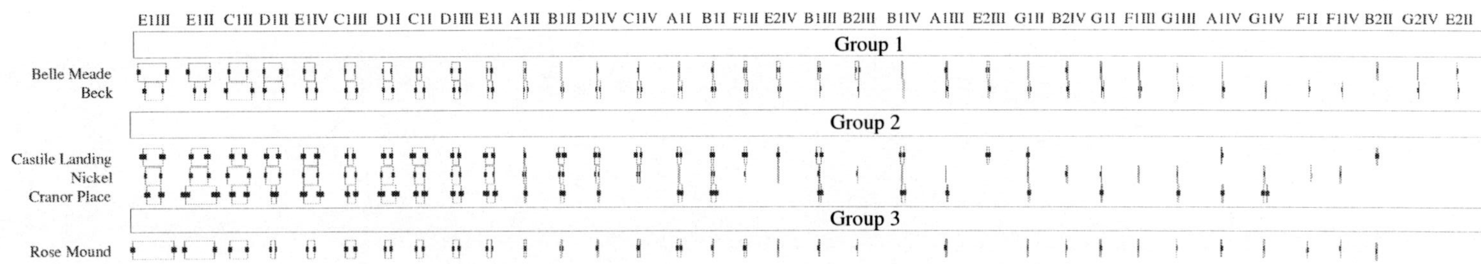

Figure 8.4. Seriation Solution for Lipo Assemblages Described by Punctate Form Attributes.

Solutions given at α=0.005 level of significance. Holden Lake is not included here as the number of punctate decorated sherds is small (*n*=27) and the error terms result in its placement anywhere in the seriations.

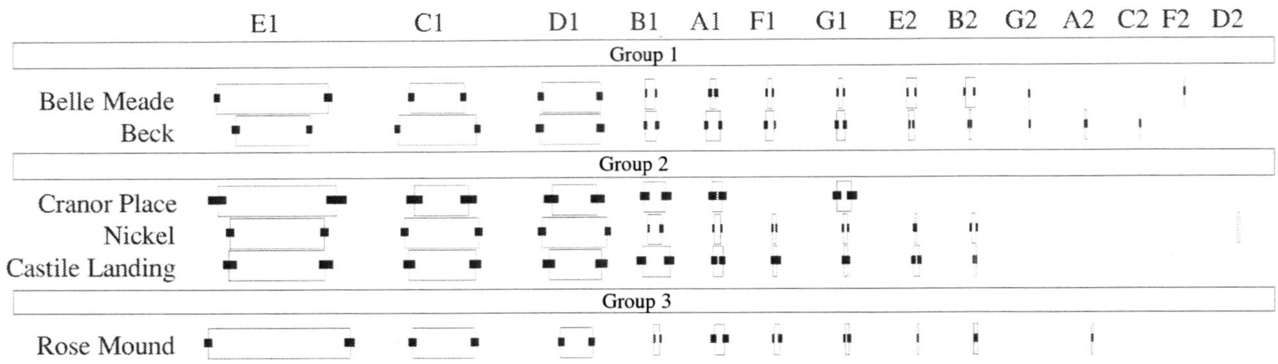

Figure 8.5. Seriation Solution for Punctate Form Types in Lipo Assemblages Using Classes Built of Two Dimensions.

Note that the order obtained here is identical to that obtained with three dimensions as shown in Figure 8.4 Solutions given at α=0.005 level of significance. Holden Lake is not included here as the number of punctate decorated sherds is small (*n*=27) results in error terms that place the assemblage anywhere in the seriation.

Table 8.7. Two-Dimension Punctate-Form Types in Lipo Assemblages.

Assemblages	E1	C1	D1	B1	A1	F1	G1	E2	B2	G2	A2	C2	F2	D2	Σ
Beck	149	162	122	23	31	18	16	11	5	1	1	2	0	0	541
Belle Meade	316	152	164	27	17	15	11	26	27	1	0	0	2	0	758
Castile Landing	58	40	31	17	6	3	2	3	1	0	0	0	0	0	161
Cranor Place	50	23	19	9	5	0	6	0	0	0	0	0	0	0	112
Holden Lake	5	9	6	0	5	1	1	0	0	0	0	0	0	0	27
Nickel	209	166	144	30	13	10	7	5	8	0	0	0	0	1	593
Rose Mound	202	89	44	8	17	9	6	1	4	0	1	0	0	0	381
Σ	989	641	530	114	94	56	49	46	45	2	2	2	2	1	2573

of PFG. Pugh (1990), who made new collections at Belle Meade, also arrived at this conclusion.

A significant contribution to the inability to generate comparable values for these types comes from PFG's definition of Barton Incised. PFG (1951:114) defined Barton Incised to consist simply as sets of lines incised "by a pointed implement applied to a moist surface with a consider amount of 'burr.'" Without any other information, it was not clear whether PFG restricted Barton Incised to sherds with oblique lines or to sherds *without* horizontal or angular lines (*i.e.*, a catch-all category somewhat like that of Parkin Punctated). Thus, in building stylistic descriptions of the combined PFG and new assemblages, I obtained the greatest resolution by aggregating these three PFG types, or rather, describing the assemblages with types constructed from dimensions that formed higher level classes (*i.e.*, dropping the orientation requirement). Like the punctate decoration types with criteria additional to those of the PFG types, it is clear that incised types built from lower level classes *could* potentially provide details about homologous similarity. However, the use of these types requires consistent information about vessel orientation, in this case, greater information than was necessary for the assignment of non-incised types.

In his type-variety system, Phillips (1970:45) recognized that Kent Incised and Barton Incised use additional attributes beyond the PFG types. Thus, Phillips proposed Barton Incised, v. Kent and Barton Incised, v. Barton as types that are distinguished by incision orientation on the vessel. In addition to Kent and Barton, Phillips (1970:45-47) also includes a number of other varieties in his Barton Incised type. These varieties include Barton Incised, v. Arcola (parallel incisions forming line filled triangles on shoulder area of subglobular jars), Barton Incised, v. Campbell (vertical parallel incision on rims of jars), Barton Incised, v. Estill (close spaced incisions forming line-filled triangles executed at a steep angle on rims of jars), and Barton Incised, v. Togo (incisions on rim adjacent to areas of punctation). Each of these varieties poses differing assignment requirements in terms of sherd size, sherd location on vessel, the ability to orient sherds as to vessel form, and information about vessel form. In addition to defining incision "varieties," Phillips also defined some of the incision variability as separate types. In the case of Mound Place Incised, for example, Phillips mixed attributes relating to vessel form into type definition. Mound Place Incised is restricted to horizontal incised lines on the rims of bowls. Consequently, the assignment of samples to Mound Place Incised requires more information than types that do not rely on vessel form (*e.g.*, Barton Incised, variety Barton). The *ad hoc* combination of attributes that comprise type-varieties without regard to assignment effects is a central reason why Phillips' (1970) varieties are have not been useful in building chronologies.

Although PFG's and Phillips' incised types are constructed inconsistently, there is clear potential to build descriptions of incisions at a lower classification level than the aggregated "Barton/Kent/Mound Placed Incised" type (*i.e.*, incisions on shell-tempered sherds without regard to orientation) I was compelled to use in Chapter 7. Useful stylistic types *can* be built using incision orientation; both my own analyses (in Figure 7.1) and PFG's seriations were successful using these classes to build seriations. That it was not possible to combine assemblages described with these types is a function of the lack of information in PFG on the formation of their types.

Taking a tack parallel to the treatment of Parkin Punctated, we can examine variability in incision patterns among the incised types. By using only variability in the relations between incised lines instead of with reference to the whole vessel, we can avoid issues of comparability that are raised with the use of Phillips' incised types. In addition, since no orientation information is needed, the relatively large number of sherds assigned to the "Barton/Kent/Mound Placed Incised" type (*n*=2,076) in my assemblages provides an opportunity to examine the effect of classification scale or level changes on type frequencies and seriation results. In their examination of sherds described as Barton, PFG noted the potential for describing incised sherds along a number of dimensions. They (PFG 1951:114) comment that:

> lines vary considerably in width, from less than 0.5 to 3.0 mm. In general, the character of the line varies considerably, but in most instances, it is irregular and carelessly executed. Where lines should abut there is a tendency for them to cross.

PFG also mention that arrangement of the lines can vary. They (PFG 1951:114) state that incised lines are arranged as "predominately groups of parallel, oblique lines slanting downward from the lip to the beginning of the shoulder area." However, despite the recognized potential for building more detail stylistic descriptions, PFG (1951:116) opted to limit their types to the level of incision presence and point out that:

> there is more variation in the handling of these simple patterns than is indicated in the foregoing description. It appears to be not so much a matter of group variations from site to site, but of individual variations within site collections. This is related to the fact that decoration in general is carelessly and haphazardly executed.

With the large sample sizes afforded by my collections, it is possible to evaluate whether decreasing the level of classification to describe stylistic variability noted by PFG could add meaningful detail to the seriation solutions. In the descriptions of my collections, I recorded information about *sets* of incised lines along three dimensions. The first dimension attends whether the incisions are (1) parallel or cross each other in a cross-hatched pattern (2). Since it was clear from the PFG descriptions that incised decoration occasionally covers large areas of vessels (such as in the case of 'Kent') or comprise patterns (such as 'triangles'), I defined the second dimension based on whether incisions are (A) distributed across the entire sherd or are (B) part of a bounded area such as a band, triangle, or trapezoid. Finally, I described the distance of incisions from each other. Incision distance was described as dense (I, distance between incisions less than the width of lines), medium (II, distance between incisions equal to width of lines [± 0.25 x the width of the lines]), or sparse (III, distance between incision greater than width of lines).

Table 8.8 presents the counts of types built with these three dimensions for my assemblages. The distribution of the frequencies for these descriptions shown in Figure 8.6 supports a hypothesis that the more detailed incised types do not provide any additional information. Unlike the case of punctate element set variability, the frequencies of incised lined types do not conform to the original seriation results. In fact, *no* seriation solutions are possible. Sherds described by combinations of these dimensions do not conform to the neutral model.

One concern using these incised types is that identifications of features along the dimension of "banded/non-banded" are heavily influenced by sherd size. Smaller sherds are less likely to include information that would allow an identification of "banded" than large sherds. Consequently, assemblages with smaller sherds would systematically have fewer numbers of banded sherds identified. In this case, the frequencies of incised types reflect the effects of post-depositional processes as well as homology. In order to assess the effect that this dimension may have had on the seriation results in Figure 8.6, I built a second set of types that substituted the shape of incision cross section for the dimension of "banded/non-banded." In this type, the shapes of line cross section was identified as (a) "v-shaped," (b) "rounded," (c) "squared," or (d) "complex." The identification of complex, in this case, refers to cross-sections of lines that appear as though they were made with a roughly broken stick or bone (*i.e.*, linear striations parallel to the length of the incision).

Table 8.9 presents the counts of sherds that meet criteria of the types created with the dimensions describing incision cross-section. As shown in Figure 8.7, the results are no better; there are significant deviations from the expectations of the neutral model. With these data, it is not possible to create spatially or temporally coherent solutions. This result confirms PFG's impression that incised types constructed at lower scales and levels only contribute noise to the analysis.

Table 8.8. Incised Types in Lipo Assemblages Using Dimensions of Alignment, Coverage, and Spacing.[a]

Assemblages	1AIII	1AII	1AI	1BII	2AIII	1BIII	2AII	1BI	2BIII	Σ
Beck	163	205	35	22	2	15	1	4	0	447
Belle Meade	247	222	37	11	13	9	2	1	0	542
Castile Landing	4	14	0	3	0	0	0	0	0	21
Cranor Place	41	22	3	10	10	11	1	1	0	99
Holden Lake	90	72	13	9	16	5	3	1	0	209
Nickel	150	101	21	8	12	8	4	1	1	306
Rose Mound	85	82	9	17	15	15	9	2	0	234
Σ	780	718	118	80	68	63	20	10	1	1858

[a]So as not to mix classification level, these counts only include sherds in which each of the dimensions could be measured.
[a]The first dimension specifies whether the incisions are parallel (1) or cross each other in a cross-hatched pattern (2). The second dimension describes whether incisions are spaced across the entire sherd (A) or are part of a bounded area (B) such as a band, triangle, or trapezoid. The third dimension describes the spacing of incisions from each other where the spacing can be dense (I, distance between incisions less than the width of lines), medium (II, distance between incisions equal to width of lines [± 0.25 x the width of the lines]), or sparse (III, distance between incision greater than width of lines).

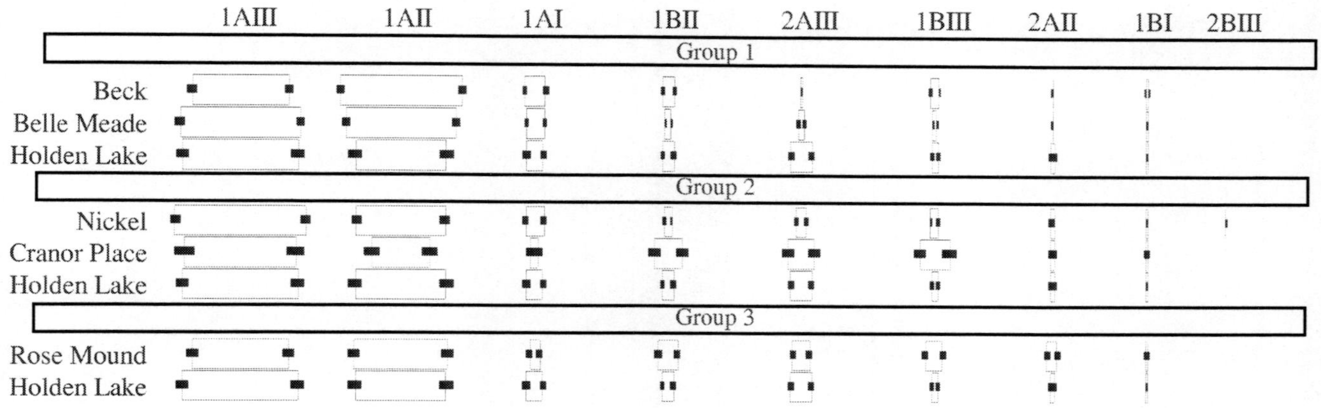

Figure 8.6. Seriation Solution for Incised Types in Lipo Assemblages Using Dimensions of Alignment, Coverage, and Spacing.

Solutions given at α=0.005 level of significance. Castile Landing is not included here as the number of incised decorated sherds is small (*n*=21) and the error terms result in its placement in any group.

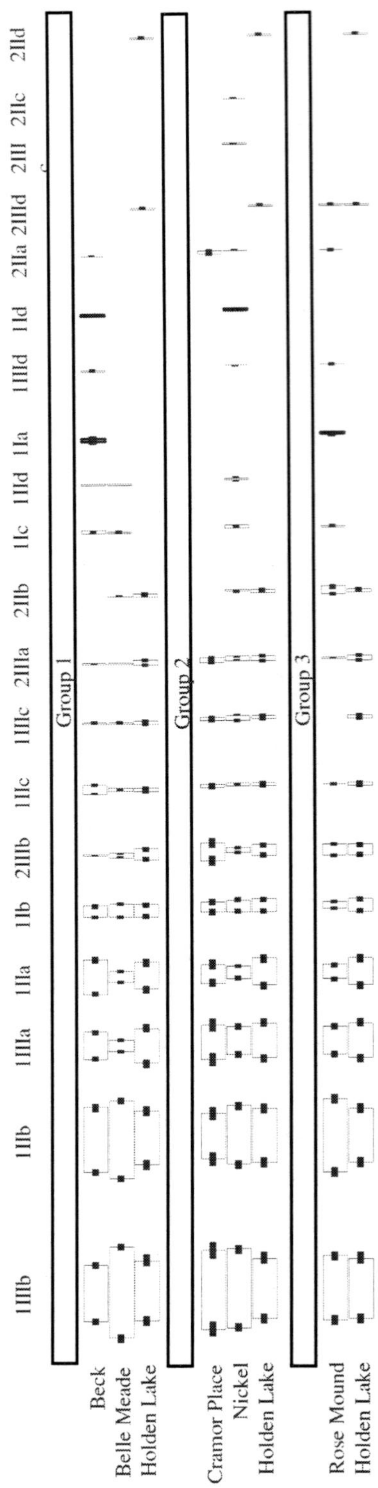

Figure 8.7. Seriation Solution for Alternate Incising Types in Lipo Assemblages.

Solutions given at a=0.005 level of significance. Castile Landing is not included here as the number of incised decorated sherds is small (n=26).

Table 8.9. Alternate Incising Types in Lipo Assemblages Using Dimensions of Alignment, Spacing, and Line Cross Section.[a]

Assemblages	1IIIb	1IIb	1IIIa	1IIa	1Ib	2IIIb	1IIc	1IIIc	2IIIa	2IIb	1Ic	1IId	1Ia	1IIId	1Id	2IIa	2IIId	2IIIc	2IIc	2IId	Σ
Beck	113	130	53	72	24	1	18	4	1	0	5	3	8	4	2	1	0	0	0	0	439
Belle Meade	225	192	26	28	34	10	6	3	3	2	3	3	0	0	0	0	0	0	0	0	535
Castile Landing	3	13	1	3	0	0	1	0	0	0	0	0	0	0	0	0	0	0	0	0	21
Cranor Place	34	20	15	8	4	8	1	1	2	0	0	0	0	0	0	1	0	0	0	0	94
Holden Lake	56	49	32	25	11	10	3	3	5	2	0	0	0	0	0	0	1	0	0	1	198
Nickel	107	80	38	17	16	7	3	7	5	2	3	4	0	1	2	1	0	1	1	0	295
Rose Mound	68	77	30	15	7	12	1	0	1	8	2	0	2	1	0	1	2	0	0	0	227
Σ	606	561	195	168	96	48	33	18	17	14	13	10	10	6	4	4	3	1	1	1	1809

[a]The first dimension describes line cross section: (a) "v-shaped," (b) "rounded," (c) "squared," or (d) "complex." The second dimension describes whether incisions are (A) spaced across the entire sherd or are (B) part of a bounded area such as a band, triangle, or trapezoid. The third dimension describes the spacing of incisions from each other where the spacing can be dense (I, distance between incisions less than the width of lines), medium (II, distance between incisions equal to width of lines [± 0.25 x the width of the lines]), or sparse (III, distance between incisions greater than width of lines).

Table 8.10. Incised-Design Types in Lipo Assemblages Using Dimensions of Incision Set Relationship and Incision Spacing.[a]

Assemblages	3II	3III	1II	1III	2III	2II	5II	3I	1I	5III	5I	4III	4I	Σ
Beck	27	11	21	13	2	0	6	5	5	1	0	0	1	92
Belle Meade	32	19	15	9	13	2	3	3	2	2	10	0	0	101
Castile Landing	0	1	4	0	0	0	0	0	0	0	0	0	0	5
Cranor Place	2	7	9	12	10	1	0	0	1	0	0	0	0	42
Holden Lake	13	14	8	5	13	2	0	0	1	0	1	0	0	57
Nickel	17	16	11	12	10	4	4	4	0	0	1	0	0	79
Rose Mound	21	18	17	14	15	7	2	1	2	1	0	1	0	99
Σ	112	86	85	65	63	16	15	13	11	4	12	1	1	475

[a] The first dimension indicates the design relationship between sets of incised lines where the sets either (1) overlap one another, are (2) adjacent and at an acute angle from one another, (3) perpendicular, (4) oppose one another, or (5) have no relationship to one another (5) as in the case of a single set of incising elements that form a shape such as a triangle. The second dimension describes the spacing of incisions from each other where the spacing can be dense (I, distance between incisions less than the width of lines), medium (II, distance between incisions equal to width of lines [± 0.25 x the width of the lines]), or sparse (III, distance between incision greater than width of lines).

Following PFG accounts of the patterns among Barton Incised sherds, I also described the new assemblages at the scale of incised line designs (*i.e.*, combinations of sets of incised lines). These descriptions are composed of measurements along three dimensions. The first dimension indicates the design relationship between sets of incised lines where the sets either (1) overlap one another, (2) are adjacent and at an acute angle from one another, (3) perpendicular, (4) oppose one another, or (5) have no relationship to one another as in the case of a single set of incising elements that form a shape such as a triangle. Table 8.10 provides descriptions of the classes constructed of this dimension and, in order to reduce the effect of chance, incision spacing where the spacing can be dense (I, distance between incisions less than the width of lines), medium (II, distance between incisions equal to width of lines [± 0.25 x the width of the lines]), or sparse (III, distance between incision greater than width of lines).

Figure 8.8 presents frequencies of the incised design types in the new assemblages. Like the types built using descriptions of incised-line elements, I could not create valid seriation solutions using measurements made with incised design types. Consequently, these descriptions do not meet the expectations of the neutral model. Unlike types built of descriptions of punctate arrangement variability, classes that incorporate features of incised-line elements and incised line element sets at classification levels and measurement scales lower than the canonical PFG type, 'Barton Incised,' do not behave as historical types.

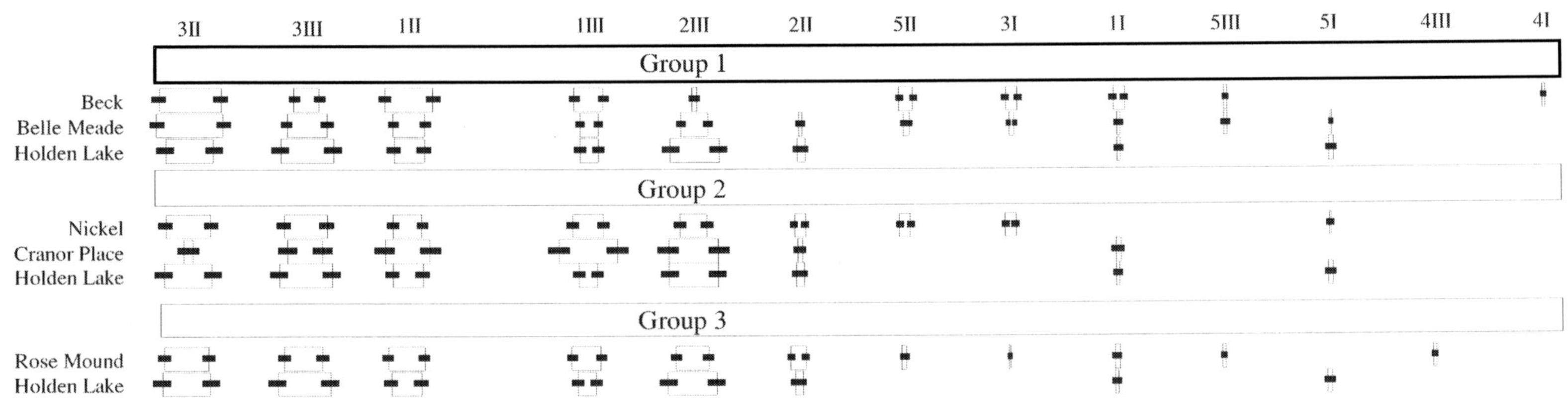

Figure 8.8. Seriation Solution for Incised Design Types in Lipo Assemblages Using Dimensions of Incision Set Relationship and Incision Spacing.

Solutions given at α=0.005 level of significance. Castile Landing is not included here as the number of incised decorated sherds is small (*n*=5).

Finally, one obvious area of decorative variability that is exploited by neither PFG nor Phillips (1970) is the inclusion of punctates with incised lines. In their discussion of Barton Incised, PFG attempt to clearly distinguish cases in which punctuates might be included with incised decoration. PFG (1951:116) point out "Barton Incised occurs as a rim decoration in combination with several types of body treatments. Sherds showing such combinations have been counted as Barton Incised on the principle that the rim decoration determines the type." However, the relationship between incising and punctation is not that clear. In their figures (PFG 1951:Figure 86a-d, Figure 95g,i-n), PFG show examples in which punctuates are mixed together with incised decoration along the rim area. In these cases, the PFG rule would fail: these sherds could be described as both Parkin Punctate and Barton Incised. Phillips (1970) specifically sought to address this confusion. Phillips (1970:46) states that:

> a slight awkwardness in the classification as published is that rim sherds of such vessels get classified as Barton Incised and body sherds as Parkin Punctated. This has been frequently pointed out by obliging critics. We were not unaware of it at the time.

This is, of course, caused by treating the type at the scale of vessel but identifying them at the scale at the potsherd. In order to address this inconsistency, Phillips (1970:46) proposed creating a variety of Barton Incised, v. Togo. The Togo variety, in Phillips' definition, consists of sherds with punctuates on the body and incisions on the rim.

However, sorting inconsistencies do not warrant creating new classification levels or measurement units at different scales. In my analyses, I was able to achieve excellent results by simply assigning *all* sherds with narrow, linear, incised lines to Barton Incised, whether or not these sherds also contained punctation. Using this rule, the types I generated behaved as historical classes. Consequently, rather than assume that Barton, v. Togo is necessarily a better unit since it captures instances of multiple decoration types, it is necessary to question whether specifying a distinct unit for sherds with both incising and punctation provides additional detail or information about the chronological and spatial relationships between the assemblages.

Table 8.11 provides a tabulation of sherds decorated with incised lines by the additional (A) presence and (B) absence of punctuates and whether (I) or not (II) the design covers the entire sherd. I included the latter dimension in order to reduce the role of chance in the frequency compositions. This is especially important since punctate presence consists of only two possibilities (punctuates/no punctuates). Figure 8.9 presents the seriation groups for the new collections described with the punctate presence/coverage types.

Table 8.11. Incised-Design Types in Lipo Assemblages Using Dimensions of Punctate Presence and Coverage.[a]

Assemblages	BI	BII	AI	AII	Σ
Beck	56	41	22	4	123
Belle Meade	80	20	28	9	137
Castile Landing	1	3	3	0	7
Cranor Place	22	23	3	1	49
Holden Lake	54	20	1	0	75
Nickel	61	15	34	2	112
Rose Mound	67	39	9	0	115
Σ	341	161	100	16	619

[a] The first dimension describes whether the decoration includes punctuations (A) or not (B). The second dimension describes if the decoration covers the entire sherd (I) or is in a limited band (II)

The expectations of the seriation model for Groups 1 and 2 are not met with the punctate presence-based descriptions of the assemblage. This result suggests that punctuate presence does not contribute to resolving details about homologous similarity between assemblages. Consequently, in this group of ceramics, there does not appear to be a reason to describe assemblages using a classification that specifies types at the scale and level of punctates/incising combinations.

SUMMARY

Large collections allowed me to evaluate the use of Phillips' (1970) varieties as well as many of PFG's classificatory decisions. PFG rejected the use of variability in punctate design for constructing types. Phillips (1970) used arrangements of punctates to form his types and varieties. Resurrecting the PFG's Castile Landing Punctated, Phillips formed Parkin Punctated, v. Castile. Phillips also defined grid arrangements of punctates on jars as Parkin Punctated, v. Harris, sherds with curved lines of punctates as Parkin Punctated, v. Transylvania and simple linear (one dimensionally aligned patterns of punctates) as Parkin Punctate, v. Parkin. Although I found that there are differences in comparability between the criteria necessary for punctate decoration type assignment, differences in the proportion of varieties are meaningful, but only at my larger sample sizes. Thus, the independent evidence presented here provides strong support for the results obtained using the PFG types.

These analyses also demonstrate that not all classes are useful for measuring homology. It is possible to build classes that consist of arrangements of incised lines as well as the form of incision. Phillips, for example, introduced varieties such as Barton Incised, v. Arcola, Barton Incised, v. Campbell, Barton Incised, v. Estill, and Barton Incised, v. Togo to account for variation he observed within incised lines patterns on shell-tempered pottery. With the evidence presented here, these classes do not contribute to studying

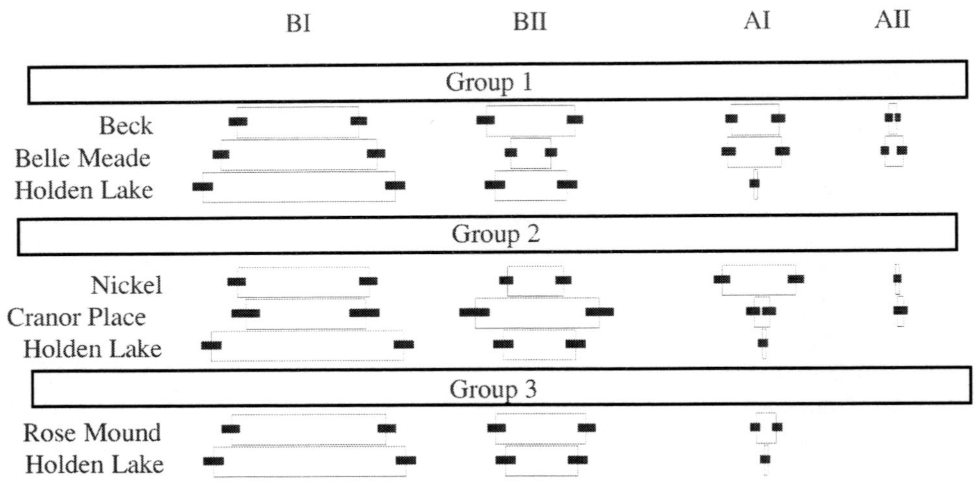

Figure 8.9. Seriation Solution for Incised Design Types in Lipo Assemblages Using Dimensions of Punctate Presence and Design Coverage.

Solutions given at α=0.005 level of significance. Castile Landing is not included here as the number of incised decorated sherds is small (n=7).

either temporal (chronology) or spatial (phases) transmission; no seriation solutions are possible with descriptions of my assemblages using these types. Differences in their proportions in assemblages are not meaningful even at my larger sample sizes. With the inclusion of Holden Lake, my samples certainly span the bulk of the duration of both types. Consequently, failure to show meaningful distributions is clear evidence of a lack of temporal meaning. Since my samples are strongly clustered (*vis a vis* the areas considered by PFG and Phillips [1970]), however, failure to detect meaningful spatial distributions does not rule out their utility at a larger scale. In this way, measurement classes embodied as Barton Incised and other PFG types contribute to our ability to study homologous similarity while classes built at higher-scales such as varieties fail.

METHODOLOGICAL ISSUES

The examination of the classification described above supports the contention that the stylistic frequencies observed in the assemblages are neither random observations nor the result of a cryptic bias in the original classifications. However, other methodological factors related to counting must be considered. In general, the methods by which abundances are generated and compared have been ignored in archaeology (Orton 1993). Since the results of the seriation analyses heavily depend on consistent descriptions of assemblages, methodological issues must be examined.

SHERD SIZE

One aspect of type assignment of which PFG were keenly aware was "error" in type frequencies due to sherd size. In order to be applicable at the physical scale of sherds, and thus useful for the surface collections, types had to be recognizable at the scale of sherds. This requirement restricts the complexity of types as a function of the relationship between sherd size and size of decorative elements. Types that include more than one element and specify relationships between elements are particularly susceptible to inconsistencies in recognition due to the patterns of sherd breakage. Although Ford mistakenly treated factors that influence type frequencies as error, in his early work he did acknowledge that the counting process produces variability. For example, Ford (1936a:17) writes:

> there is still another way in which errors are bound to occur in classification [identification, *sensu* Dunnell 1971]. In types which are formed by the combination of two or more elements, and where the isolated elements also form decoration types, there is the possibility that only one of the elements may appear on a sherd. The other element may have been present in the original decoration, but through its loss the sherd will be referred to the wrong type. For example, the combination of a number of overhanging lines drawn parallel to the vessel rim with a single row of triangular punctates just below and parallel to the lines forms a common decoration. Another type consists of the arrangement of overhanging lines without the row of punctates. It is clear that if a sherd of the first described decoration is broken so that it does not include the lower part it will be included in the second described type.

Probably because of Ford's awareness of the effects of sherd size on type assignment, most PFG types do not require extensive information about relationships between elements. If, for example, Manly Punctate is combined into Parkin Punctate (as in these analyses), the type Parkin Punctate is recognizable by the presence of just a single punctate. However, it is possible that these results incorporate variability caused by inconsistent descriptions related to other factors than the decoration attributes. Punctates that are part of Barton Incised patterns, for example, might be inconsistently assigned if the sherd is very small. Generally, PFG appreciated this problem. The original PFG sherd tabulation forms, for example, included the category of "Unclassified Shell-tempered Incised" to tabulate sherds too small to be assigned to one of the established incised types.

Subsequent researchers, however, have not been as concerned with sherd size and measurement issues as PFG. Since Phillips' (1970) interest was primarily on the documentation of empirical variability and since he treated types as empirical entities rather than measurement units, he paid very little attention to the effects of sherd size in the assignment of types. Type-varieties, therefore, are defined by the presence of decorative elements as well as a combination of vessel form dimensions and designs relationships that require substantial information on vessel orientation and shape identification. In this way, describing assemblages using varieties requires large sherds that represent large portions of vessels. Small sherds, for example, are not likely to contain all of the defining attributes of Barton Incised, v. Togo (incised lines on necks that form boundaries around areas of punctates; Phillip 1970:46). Similarly, Mound Place Incised, v. Mound Place, (horizontal lines incised below rims of bowls; Phillips 1970:135), requires sherds that are large enough to inform about vessel form as well as orientation.

Sherd size *must* influence type assignment at some point. In the case of Fortune Noded and Vernon Paul Appliqué, for example, sherds must be large enough to contain at least one full node or appliqué portion in order to be assigned to this type. Sherds with only a portion of node or appliqué may not be assigned to these types, as it is unclear whether a lump of clay on the surface is portion of a handle or some other architectural feature. Similarly, a sherd cannot be assigned to Ranch and Rhodes Incised if the sherd is particularly small since it is difficult to distinguish curvilinear lines from a small portion of hand-incised (and thus often irregularly drawn) rectilinear lines. Likewise, assigning sherds to Kent Incised and Mound Place Incised, recognized by vertical and horizontal lines, requires sherds that are large enough to be oriented relative to the vessel. Collections with small sherd sizes will generally have a preponderance of sherds assigned to types that do not require orientation or other information obtainable only from larger sherd sizes.

Additionally, sherd size also interacts with the area over which decoration occurs on a vessel. Small size sherds, for example, will result in low assignments of Hull and Walls Engraved if the decorated area involved in these designs (*i.e.*, patterns of filled lines) is only found on a small percentage of any vessel. In cases of very limited vessel decoration area, small sherds would appear "plain" more often because they will not intersect the portion of the vessel that is decorated. Similarly, it is expected that the distribution of Wallace Incised decoration on vessels will also affect the number of sherds assigned to this type. Likewise, counts of Carson Red-On-Buff require sherds large enough to recognize a portion of red slip and an area of non-red slip in order to distinguish specimens from Old Town Red.

If reliable and comparable measurements of assemblages are to be made, it is necessary to evaluate the effect that sherd size has on the assignment and tabulation of types. As shown in Equation 8.1, sherd size must be considered through three related factors: the area of sherd (S_a) depending on the classification level, the area of the decorative elements (E_a), and the area of the element sets (D_a). Large decorative elements/sets require greater sherd area for being counted than small decorative elements/sets.

Table 8.12. Counts of Decorated Shell-Tempered Sherds in Phi-Size Categories for Lipo Assemblages.

Phi Size	Beck	Belle Meade	Castile Landing	Cranor Place	Holden Lake	Nickel	Rose Mound	Σ
-6	24	40	10	4	9	25	33	145
-5.5	158	184	28	18	50	182	144	764
-5	155	210	24	37	51	173	188	838
-4	296	400	52	78	90	324	215	1455
-3.5	442	504	136	154	103	380	162	1881
Σ	1075	1338	250	291	303	1084	742	5083

As shown in Tables 8.12 and 8.13, sherd size clearly varies among my assemblages. Castile Landing, for example, has a larger proportion of the smallest sherds (≤ -3.5 phi) than the other assemblages do. Rose Mound, on the other hand, has a large proportion of large sherds (≥ -6 phi). Understanding the effects of sherd size on assemblage composition is not a matter of reconstructing pot frequencies. Rather, we are seeking to evaluate the degree to which sherd size impacts the frequency comparisons between assemblages since we know that (1) there are differences in sherd sizes between assemblages and (2) there are differences in the frequencies between decorated types. In order to determine whether differences in assemblage composition are *not* a function of sherd size, we can test two aspects of the "size-effect" hypothesis: (1) that the difference in sherd size does not predict differences in the proportion of types and (2) that differences in sherd size between assemblages do not influence the result of the seriation analyses.

Examining the relationship of type complexity with sherd size is the first way of evaluating the size effect hypothesis. The number of sherds assigned to a type decreases as sherd areas (S_a) become smaller than the area of the decorative elements (E_a) or element sets (D_a) involved in a type definition. This effect is due to the fact that, in general, the greater the number of attributes required for a type assignment, the larger the sherd size needed for type assignment. Simply, descriptions of smaller sherds are less likely to contain all of the attributes required for type assignment than larger sherds. Consequently, if sherd size has an effect on type assignment, the relative number of complex types (requiring a greater number of attributes) will be lower when measured in collections of small sherds and than when measured in the same collections but only using larger sherds.

We can evaluate the maximum significance of the size effect by comparing the frequencies of the type of greatest complexity with the simplest type and using just the largest and smallest sherds. As shown in Table 8.14, this description creates four frequencies (X_a complex type, large sherds; X_b simple type, large sherds; X_c complex type, small sherds; X_d simple type, small sherds). The size effect hypothesis is supported if we find that X_a and X_d are greater than expected by the random null since this indicates that complex types are found in a greater proportion on large sherds than expected and that the smaller sherds are generally correlated to simple types. On the other hand, if we find that the frequencies are not greater than expected, we can argue that there is no size effect. It is also possible that X_c and X_b could be greater than expected. In this case, there may be an effect relating type to size but that the effect may be due to vessel form or structural limitations relating types assignments to particular sherd sizes.

Table 8.13. Proportions of Sherds in Phi-Size Categories with 95% Confidence Interval in Lipo Assemblages.

Phi Size	Beck	Belle Meade	Castile Landing	Cranor Place	Holden Lake	Nickel	Rose Mound
-6	0.022 ± 0.009	0.030 ± 0.010	0.040 ± 0.026	0.014 ± 0.016	0.030 ± 0.021	0.023 ± 0.009	0.044 ± 0.016
-5.5	0.147 ± 0.022	0.138 ± 0.019	0.112 ± 0.041	0.062 ± 0.029	0.165 ± 0.044	0.168 ± 0.023	0.194 ± 0.029
-5	0.144 ± 0.021	0.157 ± 0.020	0.096 ± 0.039	0.127 ± 0.040	0.168 ± 0.044	0.160 ± 0.022	0.253 ± 0.032
-4	0.275 ± 0.027	0.299 ± 0.025	0.208 ± 0.052	0.268 ± 0.053	0.297 ± 0.053	0.299 ± 0.028	0.290 ± 0.033
-3.5	0.411 ± 0.030	0.377 ± 0.026	0.544 ± 0.064	0.529 ± 0.059	0.340 ± 0.055	0.351 ± 0.029	0.218 ± 0.030

Table 8.14. Four Frequencies Used in Evaluating the Sherd Size Hypothesis.

	Complex (Large Number of Attributes)	**Simple (Fewer Attributes)**
Large	X_a	X_b
Small	X_c	X_d

Table 8.15. PFG Types Ranked by Number of Defining Criteria.

PFG Type	Number of Definitional Criteria (as shown in Figure 6.1)
Barton Incised	9
Kent Incised	9
Mound Place Incised	9
Ranch Incised	8
Wallace Incised	8
Rhodes Incised	8
Walls Engraved	7
Hull Engraved	7
Parkin Punctated	5
Old Town Red	5
Fortune Noded	5
Vernon Paul Appliqué	5
Carson Red-On-Buff	5
Hollywood White	5

In order to select the "complex" and "simple" types for the comparisons, I counted the number of attributes involved in assigning a sherd to a PFG type as determined by the assignment key used in the descriptions of my collections (Figure 6.1). Table 8.15 presents a list of the PFG types ranked by the number of definitional criteria and Table 8.16 presents the number of sherds for each type and size class.

On the basis of this ranking and in order to gain the best resolution using only the smallest (-3.5 phi) and the largest (-6 phi), I chose the two types with the greatest number of sherds and with the largest differences in complexity ranking: Barton Incised and Parkin Punctate. The observed, expected, and standardized residual results from a chi-square analysis of the 2x2 matrix are shown in Table 8.17. The chi-square value (χ^2=2.430) is not significant at a 95% confidence level (p=0.119). In this case, size does not seem to have influenced type assignment.

Table 8.17. Observed (O) and Expected (E) and Standardized Residuals Values for Large and Small Sherd Sizes of Barton Incised and Parkin Punctated.

	Barton Incised	Parkin Punctated
Large (-6 phi)	O = 42 E = 50.8 Std. Residual = -1.2	O= 86 E = 77.2 Std. Residual = 1.0
Small (-3.5 phi)	O = 607 E = 598 Std. Residual = 0.4	O =900 E = 986 Std. Residual = -0.3

Table 8.16. Counts of PFG Types in Sherd Size Classes for Lipo Assemblages.[b]

Size Class (phi)[a]	Barton Incised	Kent Incised	Mound Place Incised	Ranch Incised	Rhodes Incised	Wallace Incised	Hull Engraved	Walls Engraved	Carson Red-On-Buff	Fortune Noded	Hollywood White	Old Town Red	Parkin Punctate	Vernon Paul Appliqué	Σ
-6	42	2	1	1	0	0	1	1	0	1	0	9	86	1	145
-5.5	227	6	1	7	4	13	0	8	2	18	0	34	429	3	752
-5	275	13	5	5	5	8	1	14	4	14	0	31	441	5	821
-4	438	22	10	28	4	9	0	18	3	28	2	87	769	8	1426
-3.5	607	17	8	26	14	14	3	14	3	15	3	180	900	6	1810
Σ	1589	60	25	67	27	44	5	55	12	76	5	341	2625	23	4954

[a]Size classes are measured in phi units. Smaller numbers indicate larger sizes since $x\ mm = 1/2^{y\ phi\ size}$

[b]Types are ranked from most "complex" (large number of defining criteria, 9) to "simplest" (small number of defining criteria, 5) from left to right.

As a comparison, I also examined the relationship between sherd size and the assignment of sherds to Barton Incised vs. Old Town Red (Table 8.18). Like the previous analysis, the chi-square value (χ^2=0.479) was insignificant at a 95% confidence level (p=0.489). These analyses support the hypothesis that there is negligible size effect acting on the assignment of types.

Table 8.18. Observed (O) and Expected (E) and Standardized Residuals Values for Large and Small Sherd Sizes of Barton Incised and Old Town Red.

	Barton Incised	Old Town Red
Large (-6 phi)	O = 42 E = 39.5 Std. Residual = 0.4	O= 9 E = 11.5 Std. Residual = -0.7
Small (-3.5 phi)	O = 607 E = 609.5 Std. Residual = -0.1	O =180 E = 177.5 Std. Residual = 0.2

The second means of testing the size-effect hypothesis is to simply evaluate the seriation solutions given only sherd size classes that exhibit the least amount of variability between assemblages. This analysis provides a means of evaluating whether differences in overall sherd size between assemblages impact type assignment and, therefore, the seriation results. Figure 8.10 graphically shows the distribution of sherds for my assemblages. There is considerable deviation in the proportion of sherds for each assemblage in the -6.0 phi and in the -5.0 phi size classes. In these size classes, the deviations of sherd sizes are the greatest between assemblages and potentially have the greatest impact on influencing the differences in the frequencies of types between the collections.

Consequently, an extreme evaluation of sherd size hypothesis would be to examine only the sherds that are at least –5.5 phi. These largest sherd sizes (>4.5 cm) are large enough to include multiple decorative elements. In addition, the large sherd area minimizes variability in type proportions caused by differences in the area of sherds (A_s). Table 8.19 presents the counts of the PFG types for my new assemblages for sherds that are –5.5 or –6.0 phi sized in their largest dimension. Although restricting the collections to just the largest sherd sizes increases the error term of frequency estimates, the results serve as a means for evaluating with the available data, if the overall results are biased by sherd size.

Figure 8.11 presents the results of a seriation analysis using only the descriptions of the largest sherds. Notably, there is no difference between these seriation results and the results that included sherds that were less than –5.5 phi in diameter. Although there may be errors in type assignment, *these errors are smaller than the differences in frequencies between the assemblages and are thus insignificant.* Again, we cannot falsify the null hypothesis that differences in sherd size between the assemblages are not related to the frequencies of types.

Table 8.19. Counts of PFG Types in Lipo Assemblages for –5.5 and –6.0 Phi-Sized Sherds.

Assemblage	Parkin Punctate	Barton Incised	Old Town Red	Fortune Noded	Ranch Incised	Kent Incised	Walls Engraved	Wallace Incised	Rhodes Incised	Mound Place Incised	Vernon Paul Appliqué	Carson Red On Buff	Hull Engraved	Σ
Beck A	58	23	1	3	2	1	3	2	0	0	0	0	0	93
Beck B	53	24	7	0	1	1	1	0	2	0	0	0	0	89
Belle Meade	151	55	0	3	1	3	4	3	1	0	1	1	1	224
Castile Landing	26	3	4	1	0	0	0	0	0	0	0	0	0	34
Cranor Place	5	12	3	0	0	1	0	0	0	0	0	0	0	21
Holden Lake	5	41	0	10	1	0	0	0	1	0	1	0	0	59
Nickel	112	50	23	0	1	2	1	7	0	1	2	1	0	200
Rose Mound	105	61	5	2	2	0	0	1	0	1	0	0	0	177
Σ	515	269	43	19	8	8	9	13	4	2	4	2	1	897

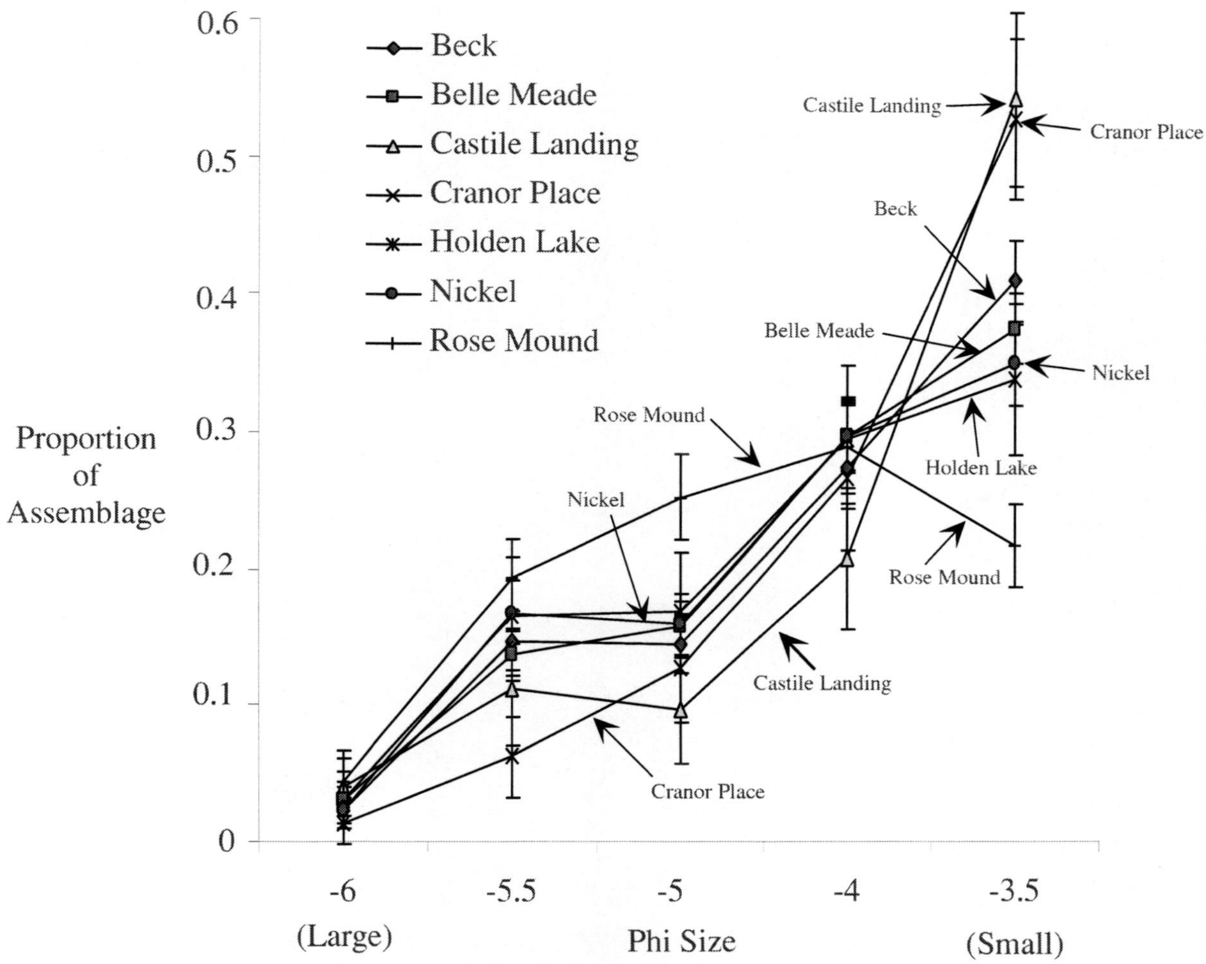

Figure 8.10. Distribution of Sherd Sizes for Lipo Assemblages.

95% confidence intervals are shown.

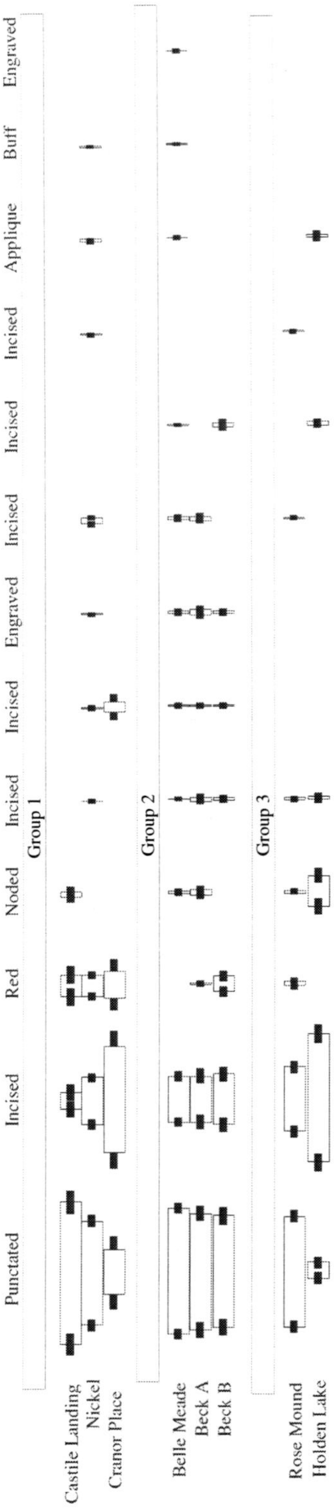

Figure 8.11. Seriation Solution For Lipo Assemblages Using Only -5.5 and -6.0 Phi Sherds.

Error terms are given for *a* =0.005 level of significance.

The types used by PFG, being based predominately on decorative technique rather than design or decoration arrangement, account for minimal influence of sherd size on type frequencies. For the most part, if a sherd is recognizable as being decorated, it can be assigned to one of PFG's cultural historical types. This aspect of the PFG classification makes it robust to problems caused by sherd size for assignment. We can also expect assignment error caused by differences in sherd size is far worse in the Phillips' varieties since sorting criteria, when available, rely on information beyond decoration technique but also decorative pattern, orientation, location on vessel, and vessel form.

THE EFFECT OF VESSEL FORM

Differences in the distribution of decorative elements across vessel forms (V_a) can also impact type frequencies. V_a is composed of two related variables: V_s, the relative size of vessels and V_f, differences in vessel forms. V_s has a relatively direct influence on type frequencies; larger vessels will mechanically contribute more to an assemblage than smaller vessels. Vessel size factors can be easily examined by measuring differences in sherd curvature and rim diameter between assemblages. However, I have not yet made these measurements on the new assemblages.

Variability in the composition of vessel forms (V_f) in each assemblage must also be considered when making the comparisons between assemblages. There is a necessary relationship between form and design composition because the application of decorations may be functionally restricted during the manufacturing process to particular forms (Krause 1972; see also Gould and Lewontin 1979). If, for example, a design element is produced only on jars and another is produced only on bowls, comparisons between assemblages with differing frequencies of bowls and jars (caused by differences in function) will necessarily produce different frequencies of design elements. These differences in frequencies of decoration indicate differences in the functional composition of the assemblage as well as style. Additionally, many decorations are restricted to areas of the neck (*e.g.*, Barton Incised, Wallace Incised, Mound Place Incised, Vernon Paul Appliqué) while others are found over just the body (*e.g.*, Parkin Punctate, Ranch Incised, Rhodes Incised, Fortune Noded, Kent Incised, Walls Engraved, Hull Engraved) and still others are found on all parts of the vessel (*e.g.*, Old Town Red). Thus, if there are differences in the proportion of vessel forms that comprise assemblages or if there are differences in the proportion of the portion of vessels that were collected (*e.g.*, rims), these differences will mechanically affect type frequencies.

To evaluate the effect of variability in vessel form in influencing the frequencies of the new assemblages, I tabulated the number of concave/plano ("neck" sherds, exterior surfaces that are plano or concave in vertical curvature) and convex ("body" sherds, exterior surfaces that are convex in vertical curvature) sherds for the decorated sherds in each assemblage and calculated the overall proportion of sherd forms in each assemblage. As is shown in Table 8.20, there is considerable variability in neck/body sherd proportions across assemblages. This variability may affect the results of the seriation analyses by biasing the proportions of types. To correct these differences, I calculated the difference in proportion for each sherd type and assemblage from the mean proportion of the sherd

Table 8.20. Sherd Forms in Lipo Assemblages.[a]

Assemblage	Number of "Neck" Sherds	Proportion of "Neck" Sherds (α=0.05)	Difference in Proportion from Mean	Number of "Body" Sherds	Proportion of "Body" Sherds	Difference in Proportion from Mean
Beck A	162	0.263 ± 0.036	0.042	404	0.657 ± 0.343	-0.026
Beck B	177	0.296 ± 0.037	0.009	398	0.666 ± 0.334	-0.035
Belle Meade	395	0.264 ± 0.022	0.041	1064	0.712 ± 0.288	-0.081
Castile Landing	38	0.172 ± 0.052	0.133	177	0.801 ± 0.199	-0.170
Cranor Place	105	0.340 ± 0.055	-0.035	188	0.608 ± 0.392	0.022
Holden Lake	180	0.496 ± 0.053	-0.191	104	0.287 ± 0.713	0.344
Nickel	316	0.273 ± 0.026	0.032	800	0.690 ± 0.310	-0.060
Rose Mound	264	0.335 ± 0.034	-0.031	491	0.624 ± 0.376	0.007
Mean Proportion for All Assemblages		0.305 ± 0.012			0.654 ± 0.346	

[a] Proportions are provided with 95% confidence intervals. Neck sherds were identified as those with concave and plano curvature. Body sherds were those with convex curvature. Sherds that included both neck and body portions were counted twice.

type for all assemblages. Using this value (difference from mean), I added the difference from the mean multiplied by the original type proportion and to the type frequencies. The values of each type were then adjusted for differences in body part frequencies and a new seriation was built.

Table 8.21 presents the corrected proportions of types for all of the new assemblages. If we use these new data to generate a seriation, as shown in Figure 8.12, we find that the results match those of the seriations for the uncorrected new assemblage samples (Figure 7.3). This result reflects the robustness of this seriation solution: despite variability in the composition in sherd forms across the assemblages, the seriation groups remain statistically valid. Of course, increasing the number of assemblages in order to increase temporal resolutions will change this result at some point.

The fact that the seriation solution is obtained despite variability in sherd form also illustrates an important point. Given the conclusions of the seriation analyses, the distributions of the proportions of "neck" and "body" sherds shown in Figure 8.13 are sorted in time. What I concluded to be the earliest assemblage, Holden Lake, has a distinctively large proportion of concave/plano sherds and the smallest proportion of convex sherds in the set of assemblages. One hypothesis that can account for this pattern is that the forms of decorated vessels are changing through time. However, differences in vessel form across assemblages invalidate an assumption made in the seriation analysis that the frequencies of types are not affected by changes in vessel form over time. Fortunately, although the differences in Holden Lake vessel forms from those of the remainder of my assemblages are large, so is its age. Consequently, the conclusions drawn from the orders and groupings in the seriation analyses remain valid but it does point to a problem that must be considered as chronologies are revised.

CONCLUSIONS

There are several general lessons to be gained from these analyses about classification and the reasons for the failure and success of certain combinations of attributes and historical types. Given a particular sample size, not all measurement scales and levels of classification are useful for describing assemblage similarity. The inclusion of punctate patterns and the arrangement of incised lines, for example, did not produce useful descriptions. This does not mean the types using these attributes were somehow "flawed." Rather, it means that though types constructed of patterns of elements (designs) are *potentially* useful for measuring historical relationships, they only serve to introduce noise in the Mississippi valley data at the sample sizes available in my collections.

In the case of the punctate design descriptions, for example, samples of particular combinations of variants in some cases were too small to add resolution between assemblages. One criterion for useful descriptions is that they produce statistically significant differences in type frequencies across the assemblages given a particular sample size. There are two related factors at work here. Classification level dictates the precision in temporal resolution. Lower-level classes afford greater resolution. However, increases in resolution also places greater burden on sample size requirements. Consequently, the "best" classification level depends upon the resolution needed for solving a particular problem and the size of the sample available.

In addition, as shown in the analysis of incised decoration, not all types describe variability in such a way to meet the expectations of the neutral model. That is, not all differences in frequencies are potentially useful for measuring homologous relations (*i.e.,* style). Differences in type composition can result from functional factors (*e.g.,* differences in temper or vessel form) or sampling error.

Table 8.21. Proportion of PFG Types in Lipo Assemblages Corrected for Differences in Sherd Form.

Assemblages	Parkin Punctate	Barton Incised	Old Town Red	Fortune Noded	Ranch Incised	Kent Incised	Walls Engraved	Wallace Incised	Rhodes Incised	Mound Place Incised	Vernon Paul Appliqué	Hull Engraved	Painted
Beck-A	0.546	0.264	0.064	0.013	0.025	0.027	0.021	0.010	0.009	0.002	0.008	0.000	0.005
Beck-B	0.460	0.395	0.045	0.005	0.011	0.025	0.011	0.006	0.013	0.004	0.006	0.002	0.002
Belle Meade	0.546	0.307	0.020	0.008	0.021	0.010	0.017	0.012	0.008	0.002	0.004	0.001	0.002
Castile Landing	0.607	0.094	0.134	0.004	0.000	0.005	0.000	0.005	0.000	0.010	0.000	0.000	0.000
Cranor Place	0.415	0.361	0.125	0.015	0.026	0.014	0.015	0.007	0.004	0.014	0.000	0.000	0.004
Holden Lake	0.119	0.616	0.013	0.105	0.009	0.008	0.000	0.005	0.004	0.011	0.008	0.000	0.018
Nickel	0.540	0.244	0.118	0.007	0.004	0.013	0.007	0.013	0.001	0.006	0.007	0.002	0.004
Rose Mound	0.537	0.339	0.075	0.024	0.004	0.003	0.000	0.004	0.001	0.004	0.001	0.000	0.000

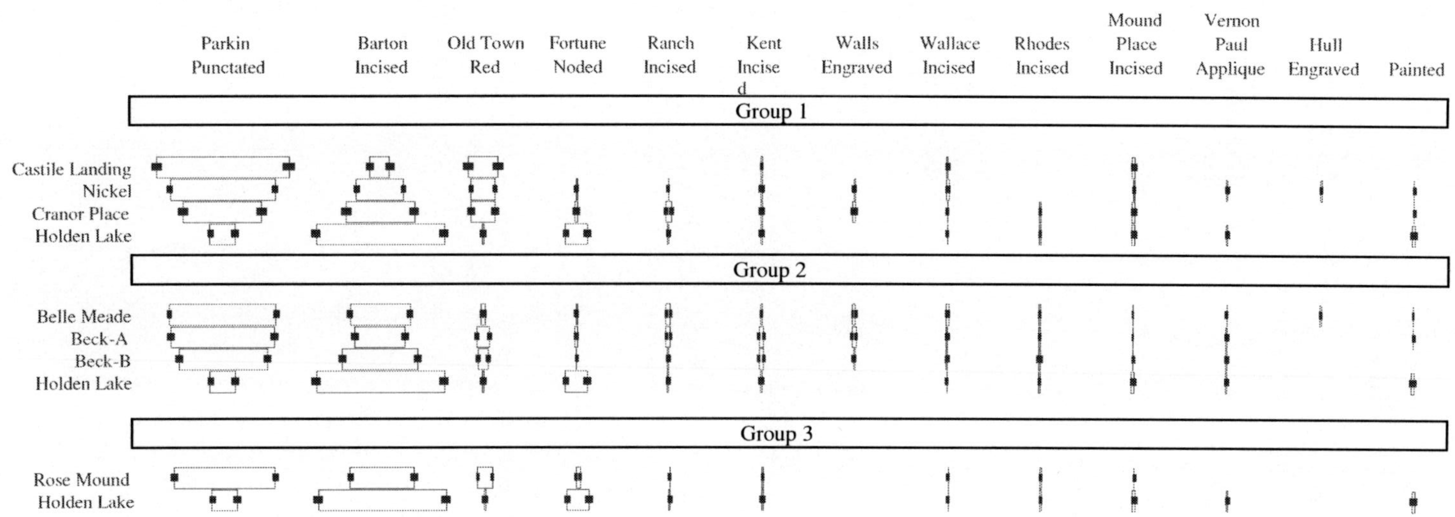

Figure 8.12. Seriation Solution for Lipo Assemblages Corrected for Differences in Proportion of Sherd Form.

Groups are formed at α=0.005 significance. Values for the cultural historical types in each assemblage were corrected by standardizing the frequencies of types for deviations from the average proportions of neck and body sherds.

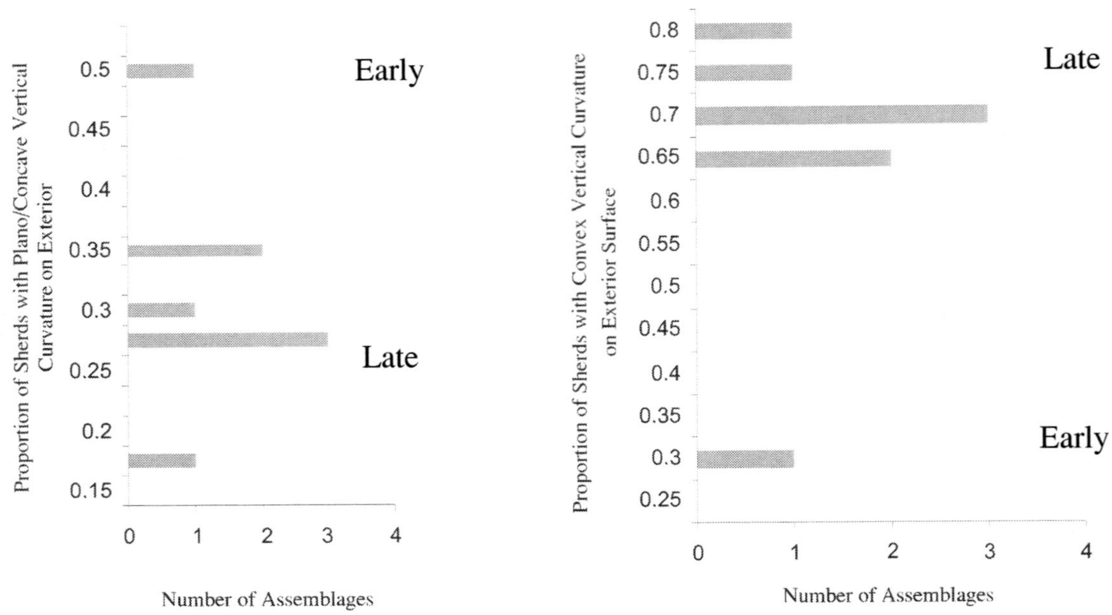

Figure 8.13. Proportions of Decorated, Shell-Tempered Plano/Convex and Convex Sherds from the Lipo Assemblage.

The distribution of the proportions of plano/concave ("neck") and convex ("body") decorated, shell-tempered sherds is potentially explained by changes in vessel form over time. The measurements include both Beck-A and Beck-B. The seriation analyses (Chapter 7) show that Holden Lake deposit is significantly earlier than the other deposits in the study. Shown here, the proportion of "neck" sherds for Holden Lake is higher and the proportion of "body" sherds is lower than all other assemblages.

Determining whether any set of attributes is effective for building descriptions in terms of stylistic similarity requires iterative evaluation of nested classifications. Finally, all measurements made with stylistic types must be comparable and made consistently across all observations. In order for stylistic classes to be effective in measuring homologous similarity, counting, and type assignment must not be driven by uncontrolled effects not related to style.

The culture historical types created by PFG meet these criteria for the most part and thus were successful in chronological construction. Successful work in the valley in terms of building chronologies and relating deposits across space have predominately relied on the PFG cultural types. These features of the types have also made them particularly useful in the interaction analyses presented here.

TYPES AND VARIETIES

It is also clear that Phillips' varieties have no utility in chronological or other analyses at the spatial and temporal scales and sample sizes in the study. Although rooted in the work of PFG, Phillips' measurement units are ontologically distinct from those used during the 1951 study. In Phillips' (1970:26) view of varieties, "intergrading is inevitable; it is in fact their very nature." Following this logic, there are few sorting criteria for Phillips' varieties. Sorting criteria changed, as new specimens were collected. He (Phillips 1970:26) states that "as data accumulate and sorting criteria become more explicit, two varieties juxtaposed in time or space naturally become more sortable." This view on the nature of sorting belies Phillips' view of what types represent. Rather than ideational units with necessary and sufficient conditions for membership, types for Phillips are extensionally defined empirical units with criteria based on their own attributes (Dunnell 1986; Ford 1954a, 1954b). Phillips viewed his types and varieties as *real* and "inherent in the specimens themselves" (Lyman *et al.* 1997:113). Although historically derived from PFG types, varieties are very different in conception. Phillips' varieties are essentialist groups rather than materialist measurement units; they are descriptions not analytic tools (Dunnell 1971, 1986).

Even if we try to form classes out of the Phillips' varieties or *ad hoc* groups, these units are not necessarily useful tools for the study of transmission in these data sets. Phillips' varieties are based primarily on the PFG definitions (*e.g.,* temper and decorative technique) with an assortment of measurement scales and levels of classes built of attributes associated with vessel form and arrangements of decorative elements (*e.g.,* designs) added. In many cases, Phillips' varieties are immediately recognizable: they consist of decorative variability abandoned by PFG for type construction since they provided no additional detail in the measurement of stylistic similarity. His "reasons" for using these combinations of attributes are entirely cryptic; he distinguishes these units simply because he can, at least conceptually.[1] To the extent that varieties are different from types, they do not further chronological or transmission studies. Thus, while they are not useful for studying homology, they have no other clear purpose either. Phillips specified no new problems for which they were suited.

Consequently, while many archaeologists discuss variability in terms of types and varieties when describing the contents of a single assemblage or when describing whole vessels, Phillips' typological system is of no use for archaeologists seeking to compare assemblages of sherds. In practice, the "type" portions of Phillips' descriptions – types made of attribute combinations defined by PFG – are the units used most often used in archaeological research across the Mississippi River valley (*e.g.*, Childress 1995; House 1990; Morse 1981; Morse and Morse 1983). The rationale that Phillips used in his classification is deeply flawed. Useful and usable types are not "norms" in any sense and varieties are not "deviations" from those norms (see also Gifford [1960]). Culture historical types are analytic tools not empirical entities. As classes, types partition variability so that it can be meaningfully quantified. The PFG types work as measurement units because PFG types are constructed to be sensitive to stylistic variability. The *ad hoc* addition of criteria for class membership such as that which has formed the varieties of Parkin Punctate and Barton Incised only adds noise to the measurements. The situation is even worse for other kinds of varieties devised by Phillips that use an assortment of functional, random, and stylistic variables to form class membership. Phillips (1970:157), for example, distinguishes Rhodes Incised, v. Horn Lake from Rhodes Incised, v. Rhodes, for example, by a combination of paste differences, incision techniques, vessel form, decoration designs. Similarly, Old Town Red is divided into six varieties that differ based on paste, temper, vessel form, surface polishing, and geographic locations. Frequently, Phillips' varieties lack a consistent set of sorting criteria.

Through these analyses, it has been possible to determine *why* the PFG types have proven useful for measuring stylistic similarity. PFG types are robust classes that generate neutral descriptions of the ceramic record. However, there is nothing particularly inherent about PFG's historical types. PFG types are not necessary for chronological analysis but they are clearly adequate. Determining the adequacy of units is, by its very nature, an iterative process. The work of PFG evaluating the distribution of types in the form of seriation insured that these units were effective for studying transmission.

---

[1] In practice, there are no sorting criteria.

CHAPTER 9: SUMMARY, CONCLUSIONS, AND FUTURE RESEARCH

> Supporters [of science] assume that the greatness and importance of a work correlates directly with its stated breadth of achievement: minor papers solve local issues, while great works claim to fathom the general and universal nature of things. But all practicing scientists know in their bones that successful studies require strict limitations. One must specify a particular problem with an accessible solution, and then find a sufficiently simple situation where attainable facts might point to a clear conclusion.
>
> -- Gould (1998:19)

SUMMARY

From the detailed efforts of culture historians in their study of the archaeological record, archaeologists have gained a fundamental framework for mapping the events of prehistory in space and time (Lyman *et al.* 1997). The work of PFG and the seriations of Ford, for example, form the basis for our understanding of the chronology of the ceramic era in the Mississippi river valley. Through a regional survey program that collected and described hundreds of thousands of pottery sherds from hundreds of sites across the valley, PFG built a dataset that allowed them to construct chronological relationships between assemblages and map the development of prehistoric cultural traditions. For the most part, their results still stand true today. Indeed, archaeologists have extended the work of PFG substantially, building elaborate ceramic typologies (*i.e.,* Phillips 1970) delineating phases, and inferring past communities that are purportedly related to ethnohistorical groups (*e.g.,* House 1991, 1993; D. Morse 1973, 1982, 1989, 1990; Morse and Morse 1983, 1996; P. Morse 1981, 1990; Smith 1990).

PFG relied on plausible, common-sense justifications for their conclusions, however. To explain the workings of seriation they claimed that the effects of popularity fads cause stylistic descriptions to have their characteristic "battleship" shape in time (PFG 1951:220-222). Plausibility notwithstanding, these reasons obscure the explanation for the success of culture-historical methods: in particular, the use of an underlying materialist metaphysic and types as measurement tools for examining stylistic descriptions of artifacts. Without an explicit appreciation for the role of materialism, over time the success of culture-historical types and seriation for building knowledge of the past became increasingly confused with their explanations. Ford's arbitrary spatial divisions ("local areas") needed for seriation, for example, came to be equated with phases and explicable as social groups (*e.g.,* Griffin 1952; House 1991, 1993, 1995; D. Morse 1973, 1982, 1989, 1990; Morse and Morse 1983, 1996; P. Morse 1981, 1990; Phillips 1970; Phillips and Willey 1953; Smith 1990; Willey and Phillips 1958). Similarly, culture-historical types became increasingly viewed as "norms," reflecting preferences of prehistoric potters rather than measurement tools designed to study chronology by intersecting continuous variability (*e.g.,* Phillips 1970; Rouse 1964; *cf.* Dunnell 1986).

These later confusions aside, the work and methods of culture historians provide an excellent starting point for building a science of archaeology. The methods employed by these archaeologists *actually worked* and have generated robust (*i.e.*, empirically tested) knowledge of the past. We know, for example, that seriation results are sensitive to transmission through space and time. We also know that culture-historical types are robust measurement units for studying interaction and that the number and kind of attributes used in building classes play a role in determining assemblage relationships. Culture historians, however, were unable to proceed beyond this point simply because of a lack of theory (Dunnell 1986). To address this lack, Dunnell has shown (1978a, 1982, 1989, 1992, 1995a; see also Teltser 1995b and papers therein) how evolutionary theory, once the theory is understood to account for cultural as well as biological change, can explain the success of culture historians by utilizing well-defined concepts such as neutrality, drift, transmission, and selection

Consequently, in order to address the needs of a scientific archaeology, my research expands on this pioneering work and takes evolutionary theory as a framework for exploiting, adopting, explaining, and modifying existing culture historical methods. To guide this effort, I defined an explanatory system rooted in evolutionary theory for studying the frequencies of decorated ceramics. Simple assumptions about cultural transmission over time and space between individuals and the effect of aggregate transmission among populations formed the basis for my explanatory framework. This theory of cultural transmission serves several purposes. First, evolutionary theory explains how and why seriation works. Second, it supplies a basis for generating and testing hypotheses about the patterns of interaction in the past that moves seriation well beyond the dating interest of the culture historians.

I investigated a central example of success for culture historians: the late prehistoric ceramics of the Mississippi river valley studied by PFG. Specifically, I outlined empirical contingencies that patterned the particular history of interaction among the late prehistoric populations in the Mississippi river valley. I found that by using a modified form of seriation, one that takes into consideration differences in sample size between assemblages, I could simultaneously evaluate the effects of space and time on stylistic (homologous) similarity of assemblages. This seriation method, based on principles derived from the neutral theory, generates groups consisting of the maximum number of assemblages that can be seriated without violating the expectations of the neutral model.

Using seriation analysis, I found that the pattern of stylistic frequencies of decorated shell-tempered types measured in collections built by PFG from the St. Francis and Memphis area were best explained as the products of locally interacting populations and not widespread communication across a single population. No single seriation group could be built and the best seriation solutions consist of assemblages that were geographically localized.

Stylistic type frequencies within the PFG assemblages, however, reflect the effects of sampling and counting practices as well as transmission processes. It was necessary, therefore, to test the hypothesis that the patterns that I observed were a function of the way in which PFG collected and described their ceramics. Without new collections, it was not possible to evaluate the degree to which the seriation results were influenced simply by chance and idiosyncratic collection methods. Consequently, I initiated a field project designed to generate new data from seven deposits in the central Mississippi river valley, six of which were originally collected by PFG. To generate these collections, I employed a research design that controlled sources of variability that influence the frequencies of culture-historical types carefully. This design took into consideration collection methods, sherd descriptions, type assignment, sherd size, sherd form, and variability in sherd distribution across deposits. Seriation analyses of the new collections and comparison with the PFG data demonstrated that the original PFG results are empirically robust and representative of the archaeological record. The impact of sampling and measurement methods contributed less variability in stylistic frequencies than the target sources of variability in the sampled archaeological populations. *Transmission* among populations through time and space is driving the composition of frequencies of stylistic types over the study area. Spatial conclusions from the seriations analyses may indeed reflect differential patterns of interaction across prehistoric populations. The similarity observed in the frequencies of decorated ceramic types supports a hypothesis that interaction among late prehistoric populations of the Mississippi river valley was spatially biased and populations were structured into lineages.

Finally, I concluded that similarity among the Parkin area deposits is potentially explained as the results of hierarchical community structuring indicative of complex organization. Overall, we can see that when factors of measurement and sampling are controlled (Chapter 4, Figures 4.5-4.13; Chapter 7, Figures 7.1-7.8), seriation provides a powerful tool for mapping patterns of prehistoric transmission. Using the results of these seriation analyses, it will be possible to pose and test further hypotheses about the evolution of interacting groups in the Mississippi river valley.

The results demonstrated the importance of a number of methodological considerations. First, large sample sizes are necessary in order to make precise distinctions between groups of seriation solutions. While it is not possible to specify a universal minimum sample size due to the effects of evenness on sample requirements, it is certainly clear that *large* samples (larger than typically acquired) are needed and that these sample size requirements must be carefully considered when building a research design. Because of the structure of historical deposits, archaeological samples are *never* too large or redundant (*cf.* Benn 1992).

Second, the collection of material in relatively small spatial units is an important step in building compositionally homogeneous samples of a deposit. Using 10x10 meter units to collect material from the surface, for example, I was able to detect compositionally distinct portions of the deposits that had been overlooked by PFG despite their concern. The sampling design also demonstrated why deposits must be collected over a large area; individual units vary significantly in composition even in the case of trash mounds, and building a representative sample of deposits requires widespread sampling.

Third, if carefully done, comparable samples can be collected by both excavation and surface collection. This requires detailed evaluation of the compositional differences, rejoining of ceramics, and theoretically-informed placement of excavation units in areas with functionally equivalent depositional histories. Fourth, factors of sherd size and vessel form must be considered when making descriptions of sherds using culture historical types. The rate of type assignment is not consistent across all sherd sizes and forms and thus must be corrected or evaluated when collections are compared. Fifth, as the culture historians found, not all attributes are useful for building types that track patterns of interaction. An iterative approach is needed to evaluate the distribution of types across assemblages; many attributes must be abandoned as inadequate. Phillips' (1970) varieties consist of *ad hoc* sets of attributes added to historical type definitions. These additions were primarily based on the observation that varieties *can* be constructed and that occasionally these units produce spatially-limited distributions. No theoretical rationale underlies their definitions and ultimately, therefore, varieties had no use here.

CONCLUSIONS

My research demonstrates that evolution can answer traditional questions in a robust, testable way, guide methodological development, and produce cumulative knowledge about the past. Evolutionary theory, however, cannot be integrated into archaeological explanations without careful consideration of the way in which the record is described. It has been long appreciated in biology that evolutionary theory is not just another theory, but a new kind of theory (Mayr 1959). The application of evolutionary theory requires *population thinking* (Dobzhansky 1951; Mayr 1977) or taking a *materialist* position in which "things are constantly in the state of become something else and categories (classes) are conventions used in description" (O'Brien 1996b:5). In a materialist ontology, cause is historically contingent and explanation necessarily historical (Gould 1986). The application of evolutionary theory to archaeology rejects the notion that types are empirical, phases are explanations, and variation in ceramic frequencies is just noise (Dunnell 1986). The works of early culture historians such as Ford provide guides for accomplishing these tasks. Once types are conceived as measurement units and variation identified as the subject for explanation, archaeological concepts become compatible with evolutionary theory. Evolutionary theory then provides the structure for building methods, testing hypotheses, and asking new questions of the archaeological record.

Building robust evolutionary explanations requires identifying sets of phenotypes that formed interacting groups from archaeological evidence alone. This task demands tremendous efforts in terms of data quantity, precision, and accuracy: requirements that archaeologists are just beginning to face. To paraphrase Lewontin (1974:8), we cannot describe the world in a way that is purely convenient, tradition-bound, or that uses common-sense attributes, and expect a scientific archaeology to emerge. Sample sizes, collection strategies, and counting all play roles in producing variability in the very things we use to estimate past interaction. Armed with the framework provided by evolutionary theory, an understanding of sampling, and careful description procedures, we now have methods available to us for defining population-scale counting units and for building scientific knowledge of the past.

PHASES

One of the primary conclusions to be drawn from this study is that phases, traditional units for describing populations, are inappropriate in an evolutionary archaeology. Archaeology based in an evolutionary framework requires detailed information about changes in how populations interact and how patterns of interaction changed through time. As shown in Chapter 7, groups of assemblages formed based on similarity are combinations of both time *and* space. One transmission component cannot fully explain similarity. For example, while spatial propinquity accounts for the grouping of most of the assemblages, intra-group relationships and the inclusion of Holden Lake into each of the seriation groups can only be explained by a strong temporal component. A simplistic space-like view of population interaction cannot account for the seriation groups and assemblage similarity as measured by stylistic (homologous) similarity.

Constructs like phases and other anthropological and intuitive units serve only to mask variability and generate potentially erroneous (and unfalsifiable) interpretations of the archaeological record. As traditionally defined, phases are inappropriate in a quantitative, scientific archaeology. Starting with a simple theory of cultural transmission and heritability, culture-historical methods such as deterministic frequency seriation provide robust methods for examining interaction history in the archaeological record.

From this model of cultural transmission, we cannot expect that transmission within structured populations will produce discretely bounded artifact distributions. Rather we should expect anomalously dense clouds of interaction within a background of constant transmission that experience convergence and divergence through time. We are left with a more complicated yet realistic model of transmission units than simple space-time blocks. Figure 9.1 presents a schematic of this view of lineages in time and space. To complicate matters further, archaeological deposits are samples of these lineages with particular time and space configurations and with varying durations. Obviously, actual archaeological samples of these deposits must be very carefully considered if one seeks to map the relationships between lineages from which they are ultimately derived.

COMPLEX SOCIETIES

The results of the style analyses of ceramic data provide an excellent empirical basis for concluding that populations in the central Mississippi river valley were structured into cultural lineages and that in at least one area these lineages are hierarchically organized in a form characteristic of complex societies (Dunnell and Lipo 1999; Dunnell and Wenke 1980). Thus, while the statistically robust seriation groups formed with the assemblages explain the outlines of units traditionally recognized as "phases," their significance is far greater. These lineage groups present a means for evaluating new hypotheses about the evolution of complexly organized social groups that are thought to have evolved in the Mississippi river valley in late prehistory. For the first time, we have the means for examining how hierarchical groups emerged within populations and studying how functionally-specialized phenotypes differentially succeeded or out competed others. These studies can lead to the evaluation of the historical constraints that drive the selection of attributes related to social organization within populations and thus build *scientific* explanations for the origins of complex societies.

Another point needs to be made. The close resemblance of the spatial distribution of seriation groupings to traditional "phase" areas might lead one to think that the conclusions drawn here are "nothing new." In addition, the conclusion that the Parkin assemblages may represent the first steps towards complex societies (as discussed in Chapter 7) might also elicit a reaction of "we already knew that." The work of Morse (1981, 1990), for example, presents an interpretation of the Parkin area deposits as the center of a settlement system and in which the major ceremonial center, the Parkin site itself, consisted of the "residence of the paramount chief" (Morse 1981:59). This is just a plausible assertion, however. Though these interpretations are compatible with the hypotheses presented in Chapter 7, it does not mean that nothing new has been learned. While many may have *believed* that Parkin assemblages are complexly organized, no one *knew* it. Even with the analyses presented here, it is still not clear if some or all Mississippian societies were complex from an evolutionary perspective, but these analyses demonstrate that it is *possible* to know and have pointed to a direction if we want to know. If it is possible to identify social complexity in the archaeological record using the methods presented here, then it is possible to identify when and why it happens where it does.

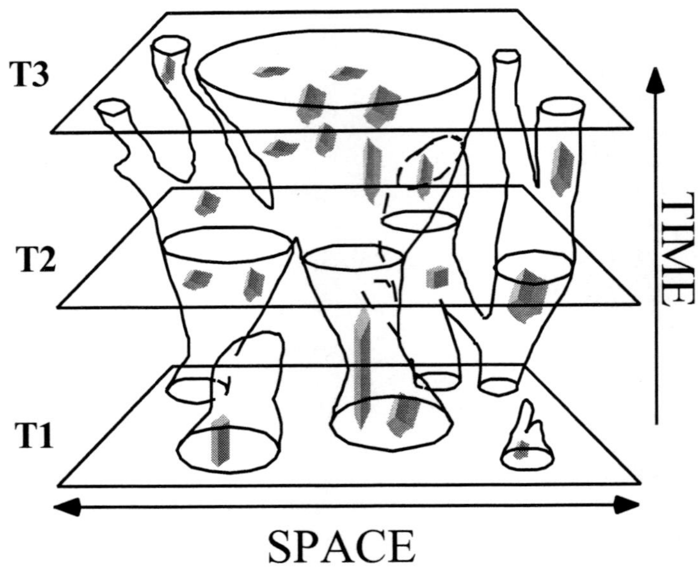

Figure 9.1. Cultural Lineages through Time and Space.

Figure is derived from (Dunnell 1982). Each branch represents an interacting population through time and within space. Over time, populations move, converge and diverge with one another. Anthropological notions of groups are equivalent to simple snapshots in time of these branches. Archaeological samples as indicated by the dark block, however, consist of samples of the remains of these lineages scattered through times.

SPATIAL ANALYSES

The focus of this study has been on generating the data necessary to evaluate claims about community patterning in late Mississippian populations. The results presented in Chapter 4 and bolstered by Chapter 7 analyses support a hypothesis that interaction among the populations was biased toward community membership. Although these data and hypotheses are framed within an evolutionary framework, no evolutionary explanation is possible. Explaining *why* the populations are structured into communities when and where they do requires examining changes in frequency data through time.

PFG were quite aware that the frequencies of their ceramic types were telling them something about change through time, though they had no theoretical or methodological tools available for examining this change in detail or for building explanations. Ford, for example, speculated about the processes driving the formation of spatial variability among the Mississippian populations over time, attributing the need for "local areas" to sedentary settlements, agriculture, and the development of political structures (PFG 1951:224). Ford pursued the study of spatial patterns among the ceramic type frequencies in his 1952 monograph. However, with only manual tabulations and no theoretical warrant, Ford was unable to explore these relationships in any rigorous detail (*e.g.,* Spaulding 1953b, 1954).

As an example of the direction we can now go given the empirical and theoretical warrant presented here, one simple means of examining the change of population structure through time is to evaluate measures of intergroup transmission. Archaeologists have long argued that differences between the degree of interaction are detected archaeologically by differences in the stylistic similarity of artifact assemblages assuming that, all other things being equal, greater stylistic similarity among assemblages indicates a greater degree of relatedness or interaction (Beals *et al.* 1945:89; Colton 1953; Englebrecht 1974; Fry and Cox 1974; Hill 1966, 1967, 1968, 1977; Hurt 1999; Kay 1975; Kidder 1924:343; Longacre 1970a:27-28; Plog 1976a, 1976b, 1980; Pollnac and Rowlett 1977; Redman 1977; Shapiro 1953; Whallon 1968). If traditional stylistic similarity is understood as homologous similarity (*sensu* Dunnell 1978a) then there is general warrant for this notion. Neiman (1995) has developed this notion in his work integrating simulation with mathematical models of drift and group interaction. In particular, Neiman (1995:26-27) suggests that the squared Euclidean distance measured as the similarity between each assemblage described with historical types and all others at a given point in time is a good measure of the amount of interaction between groups. This is accomplished by calculating the squared Euclidean distance of similarity between each pair of assemblages, i and j:

Equation 9.1. $$d_{ij}^2 = \sum_{k=1}^{n} (p_{ik} - p_{jk})^2$$

where p_k is the frequency of the *kth* type for n assemblages. Neiman demonstrates that populations with a high level of *intergroup transmission* cause low *mean-squared* distance among all assemblages, while a low level of interaction results in a high mean-squared distance. In this model of transmission, groups of individuals that interact frequently should have lower squared Euclidean distances than groups that are more isolated. Thus, using measures of squared Euclidean distance for the PFG assemblages, we should be able to get a preliminary map of how interaction between groups changed through time.

In an application of Neiman's model to the PFG data, I examined decorated culture-historical type counts from combined St. Francis and Memphis datasets. These data include clay, sand, and shell-tempered sherds since the kind of temper used in the ceramics changed over time in the LMV. In order to examine the effects of changes in temper through time, I divided the ceramics into each of four PFG "periods" (*i.e.,* those assemblages that fall within PFG's A, B, C, D designations) as determined by the original PFG seriation (*e.g.,* Figure 4.2). In order to place assemblages into one or another period, I removed all assemblages that had long durations (*i.e.,* had distributions that spanned more than one time "period"). I also included only assemblages with sufficient sample size ('A' assemblages; see Chapter 3, Figure 3.8). Then using Equation 9.1, I determined the Euclidean squared distance for assemblages from each of the "periods." The results are shown in Figure 9.2.

The results are consistent with the impression that PFG had formed for their ceramic data. Over time (from Period 'D' to 'A') there is an increase in the mean-squared Euclidean distance among the assemblages from the early to the later (Period B) assemblages. This increase in distance, following Neiman's model, potentially supports PFG's notion that populations are becoming increasingly isolated from one another through prehistory, as smaller initial populations grew and fissioned into distinct groups. Figure 9.3 shows a hypothetical scenario of how these changes in patterns of interaction and overall similarity between assemblages could occur. In the earliest period, (Period D, Figure 9.3, Model 1) population density is low and populations are able to move across the region and interact over large distances. Consequently, similarity is high and mean-squared Euclidean distance is low. One hypothesis that can be derived from Equation 9.1 and that can account for the increase in the mean-squared Euclidean distance between the early periods and Period B is that the valley filled in with individuals and population density increased. In such a scenario, overall interaction for the population would decrease as interaction regions for individual locations were reduced. Mean-squared-Euclidean distances would increase in this scenario until interaction became primarily neighbor-to-neighbor (Figure 9.3, Model 3). During the latest period (Period A), the data show a decrease in mean-squared Euclidean distance between assemblages. If sampling is not playing a major role in determining these results, then these data are consistent with a hypothesis that there was an increase in within-group interaction caused by changes in

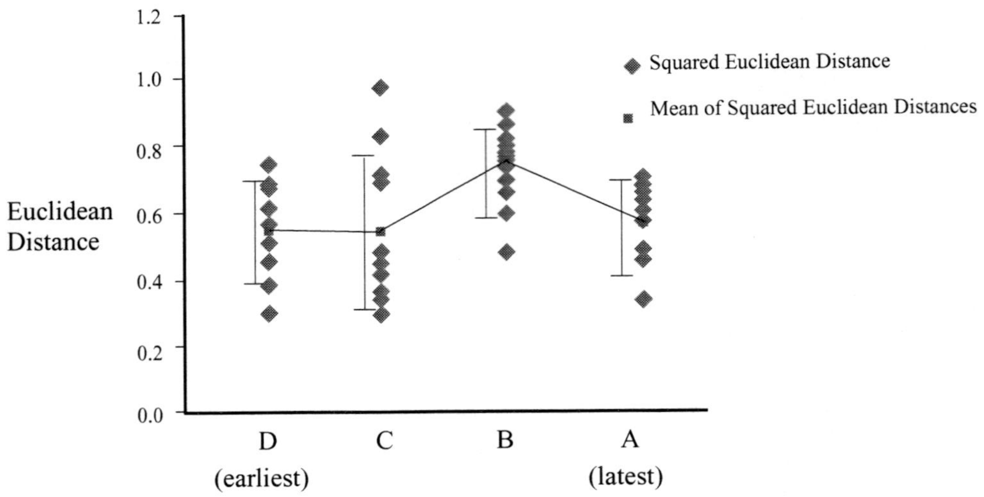

Figure 9.2. Estimates of the Squared Euclidean Distances Between the PFG Assemblages during Each of the Last Four Time "Periods" as Indicated by the PFG Seriation Results.

For each period, only the frequencies of decorated types for assemblages that fell into a single period were used. Line segments connect the mean between-assemblage distance for each time period. This line represents mean change in interassemblage distance from Baytown (D) to the Late Mississippian (A). The standard deviation for each of the mean distances is shown for each period as the vertical bars to the left of the Euclidiean Distance points for each period.

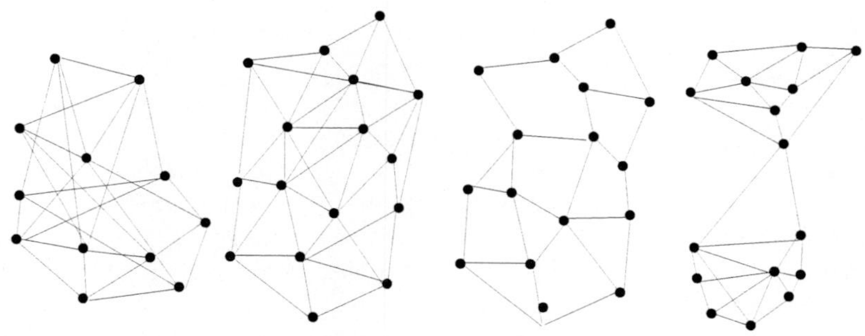

	1	2	3	4
Mean Numbers of Connections per Location:	4.61	4.53	3.00	3.67
Standard Dev.:	0.86	1.32	0.56	1.15

Figure 9.3. Four Models of Interaction.

Circles represent locations and line segment indicate paths of interaction between locations. In the first model, locations interact with each other regardless of space and the mean connectivity for each location is relatively high. Overall connectivity between locations for the second model is reduced and locations interact within a smaller area. The variance in connectedness is high in this model. In the third model, interaction occurs only to nearest neighbors and the the region over which any one location interacts is low. Finally, the fourth model describes a situation where interaction occurs at a high degree within local groups but where there is low connectivity between groups.

functional organization and the development of community groups[1]. Absolute dates for measuring chronological relationships between assemblages are needed, however, for investigating this hypothesis further. Despite the speculative nature of these conclusions, they demonstrate how the PFG data might be used to inform on not only the structure of cultural lineages in the Mississippi river valley during late prehistory (as seen in the seriation analyses), but also on the overall *evolution* of these social groups through time. Clearly, additional inquiries are warranted.

Full explanations for the evolution of social groups, however, will require additional sources of data not available from stylistic frequencies of decorated sherd samples. Temporal and geographic information will be critical for understanding rates of change and building realistic models for interaction among populations across the landscape. Independent chronological information is needed to evaluate chronological relationships of assemblages. It is also possible that differences in mean-squared distance between assemblages might be driven by differences in duration of assemblages since deposits built over a long period of time would be more likely to be susceptible to the cumulative effects of drift over time than short duration deposits. Geographic models of inter-deposit locations are also required to evaluate the effect of landscape on assemblage similarity. In addition, larger and controlled samples for the study area will be required. The need for larger sample sizes is especially important in collections earlier than the Mississippian assemblages examined here. In general, these "Baytown" and "Tchula" assemblages have lower frequencies of decorated ceramics and thus require much larger samples than those from the late Mississippian in order to make statistically meaningful comparisons. Nonetheless, the potential for these data to provide the means for exploring the evolution of social structure is tremendous and further investigations are certainly warranted.

[1] This conclusion is supported by recent work by Lyman and O'Brien (2000). In an analysis of ceramic descriptions of the lower Mississippi river valley, Lyman and O'Brien note that stylistic diversity in the area generally increased through time. They (Lyman and O'Brien 2000:67) suggest, "increasing stylistic diversity or richness may well reflect, then, a decrease in the panmictic characteristics of the human populations in the St. Francis area." In general, my research support these findings, however, further research on the chronological (and potentially spatial) nature of stylistic diversity in the assemblages of the Mississippi river valley and the theoretical framework needed to explain these patterns is required.

FUTHER RESEARCH

> We will always be mired in error. The most each generation can hope for is to reduce the error bars a little, and to add to the body of data to which error bars apply. The error bar is a pervasive, visible self assessment of the reliability of our knowledge.
> – Sagan (1996:28)

The research model presented here places enormous data requirements on efforts to detect population structure from aggregate frequency data in the archaeological record. As demonstrated by this research, however, meeting these requirements is far from impossible. With careful consideration of sample size, collection strategies, classification, and counting, I was able to statistically evaluate hypotheses regarding processes responsible for producing variability in frequencies of stylistic descriptions. The conclusions drawn here, however, point out areas that need to be evaluated. In particular, we need additional and larger samples, independent chronological data, and geographical information about the prehistoric landscape of the valley. In this way, as a scientific effort, this research is just part of a continuous and back-and-forth process. It begins where PFG left off and ends in new, future research questions, directions, and efforts.

SAMPLES

The original PFG samples are relatively robust and representative samples of the archaeological record. This is certainly a testament to the careful work of PFG, and concern of Ford, in particular, that the samples were generated in a way to insure comparability and representatively. Ford (PFG 1951:223) appreciated, for example, that large sample sizes are required for precise analyses. In addition, PFG samples were collected in ways that minimized bias by collector (using local workmen as collectors) and by splitting up "sites" into smaller collection areas when it was clear that that the deposits were chronologically mixed (PFG 1951:43).

Despite the overall quality of PFG data, however, most locations have inadequate sample sizes. PFG simply had no quantitative model with which to assess sample size and its effect on precision.[2] Larger samples permit analyses to be made with greater precision. My ability to resolve new detail about the relationship of assemblages to each other among the Parkin set, for example, demonstrates the value of larger samples detecting differences among assemblages that were formerly groups into a single set. In addition, increasing

[2] Ford (1936a:14) had tried to build a quantitative model in his early work in Louisiana and Mississippi by recollecting deposits multiple times in order to determine the degree of confidence for any particular sample. Unfortunately, this work was not continued in the PFG study.

assemblage size for sites not included in these analyses (due to small samples) will provide an opportunity to test these conclusions. If assemblages are added to the seriations yet the same configurations are obtained, increased confidence can be given to their results. For example, there are at least five Mississippian assemblages from the area of the Parkin set (grid units 11-N, 11-O, 12-N; see Figure 4.1) that would benefit from increased sample sizes and whose inclusions would help test the hypotheses generated in the analyses presented in Chapter 7.

It is also clear that larger samples are needed across the entire area. Of the 146 locations collected by PFG in the St. Francis and Memphis areas, only 22 locations have samples large enough to be included here. In several locations, the inclusion of additional assemblages into the analyses is needed in order to understand even the basics about how populations in the area are related to others. For example, in the "Nodena" area in PFG grid unit 10-P, only one assemblage (10-P-1) is large enough to have been included. Data from additional locations in this area are necessary to determine the relationship of 10-P-1 with the other assemblages (O'Brien 1996a). This need is also true of assemblages to the south (grids N-14, O-14).

Increasing the size of all assemblages in the study area will also greatly help our understanding of earlier non-Mississippian deposits. As is clear from Ford's seriation of the St. Francis area (Figure 4.2), the latest assemblages tend to have the greatest proportion of decorated ceramics. The larger the proportion of decorated to undecorated ceramics, the smaller the total sample needed to generate useful numbers of decorated sherds. Consequently, the PFG data are most effective for seriation analyses in a way that is chronologically biased towards the latest deposits. Earlier assemblages tend to have a lower proportion of decorated ceramics (often lower than 5%). Large total collections are needed in order to generate enough decorated ceramics so that these earlier collections can be included. In several cases, the overall inaccessibility of the deposits to surface collection requires recollection through excavations to be made. At Castile Landing, the time-intensive nature of the excavations resulted in a relatively small sample size. The homogeneity of the deposit and high density of ceramics, however, indicates that further work here would produce large samples comparable to surface collections. The success of the Castile Landing work in producing excavated samples comparable to surface collections suggests that excavation might also be useful in making new collections at other St. Francis-type sites that are not currently plowed or no longer accessible from the surface. Brackenseik (12-O-6)[3], for example, is a St. Francis-type site that has been greatly impacted by levee construction (Figure 9.4). Inspection of the area, however, suggests that at least some of the deposit remains intact (Figure 9.5). Careful excavations and considerations of comparability between surface assemblages and the midden deposits here could potentially produce substantial new samples that will contribute to our understanding of the populations in this area of the study region.

SPATIAL EVALUATION

Two hypotheses about spatial structure were examined in this research. The null hypothesis holds that variation is clinal such that phases are arbitrary divisions of spatial variability. An alternative hypothesis states that variation is discrete and that phases might, but not necessarily, map this structure as widely assumed (*e.g.,* Phillips 1970; D. Morse 1989; Morse and Morse 1983). In this case, units would be akin to lineages. One interpretation of the latter hypothesis is that populations are emerging as functionally independent communities. Ford (PFG 1951:224) argued, for example, that the need for using "local-areas" was due to:

> a common phenomenon for cultures at this stage of development and seems to be owing to factors such as decreased population mobility due to an increased dependence on agriculture; the establishment of more stable centers, such as ceremonial mound groups and towns; and an increase in the *cohesion of political groupings* made possible and necessitated by the improved food supply and consequent population increase; to which was added the increase availability of cultural ideas which could be combined to form "new" varieties.

Subsequent research by House (1991), D. Morse (1973, 1990), P. Morse (1981, 1990), Phillips (1970), and Williams (1954) suggests that the emergence of "phase"-scale units (developed from "local areas" of the Ford seriations) during the late Mississippian period is related to a change in social complexity. Indeed, the latest period of the archaeological record is marked by distinct shifts in settlement patterns that are represented by cohesive deposits consisting of large palisaded town sites and earthen mounds (Morse and Morse 1983).

If a change in organization were responsible for these patterns of similarity among these populations, then similarity among assemblages would be greater than similarity between lineages. Probabilities of transmission between locations would be unequal and community dependent. From the PFG data alone, however, it is not clear, whether the patterns detected by the seriation analyses reflect patterns of transmission linked to population organization or the uneven distribution of assemblages in space. In addition, one would expect that interaction between groups would be strongly affected by geography and the accessibility of populations to one another. Not all paths for interaction are of equal cost. Distance alone affects similarity by increasing or decreasing the probability (cost) of interaction.

Particular directions will be more or less costly by virtue of watercourses, currents, mountains, or swamps (Irwin 1992; Renfrew 1977:136). In the backswamps of the Mississippi river valley, one expects that geography rather than airline

---

[3] PFG (1951) misspell Brackenseik as "Brackenseed" Place.

Figure 9.4. Brackenseik at Time of PFG Visit.

Brackensiek was described by Moore (1911: 413-426; his "Rhodes place") as a flat topped mound partly washed out of shape with dwelling on top. He recovered 65 burials during excavations and on the surface numerous artifacts and earthenware sherds. Moore recovered 123 vessels from Brackenseik, many shell- tempered. PFG note during their survey that Brackenseik consists of a 120 ft sq. and 10 ft high mound with some daub showing and not much pottery. Brackenseik (12-O-6), Box 27, LMS Records 993-23, Peabody Museum Archives, Harvard University. Courtesy of the Peabody Museum of Archaeology and Ethnology, Harvard University.

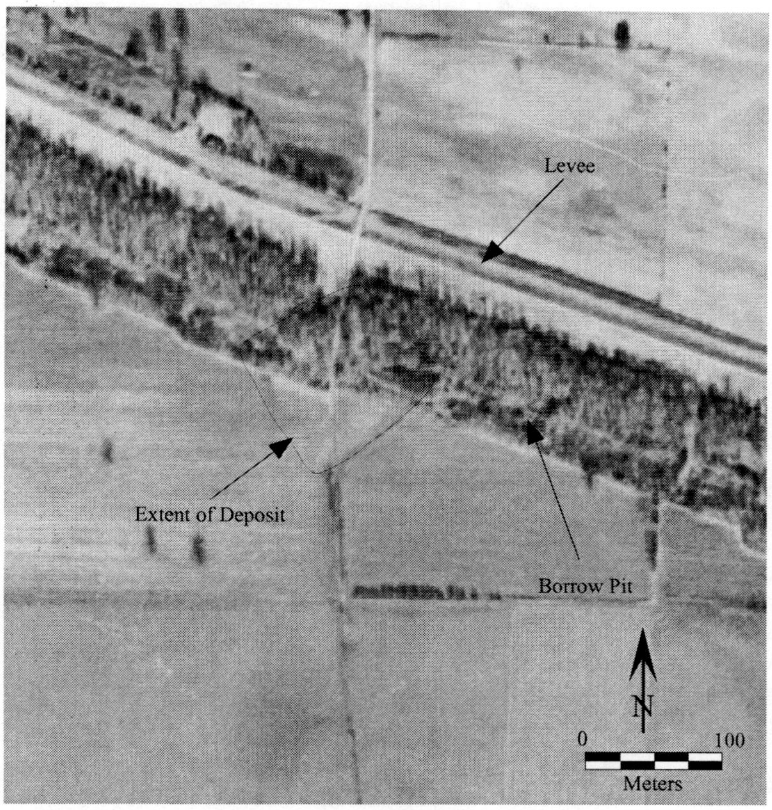

Figure 9.5. Satellite Image of the Area around Brackenseik.

The original extent of the Brackenseik deposit is visible in this satellite image as a dark square shape in the field to the south of the levee. Despite massive impacts to the area due to the construction of the levee by the Army Corp. of Engineers, inspection of the area shows that the original mound has been preserved and that portions are the midden are still extant. SPIN-2 imagery consists of digital, panchromatic, 2-meter resolution images. SPIN-2 imagery is the result of a joint venture of SOVINFORMSPUTNIK (the Interbranch Association of Russian space agencies), based in Moscow; Aerial Images, Inc. of Raleigh, North Carolina; and Central Trading Systems, Inc. of Huntington Bay, New York. SPIN-2 satellite imagery is available via http://www.terraserver.com, a site that operates as an online catalog for SPIN-2 images. This image was taken on January 4, 1989 at 17:40 GMT from an altitude of 218.6 kilometers, with a sun angle of 32.5 degrees and with a KVR-1000 camera system. The image resolution is 1.56 meters per pixel. Satellite photo used by permission of SOVINFORMSPUTNIK.

distance would be a better predictor of similarity (McCutcheon 1997). Thus, patterns of dissimilarity in this case would be predicted by the characteristics of the geographic environment rather than community membership. Air-line distances are easily measured. The critical part of analysis, therefore, is the approximation of geographic weighting; the better the geographic weighting is modeled, the closer the equation will fit the prehistoric case. McCutcheon (1997), for example, has shown that distance weighting is of sufficient importance to have major impact on correlations and that a boat-transportation model is, at least for lithic raw material, appropriate for the central Mississippi river valley.

In order to address differences in accessibility and geography across the region, GIS data coding for the hydrology, topography, and soils for the St. Francis and Memphis areas can be used. Using a model that weights distance between assemblages by the amount of "connectivity" due to water, it is possible to adjust the linear distance between assemblage locations to generate realistic estimates of interaction cost. The adjusted distances can then be used to calculate predicted similarity biased by landscape features. Patterns of similarity and dissimilarity can be hypothesized as representing the effects of interaction within spatially structured communities across varied geographies. While space structures patterns of similarity, time also plays a role that must be simultaneously considered when determining whether the observed distributions of stylistic frequencies are patterned by processes of communities structure. The greater the amount of time elapsed between depositions, the greater the differences in stylistic descriptions observed in assemblages. In this way, temporal factors confound spatial aspects of stylistic frequencies and *visa versa*. In order to conduct a spatial analysis, the temporal differences between each of the deposits must also be carefully measured. Without temporal control, it is not possible to test the hypothesis that differences not predicted by space can be explained by time or community structure. The research necessary to meet this requirement is discussed in the next section.

TEMPORAL CONTROL

> Archaeological data have an inert quality, a certain spinelessness when unaccompanied by a more or less definite chronological background.
> – (Tozzer 1926:283)

In the construction of culture histories, the temporal component of stylistic variability has been exploited to build chronological orders by minimizing the effect of spatial and social factors to transmission (*e.g.,* PFG 1951; Willey 1953; Williams 1954). In cases where one is interested in examining how community membership affects variability in stylistic descriptions, it is necessary to control differences in space *and* time so that potential variability due to social boundaries can be studied. A means of evaluating the temporal component of stylistic frequencies that is independent of the stylistic frequencies themselves is necessary to provide this information.

Generating a "date" for a group of sherds, however, is no simple task. A group of sherds can be dated in a number of ways. Each sherd has, for example, a "date" associated with its manufacture and another "date" associated with it deposition (Dunnell 1981). One might also choose to generate a "date" for a piece of charcoal found attached to a sherd assuming that it was added during use. Charcoal samples, however, may have "dates" related to contexts involving the death of the tree from which the charcoal was derived, the collection of the material as dead wood and for its ultimate transformation into charcoal during burning. There is no simple relationship between an object and a "date." Thus, when determining the "date" of a group of sherds, we must pay careful attention to which "date" we are obtaining and evaluate whether this value is related to our research goals (Dean 1978; Dunnell 2000; Dunnell and Readhead 1989). This evaluation is accomplished by examining the relationship between "dates" and events.

Events

Although it is convenient to talk about giving an object a "date," objects are not dated *per se*. Rather, events are dated. Archaeological procedures of dating consist of examinations of physical attributes of an object so that an order of these phenomena can be made either between themselves or on a fixed temporal scale (Dean 1978; Dunnell 1981, 2000; Dunnell and Feathers 1994). Events are times at which a particular configuration of physical attributes becomes a set (Dunnell 1981). Events can be arranged relative to each other as in seriation or using an independent framework (*e.g.,* calendar). Three considerations are important when building events to be dated. First, events have to be described with attributes that are related to the event of interest. In the case of the manufacturing date for a sherd, the attributes of interest are the forming and firing of the ceramic vessel. Second, the attributes must be explicable by the method chosen to date their arrangement. In the case of ceramic assemblages, the attributes of interest relate to the date of deposition, *i.e.,* when did the set of sherds come together as a group. An applicable method for dating an assemblage of sherds, therefore, is one that explains depositional attributes. Third, just as there are an infinite number of attributes for any object, there are also a large number of ways to describe the selected event attributes (Dean 1978; Dunnell 1981, 2000; Dunnell and Readhead 1989). For example, for any set of attributes, the "event" can be represented by a mean, median, or range of time at which the attributes were added (Dunnell 1981, 2000). Since each aggregation of attributes describes a different event and thus "date," the aggregation method must be the same for each assemblage for which we generate dates.

Radiocarbon (^{14}C) dating is the method most commonly used by archaeologists. Consequently, ^{14}C dating might be considered the first choice for dating ceramic assemblages. At present, only four ^{14}C dates exist for the St. Francis and Memphis area sites (Table 9.1) all of wood charcoal. While these dates suggest that the occupations for the sites overlap, their relationship to the ceramic assemblages is unspecifiable. The event dated in ^{14}C dating is the isolation of organic material from the atmospheric carbon reservoir,

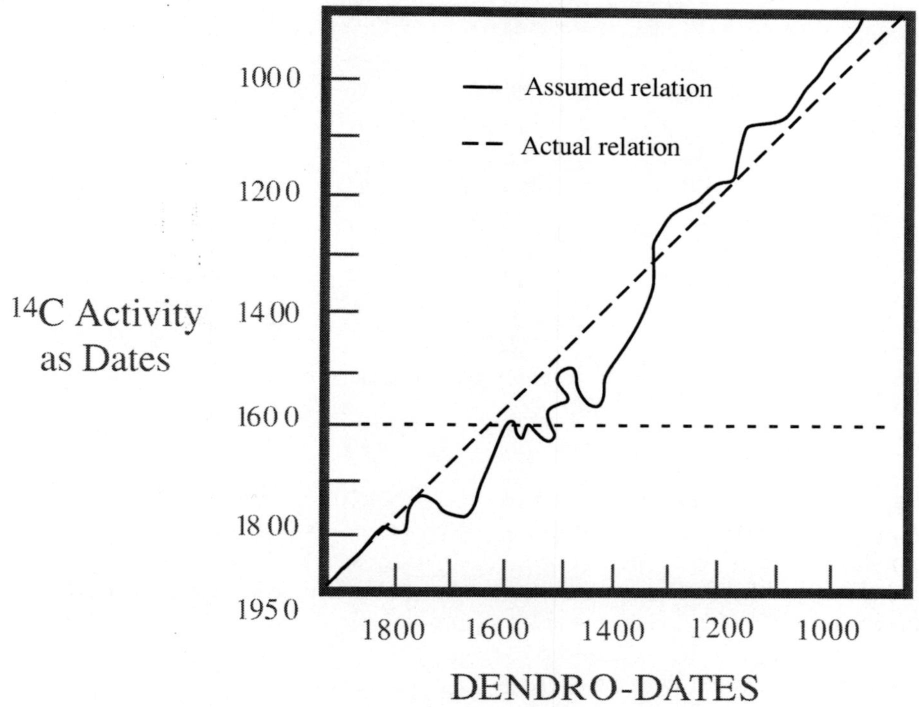

Figure 9.6. Calibration Curve for ^{14}C Dates.

Figure is derived from Stuiver *et al.* (1998). A given 14C activity may correspond to multiple dendrochronological dates. For example, a 1600 14C date corresponds to three calendar ages.

Table 9.1. Radiocarbon Dates for Memphis and St. Francis Area Deposits in the Current Study.

Site (Date)	Site #	C-14 Age B.P.	Calibrated Date (A.D)	Number of Intercepts	Calibrated Intercepts (A.D., 2 σ)[a]	Reference
Beck (TX-6844)	3CT8 13-O-5	450±60	1441	3	1334-1337 1399-1522 1571-1627	Buckner 1999
Belle Meade (GX-14768)	3CT30 13-O-7	705±195	1289	2	977-1528 1551-1633	Dye and Moore 1989
Belle Meade (SMU-2157)	3CT30 13-O-7	530±40	1414	2	1320-1351 1389-1423	Unpublished AAS files
Rose (CWRU-172)	12-N-3 3CS27	900±65	1158	2	1003-1010 1016-1275	Brose 1976

[a]Calculated using Version 4.1 of the University of Washington Quaternary Isotope Lab Radiocarbon Calibration Program (Stuiver *et al.* 1998).

usually death in plants and animals. House timbers may have been salvaged from earlier buildings one or more times, before being used in the building in which they are found or turned into firewood. The wood may have been acquired as deadwood in the first place (Dunnell 2000; Schiffer 1987). In addition, wood can vary in age based on the location from which the sample is derived; the center portion of a tree, for example, can be centuries older than the outer areas. With ^{14}C the temporal relationship between the dated event and the deposition of the ceramics (the event of interest) is unknown. Determining this relationship requires independent data.

The correlation between ^{14}C dates and depositional events is particularly elusive in late prehistoric deposits of the Mississippi river valley for several reasons. First, the nature of the deposit makes it particularly difficult to confidently assert an association between individual objects. St. Francis-type sites consist of large amounts of heavily modified and reshuffled midden material. While often stratified, the stratification orders midden management events not artifact manufacture or deposition (Dunnell 2000). Proximity is not a reliable indicator of temporal association. Dates generated on carbon samples, for example, can differ by hundreds or even thousands of years from the date of deposition of a nearby sherd. More recent charcoal may have been introduced from the surface or older charcoal been moved up from deeper in the deposit. A more direct means of dating the events related to the formation of ceramic assemblages is needed.

In addition to these problems, correctly calibrating ^{14}C dates is extremely difficult for the most recent several hundred years due to the changes in the ^{14}C reservoir (Stuiver *et al.*1998). The calibration for ^{14}C dates for the past several hundred years is presented in Figure 9.6. For more recent ^{14}C dates (*i.e.,* A.D. 1400-A.D. 1800), more than one calendrical date is associated with each level of measurement of ^{14}C. A ^{14}C date of 1800, for example, corresponds to three potential calendrical dates ranging from A.D. 1680 to 1850 (Ramenofsky 1987; Renfrew and Clark 1974). During the late prehistoric and historic period, the variability in ^{14}C in atmospheric CO_2 is significant (Browman 1981). Equal amounts of ^{14}C may represent several calendrical dates. In the case of the St. Francis and Memphis area ^{14}C dates, there are as many as three calendrical dates for each calibrated age (Table 9.1). Further, the 95% confidence intervals for these three dates are statistically indistinguishable.

Other than being able to say that each sample dates to the late prehistoric period, no temporal conclusions can be drawn from these ^{14}C dates. The inability to relate the dated event with the event of interest and the inability to statistically discriminate dates from late prehistory make the ^{14}C dating method inappropriate for dating ceramic assemblages from late Mississippian, St. Francis-type sites.

Luminescence Dating

Since we are interested in the formation of a ceramic assemblage, methods that date the manufacturing and use of the ceramic objects provide a closer relationship between the target event and the dated event. Luminescence is particularly well suited for the dating of ceramics since the method measures the time elapsed since vessels were heated during manufacturing and/or use (Aitken 1985, 1989; Dunnell 2000; Dunnell and Feathers 1994; Dunnell and Readhead 1989). Luminescence dating is a well established, if little used (in the United States), dating method that avoids the problems of associating dated events and the events of interests for sherds (Dunnell and Feathers 1994) and thus is especially relevant to dating St. Francis area deposits. It is also immune to the problems related to variance in ^{14}C production.

In order to obtain a measurement, luminescence dating requires that one estimate the dose of radiation derived from the internal matrix of the sherd and the environmental dose to which the sample was subjected. Environmental radiation can be measured by evaluating samples of sediment matrix surrounding the sherd for amounts of potassium, uranium, and thorium. Since the selection of the sample for dating depends on subsequent stylistic analysis, it was not possible to collect sediments for each sherd *a priori*. Instead, I collected a minimum of five sediment samples from locations across each deposit in the study to be used for evaluating radioactivity. While the homogeneous composition of St. Francis-type deposits suggests that variability in sediment radioactivity is minimized, this hypothesis can be evaluated by examining whether differences exist between samples. If no significant differences are found to exist across the deposit, then samples can be evaluated using the closest available sample. In the case of significant differences, sherd samples will have to be limited to only those immediately proximate to the sediment sample.

There are two sets of measurements involved in luminescence dating: the radioactivity of the sherd and the sample from which is derived the dose rate and the luminescence signal, and its sensitivity, from which is derived the equivalent dose. The equivalent dose is the amount of accumulated dose in the sherd through time and is determined by calibrating the natural luminescence signals against signals resulting from laboratory irradiations. Measuring the equivalent dose is probably the largest source of random error, although it contributes systematic error as well. The dose rate, on the other hand, is more prone to systematic error.

Because luminescence dating requires measurement of several variables, precision is usually poorer than radiocarbon. Errors are typically 8-10 percent. Precision can be improved by interval scale dating where the interest is not in the absolute date but in the differences in ages among sherds. This is accomplished by eliminating systematic error, where this can be assumed to be of the same magnitude and direction for all sherds (Dunnell and Feathers 1994; Feathers 1997:52). If, for example, one could assume that the environmental dose rate was the same for all sherds in a deposit, associated error would cancel out. By systematically assigning variability in radiation calibration across all samples, the error term associated with dates is reduced because error is applied to the *difference* rather than absolute amounts. A 1.0% error in radiation calibration translates into an error term of ±6 years on 600-800 year old ceramics.[1] However, in the comparison of 600 and 800-year dates the same error term is reduced to only ±2 years.[2] Thus, the use of luminescence interval scale dating can aid in making detailed comparisons between assemblages.

Generating Luminescence Dates for Assemblage Events

Generating dates for ceramic assemblages, however, still requires understanding the relationship between the dated events and the target events. Luminescence dates the manufacturing/use events of the ceramic vessels from which the sherds are derived. Although the relationship between the luminescence-dated event (exposure to ≥ 450°C) and the target event (the formation of the ceramic assemblage) is more direct than in the case of ^{14}C dating, there is still an unknown amount of time separating each event. First, assemblages are a collection of objects and the attributes for the assemblage deposition event consist of the aggregation of all sherds into the assemblage. Assemblages, like individual objects, have a temporal range as well as a mean date that is determined by the rate of individual object deposition (Figure 9.7; Dunnell 1981). The luminescence dating of a single sherd in a ceramic assemblage, therefore, will not necessarily produce an accurate estimate for the assemblage deposition event any more than a single ^{14}C date. A large sample of luminescence dates is needed to generate the temporal distribution of each assemblage. However, the temporal characterization of all culture-historical types in a large assemblage potentially requires hundreds of luminescence dates. The absolute number depends on the characteristics of the rate of deposition and how quickly a statistical approximation of the temporal distribution can be generated.

Second, the amount of time from the manufacturing/use event that sets the luminescence clock for any individual sherd to its addition into an archaeological deposit can vary. If one kind of vessel, for example, was systematically in use longer before being added to a deposit than another kind of vessel, then luminescence dates obtained from sherds of the first vessel will be early relative to the dates of the second vessel even though the depositional events are contemporaneous. In general, however, pots are extremely fragility and differences in use-life are much smaller than the differences in age of interest. Thus, one can eliminate this source of variability by considering the general function, thickness, and form of the vessel from which the sherds are derived. Sherds should be derived from vessels that have equivalent use and are similar in terms of fragileness. Finally, one must carefully select the sherds to be dated and consider the mean and range of age distributions for historical types. For example, in order to control temporal factors in type frequencies it is necessary to generate dates from only decorated sherds that are assigned to a single culture-historical type and that have a distribution that is narrow in temporal range. Sherds of culture-historical types with short durations are good choices for controlling for temporal variability since a statistical description can be

---

[1] In this case, 1.0% error multiplied by 600 years = ±6 years.

[2] If we interested in comparing two dates (*e.g.,* 800 and 600 years), then a 1% error is only ± 2 years since 800 – 600 = 200 * 1.0% = 2 years.

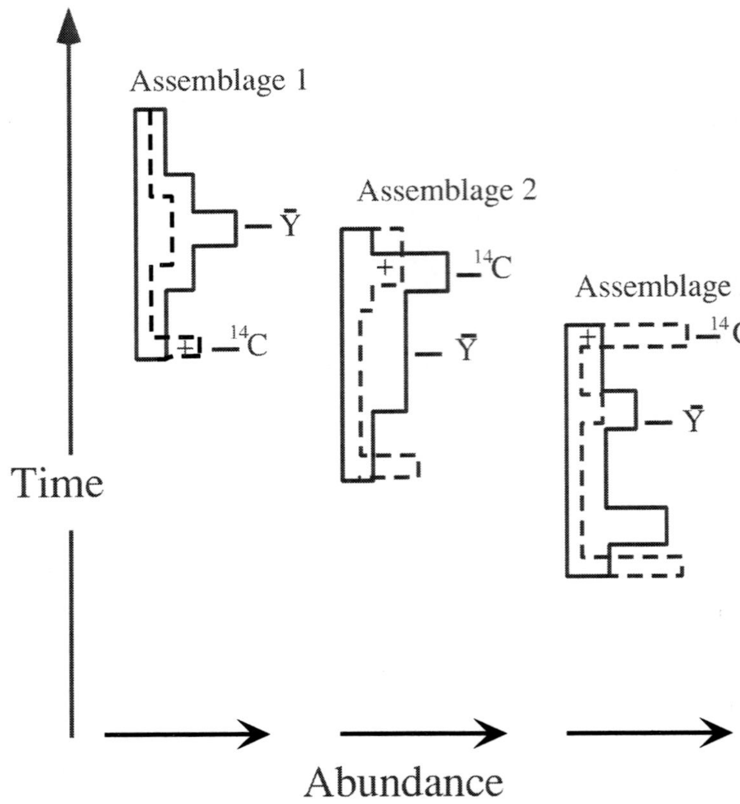

Figure 9.7. Comparison of Frequency Seriation and Radiocarbon Orders.

Figure is derived from Dunnell (1981:89). Accumulation histories for three assemblages (1 - 3) for ceramics (solid lines) and carbon (dotted lines). The actual carbon sample dated is indicated with a "+". Frequency seriation would provide an order of 1, 2, 3; radiocarbon dating would arrange them 1, 3, 2. Since radiocarbon dating treats a discrete element, the actual sample could have originated anytime during the events in question and thus a different set of samples might yield a different order. Averaging a large set of dates would still fail to produce identical orders because the underlying accumulation histories differ.

produced of a very short duration with fewer samples than one with a long duration.

Thus, the culture-historical type that should be dated will minimize variability in time between manufacture and deposition and have a narrow temporal range. These conditions are met by sherds from a single culture-historical type that has a relatively narrow distribution in time and that is associated with a single vessel size and form. PFG types that meet these conditions include Walls Engraved and Fortune Noded. Sherds of each of these types are found in relatively high abundance at each of the study assemblages, are easily and unambiguously recognizable, appear to be limited in terms of temper composition and vessel form, and have a brief period of occurrence late in prehistory.

Comparing assemblages involves systematically examining the simultaneous effects of time, space, and geography. First, we examine the correlation between determined age and type frequency. Samples of the same age will have no significant correlation between age and frequency. Samples that are chronologically distant, on the other hand, will have a significant correlation between age and frequency. In this way, we can evaluate chronological hypotheses about the relationships between assemblages. For example, if differences in Fortune Noded abundance in two assemblages cannot be found accountable by differences in time between each assemblage then we must account for this variability as transmission across space. Assemblage similarity must then be considered as the effects of space and geography on transmission patterns.

Preliminary Luminescence Dating Results

With the assistance of Dr. James Feathers at the University of Washington Luminescence Laboratory, I have initiated luminescence dating of Mississippi valley ceramics in the St. Francis and Memphis areas. Four dates have been generated so far: one each for Beck-A, Belle Meade, Holden Lake, and Rose Mound (Table 9.2). The dates for Belle Meade and Beck-A (both on Walls Engraved sherds) are, for all practical purposes, contemporaneous and date to late prehistory (just overlapping Hernando de Soto's arrival into the valley in AD 1542). The Rose Mound sherd, on the other hand, is very late (*i.e.,* AD 1709) and may have been impacted by factors related to post-manufacturing burning, dose fading that was not detected in the luminescence measurements, disequilibrium, or some other systematic error in the dose rate (J. Feathers, personal communication). Holden Lake, as suggested by the seriation results, is early. More dates, however, are needed to evaluate the significance of these assessments.

The luminescence dates, however, help resolve the ambiguity of ^{14}C samples. Beck is best represented chronologically by the 16th century intercept (the A.D. 1571-1627 date). Belle Meade is also best represented by the later intercept (A.D. 1551-1633). The Rose Mound chronology, however, is still unclear. Further luminescence dates will help resolve this problem.

A FINAL NOTE

The magnitude of the empirical, technical, methodological, and theoretical problems facing archaeologists working to build a historical science is daunting. Some archaeologists and philosophers of science have thrown up their hands and have declared that the effort is too complex for contemporary imagination, and may forever lie beyond our reach. The post-processualists, for example, fueled by the rather spectacular failures of the new archaeology to provide an empirical means by which to evaluate conclusions, argue against science altogether as a means evaluating knowledge of the past (*e.g.,* Hodder 1984, 1986, 1989; Hodder *et al.* 1995; Shanks and Hodder 1995; Shanks and Tilley 1987). The rejection of science currently threatens to render all of the social sciences irrelevant. If no empirical standard is possible, why should archaeologists be paid to tell stories about the past? If anything, the post-processualists' critique

Table 9.2. Luminescence Dates for Lipo Assemblages.

Assemblage	Age ± 1 σ Error	Culture-Historical Type
Beck - A	AD 1539 – 1639	Walls Engraved
Belle Meade	AD 1462 – 1594	Walls Engraved
Holden Lake	AD 1055 – 1235	Fortune Noded[1]
Rose Mound	AD 1637 – 1781	Fortune Noded

[1]The first luminescence dates (Beck and Belle Meade) were made prior to the completion of the initial ceramic analyses. In Beck and Belle Meade Walls Engraved sherds were dated. Walls Engraved was ubiquitous in PFG assemblages and it was assumed it would be in my samples. In the Holden Lake and Rose Mound samples, however, no sherds were assigned to Walls Engraved. Consequently, these dates were made from Fortune Noded sherds. All of the Lipo assemblages have sherds assigned to Fortune Noded; future luminescence dates will focus on this type.

warns of the urgency of resolving archaeology's central dilemma: is a scientific archaeology possible?

One primary goal of this research has been to demonstrate that if we are to build a cumulative body of knowledge about the past, it is not only feasible but also *necessary* to integrate the theoretical and empirical aspects of scientific inquiry to archaeology. This research also demonstrates that a scientific archaeology demands more of archaeological research than traditional wisdom dictates. Telling stories is a creative effort that requires nothing more than a comfortable chair and a good cup of tea. Building scientific knowledge, on the other hand, takes not only tremendous creativity but also a good deal of *hard work:* time, effort, attention to rigor, perseverance, daring, knowledge, and obsession for accuracy and precision. For some, what is gained in certainty of conclusions will seem to come as a loss to those pondering diverse or "interesting" questions about the past (*e.g.*, Wylie 1995:208-209). However, without a scientific archaeology as a system to build knowledge about the past, the product of archaeology will be reduced to "just-so stories" (Gould and Lewontin 1979). As Fox (1996:341) has eloquently described:

> we are here accepting the "spirit of science" as much as we quantify and use statistics. These are only relevant to certain kinds of hypotheses. If you are indeed interested in the "truth" – and whatever you might think, you really are – then you *must* use the scientific mode, at however low a level, to arrive at it. Evolution would not work otherwise. We would not be here to discuss the matter today. In our everyday thinking we are constantly testing and confirming and falsifying hypotheses; this, more than "conditioning" explains how we behave as we do. We are natural *natural* scientists; we have no other choice.

Science *is* a means by which we gather information about the way world works. Rejecting science as a goal for archaeology dooms the discipline to a quagmire of relativistic, individualistic, and non-cumulative stories about the past. It is entirely possible that the more pessimistic critics will be right about archaeology. However, there is only one way to find out. If archaeology is to become a science, we must be willing to persevere and to take on the required hard work. But, as Wilson (1998:209) points out, "the more forbidding the task, the greater the prize for those who dare to undertake it.

REFERENCES CITED

Abbott, D. N.
1972 The Utility of the Concept of Phase in the Archaeology of the Southern Northwest Coast. *Syesis* 5:267-278.

Ahouse, J.
1998 The Tragedy of A Priori Selectionism: Dennett and Gould on Adaptationism. *Biology and Philosophy* 13:359-391.

Aitken, M. J.
1985 *Thermoluminescence Dating*. Academic Press, London.
1989 Luminescence Dating: A Guide for Non-Specialists. *Archaeometry* 31:147-159.

Aldenderfer, M. S. and R. K. Blashfield
1984 *Cluster Analysis*. Sage Publications, Beverly Hills.

Allen, M. S.
1996 Style and Function in East Polynesian Fish-Hooks. *Antiquity* 70:97-116.

Ascher, R.
1961 Analogy in Archaeological Interpretation. *Southwestern Journal of Anthropology* 17:317-325.

Autin, W. J., S. F. Burns, B. J. Miller, R. T. Saucier and J. I. Snead
1991 Quaternary Geology of the Lower Mississippi Valley. In *Quaternary Nonglacial Geology; Coterminous U.S.*, edited by R. B. Morrison, pp. 547-582. The Geology of North America, Volume K-2. Geological Society of America, Boulder.

Barton, C. M. and G. A. Clark (editors)
1997 *Rediscovering Darwin: Evolutionary Theory and Archeological Explanation*. Archeological Papers of the American Anthropological Association, No. 7. American Anthropological Association, Washington.

Beals, R. L., G. W. Brainerd and W. Smith
1945 Archaeological Studies in Northeast Arizona. A Report on the Archaeological Work of the Rainbow Bridge-Monument Valley Expedition. *University of California Publications in American Archaeology and Ethnology,* Volume 4:1-236.

Beck, C.
1995 Style, Function, Technology, and Chronological Units. Paper presented at the 60th Annual Meeting of the Society for American Archaeology, Minneapolis.

Benn, D. W. (editor)
1992 *Excavations at the Moon Site (3PO488), A Middle Mississippian Village in Northeastern Arkansas*. Center for Archaeological Research, Southwest Missouri State University, Springfield.

Berry, K. J., K. L. Kvamme and P. W. J. Mielke
1980 Permutation Techniques for the Spatial Analysis of the Distribution of Artifacts into Classes. *American Antiquity* 45:55-59.
1983 Improvements in the Permutation Test for the Spatial Analysis of the Distribution of Artifacts into Classes. *American Antiquity* 48:547-553.

Best, M. B.
1997 Models for Interacting Populations of Memes: Competition and Niche Behavior. In *Proceedings of the Fourth European Conference on Artificial Life*, edited by P. Husbands and I. Harvey, pp. 154-162. MIT Press/Bradford Books, Cambridge.

Best, M. L. and R. Pocklington
1999 Meaning as Use: Transmission Fidelity and Evolution in NetNews. *Journal of Theoretical Biology* 196:389-395.

Bettinger, R. L.
1991 *Hunter-Gatherers: Archaeological and Evolutionary Theory*. Plenum, New York.

Binford, L.
1962 Archaeology as Anthropology. *American Antiquity* 28:217-225.
1965 Archaeological Systematics and the Study of Culture Process. *American Antiquity* 31:203-210.
1967 Smudge Pits and Hide Smoking: The Use of Analogy in Archaeological Reasoning. *American Antiquity* 32:1-12.
1968a Archeological Perspectives. In *New Perspectives in Archaeology*, edited by L. Binford, pp. 5-32. Academic Press, New York.
1968b Methodological Considerations of the Archaeological Use of Ethnographic Data. In *Man the Hunter*, edited by R. Lee and I. DeVore, pp. 268-273. Aldine, Chicago.
1977 General Introduction. In *For Theory Building in Archaeology*, edited by L. R. Binford, pp. 1-10. Academic Press, New York.
1978 *Nunamiut Ethnoarchaeology*. Academic Press, New York.
1980 Willow Smoke and Dog's Tails: Hunter-Gatherer Settlement Systems and Archaeological Site Formation. *American Antiquity* 45:4-20.
1981 Behavioral Archaeology and the 'Pompeii Premise.' *Journal of Anthropological Research* 37:195-208.

1986 An Alyawara Day: Making Men's Knives and Beyond. *American Antiquity* 51: 547-562.

Blackmore, S.
1999 *The Meme Machine*. Oxford University Press, New York.

Blute, M.
1979 Sociocultural Evolutionism. *Behavioral Science* 24:46-59.

Boas, F.
1896 The Limitations of the Comparative Method in Anthropology. *Science* 4:901-907.
1902 Some Problems in North American Archaeology. *American Journal of Archaeology* 6:1-6.
1909 *Anthropology*. Columbia University Press, New York.

Bonner, J. T.
1980 *The Evolution of Culture in Animals*. Princeton University Press, Princeton.

Boone, J. and E. A. Smith
1998 Is it Evolution Yet? A Critique of Evolutionary Archaeology. *Current Anthropology* 39:S141-S174.

Bordes, F.
1952 Stratigraphic du Loess et Evolution des Industries Paleolithiques Dans l'Ouest du Basin de Paris. *L'Anthropologie* 56:405-452.

Boyd, R. and P. J. Richerson
1985 *Culture and the Evolutionary Process*. University of Chicago Press, Chicago.

Brackenridge, H.
1818 On the Population and Tumuli of the Aborigines of North America. In a Letter from H. H. Brackenridge, Esq. to Thomas Jefferson, Read Oct. 1, 1813. *Transactions of the American Philosophical Society* 1:151-159.

Brain, J. P.
1978 The Archaeological Phase: Ethnographic Fact or Fancy? In *Archaeological Essays in Honor of Irving B. Rouse*, edited by R. C. Dunnell and E. S. Hall, pp. 311-318. Mouton, The Hague.
1985 Introduction: Update of De Soto Studies Since the United States De Soto Expedition Commission Report. In *Final Report of the United States De Soto Commission*, edited by J. R. Swanton, pp. xi-xix. Smithsonian Institution Press, Washington.
1989 *Winterville: Late Prehistoric Culture Contact in the Lower Mississippi Valley*. Mississippi Department of Archives and History Archaeological Report 23, Jackson.

Brauchli, K., T. Killingback and M. Doebeli
1999 Evolution of Cooperation in Spatially Structured Populations. *Journal of Theoretical Biology* 200:405-417.

Brose, D.
1976 Some Mississippian Ceramics from Arkansas, in the Cleveland Museum of Natural History. *Kirtlandia* 21:1-11.
1980 How Capt. Riggs Hunts for Mound Builders' Relics: an Historical Investigation of Some Influences on C.B. Moore. *Southeastern Archaeological Conference, Bulletin* 22:145-52.

Broughton, J. M. and J. F. O'Connell
1999 On Evolutionary Ecology, Selectionist Archaeology, and Behavioral Archaeology. *American Antiquity* 64:153-165.

Browman, D. L.
1981 Isotopic Discrimination and Correction Factors in Radiocarbon Dating. In *Advances in Archaeological Method and Theory 4*:241-295.

Brown, C.
1926 *Archeology of Mississippi*. University of Mississippi, Mississippi Geological Survey, Oxford.

Brown, I. W.
1998 *Decorated Pottery of the Lower Mississippi Valley: A Sorting Manual*. Alabama Museum of Natural History, Tuscaloosa.

Buckner, C. A.
1999 *Final Report: Three Buried Mississippian Sites on the Kochititzky Ditch No. 1: Data Recovery Excavations at the Kotchtitzky Ditch (3MS599), Perry Dixon (3MS600), & John's Lake (3MS601) Sites, Mississippi County, Arkansas*. Panamerican Consultants, Memphis.

Butzer, K.
1971 *Environment and Archaeology: An Ecological Approach to Prehistory*. Aldine, Atherton, NY.

Cavalli-Sforza, L. L. and M. W. Feldman
1981 *Cultural Transmission and Evolution*. Princeton University Press, Princeton.

Chang, K.-C.
1967 *Rethinking Archaeology*. Random House, New York.

Chapman, C. H.
1989 *The Archaeology of Missouri, Volume II*. University of Missouri Press, Columbia.

Childress, M. R.
1992 Mortuary Vessels and Comparative Ceramic Analysis: An Example from the Chucalissa Site. *Southeastern Archaeology* 11:31-50.

Clarke, D. L.
1968 *Analytical Archaeology*. Methuen, London.

Close, A.
1978 The Identification of Style in Lithic Artefacts. *World Archaeology* 10:223-237.

Cochran, W. G.
1975 *Sampling Techniques*. Wiley, New York.

Collins, H. B., Jr.
1927a Archeological Work in Louisiana and Mississippi. *Smithsonian Miscellaneous Collections* 78:200-207.
1927b Potsherds from Choctaw village sites in Mississippi. *Journal of the Washington Academy of Sciences* 17:259-263.

Colton, H. S.
1953 *Potsherds: An Introduction to the Study of Prehistoric Southwestern Ceramics and Their Use in Historic Reconstruction*. Bulletin, no. 25, Museum of Northern Arizona, Flagstaff.

Colton, H. S. and L. L. Hargrave
1937 Handbook of Northern Arizona Pottery Wares. *Museum of Northern Arizona, Bulletin* 11.

Conant, A. J.
1878 Archaeology of Missouri. *Transactions of the Academy of Science of St. Louis* 3:353-368.
1879 *Foot-prints of Vanished Races in the Mississippi Valley; Being an Account of Some of the Monuments and Relics of Pre-Historic Races Scattered over its Surfaces, with Suggests as to Their Origin and Uses*. St. Louis.

Conkey, M. W. and C. A. Hastorf
1990 Experimenting with Style in Archaeology: Some Historical and Theoretical Issues. In *The Uses of Style in Archaeology*, edited by M. W. Conkey and C. A. Hastorf, pp. 5-17. Cambridge University Press, Cambridge.

Cordell, L. and F. Plog
1979 Escaping the Confines of Normative Thought: A Reevaluation of Puebloan Prehistory. *American Antiquity* 44:405-429.

Cordell, L. S. and V. J. Yannie
1991 Ethnicity, Ethnogenesis, and the Individual: a Processual Approach toward Dialogue. In *Processual and Postprocessual Archaeologies: Multiple Ways of Knowing the Past*, edited by R. Preucel, pp. 96-107. University of Southern Illinois Press, Carbondale.

Crow, J. F. and M. Kimura
1970 *An Introduction to Population Genetics Theory*. Harper Row, New York.

Daniel, P.
1977 *Deep'N As It Come: The 1927 Mississippi River Flood*. Oxford University Press, New York.

Darwin, C.
1859 *On the Origin of Species By Means of Natural Selection, or the Preservation of Favoured Races in the Struggle for Life*. John Murray, London.

Davis, D. D.
1981 Ceramic Classification and Temporal Discrimination: a Consideration of Later Prehistoric Stylistic Change in the Mississippi River Delta. *Midcontinental Journal of Archaeology* 6:53-89.

Davis, M. B.
1976 Pleistocene Biogeography of Temperate Deciduous Forests. *Geoscience and Man* XIII:13-26.
1985 Climatic Instability, Time Lags, and Community Disequilibrium. In *Community Ecology*, edited by J. Diamond and T. J. Case, pp. 269-284. Harper and Row, New York.

Dawkins, R.
1976 *The Selfish Gene*. Oxford University Press, New York.

Deacon, T. W.
1997 *The Symbolic Species: The Co-evolution of Language and The Brain*. W.W. Norton and Company, New York.

Dean, J. S.
1978 Independent Dating in Archaeological Analysis. *Advances in Archaeological Method and Theory 1*: 223-255.

DeBoer, W. R.
1990 Interaction, Imitation, and Communication as Expressed in Style. In *The Uses of Style in Archaeology*, edited by M. W. Conkey and C. A. Hastorf, pp. 82-104. Cambridge University Press, Cambridge.

Deetz, J.
1965 *The Dynamics of Stylistic Change in Arkira Ceramics*. Illinois Studies in Anthropology, Urbana.

Deetz, J. and E. N. Dethlefsen
1965 The Doppler-effect and Archaeology: a Consideration of the Spatial Aspects of Seriation. *Southwestern Journal of Anthropology* 21:196-206.
1971 Some Social Aspects of New England Colonial Mortuary Art. *Memoirs of the Society for American Archaeology* 25:30-38.

Dellinger, S. C. and S. D. Dickinson
1940 Possible Antecedents of the Middle Mississippian Ceramic Complex in Northeastern Arkansas. *American Antiquity* 6:133-47.

Dennett, D. C.
1995 Darwin's Dangerous Idea. *The Sciences* 35:34-40.

Dickinson, S. D. and S. C. Dellinger
1940 A Survey of the Historic Earthenware of the Lower Arkansas Valley. *Texas Archeological and Paleontological Society, Bulletin* 12:76-96.

Dobzhansky, T.
1951 Mendelian Populations and Their Evolution. In *Genetics in the 20th Century*, edited by L. C. Dunn, pp. 573-589. Macmillian, New York.

Dougherty, R. P.
1929 *Nabonidus and Belshazzar: a Study of the Closing Events of the Neo-Babylonian Empire.* Yale University Press, New Haven.

Dunnell, R. C.
1970 Seriation Method and its Evaluation. *American Antiquity* 35:305-319.
1971 *Systematics in Prehistory*. Free Press, New York.
1978a Style and Function: A Fundamental Dichotomy. *American Antiquity* 43:192-202.
1978b Archaeological Potential of Anthropological and Scientific Models of Function. In *Archaeological Essays in Honor of Irving Benjamin Rouse*, edited by R. C. Dunnell and E. S. Hall, pp. 41-73. Mouton, The Hague.
1980 Evolutionary Theory and Archaeology. *Advances in Archaeological Method and Theory 3*:35-99.
1981 Seriation, Groups, and Measurement. In *Manejos de Datos Y Methods Mathematicos de Arqualogia*, edited by G. L. Cowgill, R. Whallon and B. S. Ottaway, pp. 67-90. Union Internacional de Ciences Prehistoricas y Protohistoricas, Mexico DF.
1982 Science, Social Science and Common Sense: The Agonizing Dilemma of Modern Archaeology. *Journal of Anthropological Research* 38:1-25.
1983 Aspects of the Spatial Structure of the Mayo Site (15-JO-14) Johnson County, Kentucky. In *Lulu Linear Punctated*, edited by R. Dunnell and D. Grayson, pp. 109-165. Museum of Anthropology, University of Michigan, Ann Arbor.
1985a Archaeological Survey in the Lower Mississippi Alluvial Valley, 1940-1947: A Landmark Study in American Archaeology. *American Antiquity* 50:297-300.
1985b Methodological Issues in Contemporary Americanist Archaeology. In *PSA 1984, Volume 2*, edited by P. D. Asquith and P. Ketcher, pp. 717-744. Philosophy of Science Association, East Lansing.
1986 Methodological Issues in Americanist Artifact Classification. *Advances in Archaeological Method and Theory 9*: 149-207.
1989 Aspects of the Application of Evolutionary Theory in Archaeology. In *Archaeological Thought in America*, edited by C. C. Lamberg-Karlovsky, pp. 35-99. Cambridge University Press, Cambridge.
1992 Archaeology and Evolutionary Science. In *Quandaries and Quests: Visions of Archaeology's Future*, edited by L. Wandsnider, pp. 209-224. Center for Archaeological Investigations, Occasional Paper No. 20. Southern Illinois University Press, Carbondale.
1995a What Is It That Actually Evolves? In *Evolutionary Archaeology*, edited by P. A. Teltser, pp. 33-50. University of Arizona Press, Tucson.
1995b Style: Concepts and Practice. Paper presented in the Symposium "Expressions of Ethnicity: Evolutionary and Historical Perspectives on Plains Indian Art," McClung Museum, University of Tennessee, 11 November 1995.
1996a Griffin's 'Cultural Change and Continuity'; the Synthesis of Time and Space. Paper presented at the First Annual G. R. Willey Symposium on the History of Archaeology, 61st Annual Meeting of the Society for American Archaeology, April 1996. New Orleans.
1996b Style: The Development of a Concept. Paper presented at the Department of Anthropology, University of Hawai'i, Honolulu, 19 March 1996.
2000 Dating. In *Archaeological Method and Theory: An Encyclopedia*, edited by L. Ellias, pp. 143-150. Garland, New York.

Dunnell, R. C. and S. K. Campbell
1977 *Aboriginal Occupation of Hamilton Island, Washington.* University of Washington, Department of Anthropology, Reports in Archaeology No. 4, Seattle.

Dunnell, R. C. and W. S. Dancey
1983 The Siteless Survey: A Regional Scale Data Collection Strategy. In *Advances in Archaeological Method and Theory 6*: 267-287.

Dunnell, R. C. and J. K. Feathers
1991 Late Woodland Manifestations of the Malden Plain, Southeast Missouri. In *Stability, Transformation, and Variation: the Late Woodland Southeast*, edited by M. S. Nassaney and C. R. Cobb. pp. 21-45. Plenum Press, New York.
1994 Thermoluminescence Dating of Surficial Archaeological Material. In *Dating in Exposed and Surface Contexts*, edited by C. Beck, pp. 115-137. University of New Mexico Press, Albuquerque.

Dunnell, R. C., L. H. Hansen, Jr. and D. L. Hardesty
1971 *The Woodside Component of the Slone Site, Pike County, Kentucky*. Southeastern Archaeological Conference, Bulletin 14, Morgantown.

Dunnell, R. C. and C. P. Lipo
1999 Complexity in Evolutionary Perspective. Paper Presented at the 1999 Annual Meeting of the Southeastern Archaeological Conference, Pensacola, Florida.

Dunnell, R. C. and M. L. Readhead
1989 The Relation of Dating and Chronology: Comments on Chatters and Hoover (1986) and Butler and Stein (1988). *Quaternary Research* 39:232-233.

Dunnell, R. C. and J. F. Simek
1995 Artifact Size and Plowzone Processes. *Journal of Field Archaeology* 22:305-319.

Dunnell, R. C., P. A. Teltser and R. Vercruysse
1986 Efficient Error Reduction in Large Data Sets. *Advances in Computer Archaeology* 3:22-37.

Dunnell, R. C. and R. J. Wenke
1980 An Evolutionary Model of the Development of Complex Societies. Paper Presented at the 1980 Annual Meeting of the American Association for the Advancement of Science, San Francisco, January.

Durham, W. H.
1976 The Adaptive Significance of Cultural Behavior. *Human Ecology* 4:89-121.
1979 Toward a Coevolutionary Theory of Human Biology and Culture. In *Evolutionary Biology and Human Social Behavior: An Anthropological Perspective*, edited by N. Chagnon and W. Irons, pp. 39-58. Duxbury Press, North Scituate.
1990 Advances In Evolutionary Culture Theory. *Annual Review of Anthropology* 19:187-210.
1991 *Coevolution: Genes, Culture, and Human Diversity*. Stanford University Press, Stanford.
1992 Applications of Evolutionary Culture Theory. *Annual Review of Anthropology* 21:331-355.

Dye, D. H.
1998 An Overview of Walls Engraved Pottery in the Central Mississippi Valley. In *Changing Perspectives on the Archaeology of the Central Mississippi Valley*, edited by M. J. O'Brien and R. C. Dunnell, pp. 80-98. The University of Alabama Press, Tuscaloosa.

Dye, D. H. and S. D. Moore
1989 A Domestic Structure from the Belle Meade Site (3CT30), Crittenden County Arkansas. Paper Presented at the 24th Annual Meeting of the Southern Anthropological Society, Memphis.

Eddington, A.
1953 *The Nature of the Physical World*. University of Michigan Press, Ann Arbor.

Efron, B.
1982 *The Jackknife, the Bootstrap and Other Resampling Plans*. SIAM, New York.

Efron, B. and R. Tibshirani
1993 *An Introduction to the Bootstrap*. Chapman and Hall, New York.

Eighmy, J. L. and J. M. LaBelle
1996 Radiocarbon Dating of Twenty-Seven Plains Complexes and Phases. *Plains Anthropologist* 41:53-69.

Engelbrecht, W.
1974 The Iroquois: Archaeological Patterning on the Tribal Level. *World Archaeology* 6:52-65.

Evers, E.
1880 *The Ancient Pottery of Southeastern Missouri. Contributions to the Archaeology of Missouri by the Archaeological Section of the St. Louis Academy of Science Part I*, pp. 21-30. G.A Bates, Salem.

Ewald, P. W.
1994 *Evolution of Infectious Disease*. Oxford University Press, New York.

Ewens, W. J.
1972 The Sampling Theory of Selectively Neutral Alleles. *Theoretical Population Biology* 3:87-112.

Feathers, J. K.
1990 *Explaining the Evolution of Ceramics in Southeast Missouri*. Unpublished Ph.D. Dissertation, Department of Anthropology, University of Washington, Seattle.
1997 The Application of Luminescence Dating in American Archaeology. *Journal of Archaeological Method and Theory* 4:1-66.

Fenneman, N. M.
1938 *The Physiography of the Eastern United States*. McGraw-Hill, New York.

Fisher, R. A.
1936 *Statistical Methods for Research Workers*. Oliver and Boyd, Edinburgh.

Fisk, H. N.
1944 *Geological Investigation of the Alluvial Valley of the Lower Mississippi River*. Mississippi River Commission Publications. No. 52. Mississippi River Commission, USACE, Waterways Experimental Station, Vicksburg.

Fitzhugh, W. W.
1987 Archaeological Ethnicity and the Prehistory of Labrador. In *Ethnicity and Culture*, edited by R. Auger, M. F. Glass, S. MacEachern and P. H. McCartney, pp. 141-153. Archaeological Association, University of Calgary.

Flannery, K. V.
1967 Culture History v. Cultural process: A Debate in American Archaeology. *Scientific American* 217:119-122.
1976 The Trouble with Regional Sampling. In *The Early Mesoamerican Village,* edited by K. Flannery, pp. 159-160. Academic Press, New York.
1986 A Visit to the Master. *Guila Naquitz: Archaic Foraging and Early Agriculture in Oaxaca, Mexico*, edited by K. V. Flannery, pp. 511-519. Academic Press, Orlando.

Ford, J. A.
1935 *Ceramic Decoration Sequence at an Old Indian Village Site, Near Sicily Island, Louisiana.* Department of Conservation, Louisiana State Geological Survey, Anthropological Study 1, New Orleans.
1936a *Analysis of Indian Village Site Collections from Louisiana and Mississippi.* Department of Conservation, Louisiana State Geological Survey, Anthropological Study 2. New Orleans.
1936b Archaeological Methods Applicable to Louisiana. *Proceedings of the Louisiana Academy of Sciences* 3:102-105.
1938a A Chronological Method Applicable to the Southeast. *American Antiquity* 3:260-264.
1938b *An Examination of Some Theories and Methods of Ceramic Analysis.* Unpublished Masters Thesis, University of Michigan, Ann Arbor.
1939 Archaeological Exploration in Louisiana During 1938. *Louisiana Conservation Review* 7:15-17.
1949 Cultural Dating of Prehistoric Sites in Virú Valley Peru. *American Museum of Natural History, Anthropological Papers* 5:263-266.
1951 *Greenhouse: A Troyville-Coles Creek Period site in Avoyelles Parish, Louisiana.* American Museum of Natural History, Anthropological Papers 44, Pt. 1. New York.
1952 *Measurements of Some Prehistoric Design Elements in the Southeastern States* Anthropological Papers of the American Museum of Natural History 44, Pt. 3. New York.
1954a Comment on A. C. Spaulding, "Statistical Techniques for the Discovery of Artifacts Types" *American Antiquity* 19:390-391.
1954b On the Concept of Types. *American Anthropologist* 56:42-54.
1961 In Favor of a Simple Typology. *American Antiquity* 27:113-114.
1962 *A Quantitative Method for Deriving Cultural Chronology.* Technical Manual, No. 1. Pan American Union, Washington.
1963 *Hopewell Culture Burial Mounds Near Helena, Arkansas.* American Museum of Natural History Anthropological Papers, American Museum of Natural History, 50, Pt. 1. New York.

Ford, J. A. and J. Griffin
1938 Report of the Conference on the Southeastern Pottery Typology. *Newsletter of the Southeastern Archaeological Conference* 7:10-22.

Ford, J. A. and G. I. Quimby
1945 *The Tchefuncte Culture, an Early Occupation of the Lower Mississippi Valley.* Society of American Archaeology, Memoirs No. 2. Menasha.

Ford, J. A. and G. R. Willey
1940 *Crooks Site, a Marksville Period Burial Mound in La Salle Parish Louisiana.* Department of Conservation, Louisiana State Geological Survey, Anthropological Study 3. New Orleans.
1941 An Interpretation of the Prehistory of the Eastern United States. *American Anthropologist* 43:325-363.

Fox, G. L.
1992 *A Critical Evaluation of the Interpretive Framework of the Mississippi Period in Southeast Missouri.* Unpublished Ph.D. Dissertation, Department of Anthropology, University of Missouri, Columbia.
1998 An Examination of Mississippian-Period Phases in Southeastern Missouri. In *The Archaeology of the Central Mississippi River Valley*, edited by R.C. Dunnell and M. J. O'Brien, pp. 31-58. University of Alabama Press, Tuscaloosa.

Fox, R.
1996 Art/Science in Anthropology. In *The Flight From Science and Reason*, edited by P. Gross, N. Levitt and M. Lewis, pp. 327-345. New York Academy of Science, New York.

Fry, R. E. and S. C. Cox
1974 The Structure of Ceramic Exchange at Tikal, Guatemala. *World Archeology* 6:209-225.

Fuller, J. W.
1981 *Developmental Change in Prehistoric Community Patterns: The Development of Nucleated Village Communities in Northern West Virginia.* Unpublished Ph.D. Dissertation, Department of Anthropology, University of Washington, Seattle.

Gabora, L.
1996 A Day in the Life of a Meme. *Philosophica* 57: 901-938.

Gifford, J. C.
1960 Type-Variety Method. *American Antiquity* 9:29-44.

Gillespie, J. H.
1991 *The Causes of Molecular Evolution.* Oxford University Press, New York.

Good, P.
1994 *Permutation Test: A Practical Guide to Resampling Methods for Testing Hypotheses.* Springer-Verlag, New York.

Gould, R.
1968a Living Archaeology: the Ngatatjara of Western Australia. *Southwestern Journal of Anthropology* 24:101-122.
1968b Chipping Stone in the Outback. *Natural History* 77:42-49.
1969 *Yiwara: Foragers of the Australian Desert.* Scribner, New York.
1978 Beyond Analogy in Ethnoarchaeology. In *Explorations in Ethnoarchaeology*, edited by R. Gould, pp. 249-293. University of New Mexico Press, Albuquerque.

Gould, R. A. and P. J. Watson
1982 A Dialogue on the Meaning and Use of Analogy in Ethnoarchaeological Reasoning. *Journal of Anthropological Archaeology* 1:355-381.

Gould, S. J.
1977 Darwin's Untimely Burial. In *Ever Since Darwin: Reflections in Natural History*, edited by S. J. Gould, pp. 39-48. Norton, New York.
1986 Evolution and the Triumph of Homology, or Why History Matters. *American Scientist* 74:60-69.
1989 *Wonderful Life: The Burgess Shale and the Nature of History*. W.W. Norton, New York.
1998 Writing in the Margins. *Natural History* 107:16-20.

Gould, S. J. and R. C. Lewontin
1979 The Spandrels of San Marco and the Panglossian Paradigm: A Critique of the Adaptationist Programme. *Proceedings of the Royal Society, London, B* 205:581-598.

Gould, S. J., D. M. Raup, J. J. Sepkoski, T. J. M. Schopf and D. S. Simberloff
1977 The Shape of Evolution: a Comparison of Real and Random Clades. *Paleobiology* 3:23-40.

Graves, M. B.
1994a Community Boundaries in Late Prehistoric Puebloan Society: Kalinga Ethnoarchaeology as a Model for the Southwestern Production and Exchange of Pottery. In *The Ancient Southwestern Community: Models and Methods for the Study of Prehistoric Social Organization*, edited by W. H. Wills and R. D. Leonard, pp. 149-170. University of New Mexico Press, Albuquerque.
1994b Kalinga Social and Material Cultural Boundaries. In *Kalinga Ethnoarchaeology*, edited by W. A. Longacre and J. M. Skibo, pp. 14-49. Smithsonian Institution Press, Washington.

Graves-Brown, P.
1996 In Search of the Watchmaker: Attribution of Agency in Natural and Cultural Selection. In *Darwinian Archaeologies*, edited by H. D. G. Maschner, pp. 165-179. Plenum Press, New York.

Grayson, D. K.
1989 Sample Size and Relative Abundance in Archaeological Analysis: Illustrations from Spiral Fractures and Seriation. In *Quantifying Diversity in Archaeology*, edited by R. D. Leonard and G. T. Jones, pp. 79-84. Cambridge University Press, Cambridge.

Griffin, J. B.
1938 The Ceramic Remains from Norris Basin, Tennessee. In *An Archaeological Survey of the Norris Basin in Eastern Tennessee*, edited by W. S. Webb, pp. 253-259. Bulletin 118. Bureau of American Ethnology, Washington.
1939 Report on the Ceramics of Wheeler Basin. In An Archaeological Survey of Wheeler Basin on the Tennessee River in Northern Alabama, edited by W. S. Webb, pp. 127-165. Bulletin 122. Bureau of American Ethnology, Washington.
1943 *The Fort Ancient Aspect in Cultural and Chronological Position in Mississippi Valley Archaeology*. University of Michigan Press, Ann Arbor.
1946 Cultural Change and Continuity in Eastern United States Archaeology. In *Man in Northeastern North America 3*, edited by F. Johnson, pp. 37-95. Peabody Foundation for Archaeology, Cambridge.
1951 Essays on Archaeological Methods: Proceedings of a Conference Held Under Auspices of the Viking Fund. University of Michigan Press, Ann Arbor.
1952 Prehistoric Cultures of the Central Mississippi Valley. In *Archeology of Eastern United States*, edited by J. B. Griffin, pp. 226-238. University of Chicago Press, Chicago.
1981 The Acquisition of a Little-Known Pottery Haul From the Lower Mississippi Valley. *Geoscience and Man* XXII:51-55.

Gulick, J. T.
1872 On Diversity of Evolution Under One Set of External Conditions. *Journal of the Linnean Society (Zoology)* 11:496-505.
1888 Divergent Evolution through Cumulative Segregation. *Journal of the Linnaean Society of London (Zoology)* 20:189-274.
1905 *Evolution, Racial and Habitudinal.* Carnegie Institution, Washington.

Hägerstrand, T.
1952 *The Propagation of Innovation Waves.* Department of Geography, Royal University of Lund, London.

Hallpike, C.
1986 *The Principles of Social Evolution.* Claredon Press, Oxford.

Hammond, N.
1972 Location Models and the Site of Lubaantun: A Classic Maya Center. In *Models in Archaeology*, edited by D. L. Clarke, pp. 257-300. Methuen, London.

Heizer, R. F. (editor)
1949 *A Manual of Archaeological Field Methods.* National Press, Palo Alto.
1950 *A Manual of Archaeological Field Methods, Revised Edition.* National Press, Palo Alto.

Hill, J. N.
1966 A Prehistoric Community in Eastern Arizona. *Southwestern Journal of Anthropology* 22:9-30.
1967 Structure, Function, and Change at Broken K Pueblo. *Fieldiana: Anthropology 59*:158-167.
1968 Broken K Pueblo: Patterns of Form and Function. In *New Perspectives in Archeology*, edited by S. R. Binford and L. R. Binford, pp. 103-142. Aldine, Chicago.
1977 Individual Variability in Ceramics and the Study of Prehistoric social Organization. In *The Individual in Prehistory*, edited by J. N. Hill and J. Gunn, pp. 55-108. Academic Press, New York.

Hodder, I.
1984 Burials, Houses, Women and Men in European Neolithic. In *Ideology, Power, and Prehistory*, edited by D. Miller and C. Tilley, pp. 51-68. Cambridge University Press, Cambridge
1986 *Reading the Past.* Cambridge University Press, Cambridge.
1989 *The Meanings of Things.* Unwin Hyman, London.

Hodder, I., M. Shanks, A. Alexandri, V. Buchli, J. Carman, J. Last and G. Lucas
1995 *Interpreting Archaeology: Finding Meaning in the Past.* Routledge, New York.

Holmes, W. H.
1886a Ancient Pottery of the Mississippi Valley. In *Fourth Annual Report of the Bureau of Ethnology for 1882-83*, pp. 367-436. Smithsonian Institution, Washington.
1886b Origin and Development of Form and Decoration in Ceramic Art. In *Fourth Annual Report of the Bureau of Ethnology for 1882-83*, pp. 437-465. Smithsonian Intitution, Washington.
1892 Modern Quarry Refuse and the Paleolithic Theory. *Science* 20:295-297.
1903 *Aboriginal Pottery of the Eastern United States.* Bureau of American Ethnology, 20th Annual Report. Smithsonian Institution, Washington.

House, J.
1991 Monitoring Mississippian Dynamics: Time, Settlement, and Ceramic Variation in the Kent Phase, Eastern Arkansas. Unpublished Ph.D. Dissertation, Department of Anthropology, Southern Illinois University, Carbondale.
1993 Dating the Kent Phase. *Southeastern Archaeology* 12:21-32.
1995 Investigating Mississippian Settlement Variability in the Lower St. Francis Basin, Eastern Arkansas. Paper presented at the 52nd Southeastern Archaeological Conference, Knoxville, Tennessee. November 9, 1995.

Hudson, C.
1985 De Soto in Arkansas: a Brief Synopsis. *Arkansas Archeological Society Field Notes* 205:3-12.

Hull, D.
1980 Individuality and Selection. *Annual Review of Ecology and Systematics* 11:311-332.
1988 *Science as Process.* University of Chicago Press, Chicago.

Hurt, J. D.
1999 Stylistic Variation as an Indicator of Interaction and Social Distance Among Kayenta Anasazi Groups. In *Explanations of Change: Case Studies in Evolutionary Archaeology*, edited by R. C. Dunnell and R. D. Leonard. University of Utah Press, Salt Lake City (submitted).

Irwin, G.
1992 *The Prehistoric Exploration and Colonization of the Pacific.* Cambridge University Press, New York.

Jacob, F.
1998 *Of Flies, Mice, and Men.* Harvard University Press, Cambridge.

Jakob, P. L.
1990 *Visions of a Flying Machine.* Smithsonian Institution Press, Washington.

Jelinek, A. J.
1976 Form, Function and Style in Lithic Analysis. In *Cultural Change and Continuity: Essays in Honor of James Bennett Griffin*, edited by C. E. Cleland, pp. 19-34. Academic Press, New York.

Jenkins, P. F. and A. J. Baker
1984 Mechanisms of Song Differentiation in Introduced Populations of the Chaffinches *Fringilla coelebs* in New Zealand. *Ibis* 126:510-524.

Johnson, K. W. and B. C. Nelson
1990 The Utina: Seriations and Chronology. *The Florida Anthropologist* 43:48-62.

Jones, G. T., D. K. Grayson and C. Beck
1983 Artifact Class Richness and Sample Size in Archaeological Surface Assemblages. In *Lulu Linear Punctated: Essays in Honor of George Irving Quimby*, edited by R. C. Dunnell and D. K. Grayson, pp. 55-73. Museum of Anthropology, University of Michigan, Ann Arbor.

Kaufmann, S.
1993 *Origins of Order: Self-Organization, Evolution and Natural Selection*. Oxford University Press, Oxford.

Kay, M.
1975 Social Distance among Central Missouri Hopewell Settlements: a First Approximation. *American Antiquity* 40:64-71.

Kidder, A. V.
1915 Pottery of the Pajarito Plateau and Some Adjacent Regions in New Mexico. *American Anthropological Association, Memoir* 2:407-462.
1917 A Design-Sequence from New Mexico. *National Academy of Sciences, Proceedings* 3:369-370.
1924 *An Introduction to the Study of Southwestern Archaeology with a Preliminary Account of the Excavations at Pecos*. Yale University Press, New Haven.

Kidder, M. A. and A. V. Kidder
1917 Notes on the Pottery of Pecos. *American Anthropologist* 19:325-360.

Kimura, M.
1968 Evolutionary Rate at the Molecular Level. *Nature* 217:624-626.
1977 Preponderance of Synonymous Changes as Evidence for the Neutral Theory of Evolution. *Nature* 267:275-276.
1983 *The Neutral Theory of Molecular Evolution*. Cambridge University Press, Cambridge.

Kimura, M. and F. Crow
1964 The Number of Alleles that Can Be Maintained in a Finite Population. *Genetics* 49:725-738.

Kimura, M. and T. Ohta
1971 *Theoretical Aspects of Population Genetics*. University of Princeton Press, Princeton, New Jersey.

King, A. and J. Freer
1995 The Mississippian Southeast: A World-Systems Perspective. In *Native American Interactions: Multiscalar Analyses and Interpretations in the Eastern Woodlands*, edited by M. S. Nassaney and K. E. Sassaman, pp. 266-288. The University of Tennessee Press, Knoxville.

King, J. L. and T. H. Jukes
1969 Non-Darwinian Evolution: Random Fixation of Selectively Neutral Mutations. *Science* 164:788-798.

Kintigh, K. W.
1984 Measuring Archaeological Diversity by Comparison with Simulated Assemblages. *American Antiquity* 49:44-54.

Klejn, L. S.
1973 On Major Aspects of the Interrelationship of Archaeology and Ethnology. *Current Anthropology* 14:311-320.

Kornbacher, K.
2000 Explaining Groundstone Tool Technology, in *Posing Questions for a Scientific Archaeology*, edited by T. Hunt, C.P. Lipo and S. Sterling. Greenwood Press, Westport (in press).

Krause, R. A.
1972 Toward a Formal Account of Bantu Ceramic Manufacture. In *Archaeological Essays in Honor of Irving B. Rouse*, edited by R. C. Dunnell and E. S. Hall, pp. 87-120. Mouton Publishers, New York.
1977 Taxonomic Practice and Middle Missouri Prehistory: A Perspective on Donald J. Lehmer's Contributions. *Plains Anthropologist, Memoir* 13:5-13.
1990 Ceramic Practice and Semantic Space - An Ethnoarchaeological Inquiry into the Logic of Bantu Potting. *Antiquity* 64:711-726.

Kreitman, M.
1996 The Neutral Theory is Dead. Long Live the Neutral Theory. *BioEssays* 18:678-682.

Krieger, A. D.
1944 The Typological Concept. *American Antiquity* 9:271-288.

Kroeber, A. L.
1916 Zuñi *Potsherds*. Anthropological Paper 18(1). American Museum of Natural History, New York.
1919 On the Principle of Order in Civilization as Exemplified by Changes of Fashion. *American Anthropologist* 21:235-263.

Lang, A. L.
1995 *Cultural Evolution in the Song of the Introduced Eurasian Tree Sparrow*. Unpublished PhD. Dissertation, Department of Zoology, University of Toronto, Toronto.

Lang, A. L. and J. C. Barlow
1997 Cultural Evolution in the Eurasian Tree Sparrow: Divergence Between Introduced and Ancestral Populations. *Condor* 99:413-423.

Lehmer, D. J.
1966 *Fire Heart Creek Site*. Publications in Salvage Archeology 1. Smithsonian Institution, River Basin Surveys, Lincoln.
1971 *Introduction to Middle Missouri Archeology*. Anthropological Papers 1. National Park Service, Washington.

Lemley, H. J. and S. D. Dickinson
1937 Archaeological Investigations on Bayou Macon in Arkansas. *Texas Archeological and Paleontological Society, Bulletin* 9:11-47.

Leonard, R. D. and G. T. Jones
1987 Elements of an Inclusive Evolutionary Model for Archaeology. *Journal of Anthropological Archaeology* 6:199-219.
1989 (editors) *Quantifying Diversity in Archaeology*. Cambridge University Press, Cambridge.

Lewontin, R. C.
1974 *The Genetic Basis of Evolutionary Change*. Columbia University Press, New York.

Lipo, C. P. and M. E. Madsen
1996 Problems and Solutions in the Study of Social Organization. Paper Presented at the 61st Annual Meetings for the Society for American Archaeology, New Orleans.
1998 Comment to Lyman and O'Brien: The Goals of Evolutionary Archaeology: History and Explanation. *Current Anthropology* 39:615-652.
1999 The Evolutionary Biology of Ourselves: Unit Requirements and Organizational Change in United States History. In *Evolutionary Explanations of the Archaeological Record*, edited by R.C. Dunnell and R. Leonard. University of Utah Press, Salt Lake City (submitted).
2000 Neutrality, Style and Drift: Building Methods for Studying Cultural Transmission in the Archaeological Record. In *Style and Function in Archaeology*, edited by G. Rakita and T. Hurt. Greenwood Press, Westport (in press).

Lipo, C. P., M. E. Madsen, T. Hunt and R. C. Dunnell
1997 Population Structure, Cultural Transmission and Frequency Seriation. *Journal of Anthropological Archaeology* 16:301-333.

Longacre, W. A.
1964 Sociological Implications of the Ceramic Analysis. In *Chapters in the Prehistory of Eastern Arizona Volume II*, pp. 155-167. Field Museum of Natural History, Chicago.
1970a *Archaeology as Anthropology: A Case Study*. Anthropological Papers of the University of Arizona No. 17. Tucson.
1970b Current Thinking in American Archaeology. *Bulletin of the American Anthropological Association* 3:126.

Longacre, W. A. and M. T. Stark
1992 Ceramics, Kinship and Space: A Kalinga Example. *Journal of Anthropological Archaeology* 11:125-136.

Love, M. W.
1993 Ceramic Chronology and Chronometric Dating: Stratigraphy and Seriation at La Blanca Guatemala. *Ancient Mesoamerica* 4:17-29.

Lowen, C. B. and R. I. M. Dunbar
1996 Cultural Evolution in a Structured Population. Unpublished Manuscript in Possession of Author.

Lumb, L. C. and C. H. McNutt
1988 *Chucalissa: Excavation in Units 2 and 6, 1957-1967*. Memphis State University, Archaeological Research Center, Occasional Paper No. 15, Memphis.

Lyman, R. L. and M. J. O'Brien
1998 The Goals of Evolutionary Archaeology: History and Explanation. *Current Anthropology* 39:615-652.
2000 Change and Diversity Diagrams. *Journal of Anthropological Archaeology*. 19:39-74.

Lyman, R. L., M. J. O'Brien and R. C. Dunnell
1997 *The Rise and Fall of Culture History*. Plenum Press, New York.

Lynch, A(aron).
1996 *Thought Contagion: How Belief Spreads Through Society*. Basic Books, New York.

Lynch, A(lejandro).
1996 The Population Memetics of Birdsong. In *Ecology and Evolution of Acoustic Communication in Birds*, edited by D. E. Kroodsma and E. H. Miller, pp. 181-197. Cornell University Press, Ithaca.

Lynch, A. and A. J. Baker
1993 A Population Memetics Approach to Cultural Evolution in Chaffinch Song: Meme Diversity within Populations. *American Naturalist* 141:597-620.
1994 A Population Memetics Approach to Cultural Evolution in Chaffinch Song: Differentiation among Populations. *Evolution* 48:351-359.

Lynch, A., G. M. Plunkett, A. J. Baker and P. F. Jenkins
1989 A Model of Cultural Evolution of Chaffinch Song Derived with the Meme Concept. *American Naturalist* 133:634-653.

Madsen, M. and C. Lipo
1993 Units of Evolution within a Darwinian Approach to Anthropology: Definitions and Implications for the Subfields. Paper Presented at the 92nd

Meeting of the American Anthropological Association, Washington.

Mainfort, R. C.
1995 Cluster Analysis and Archaeological Phases: an Example from the Central Mississippi Valley. Paper Presented at the Midwest Archaeological Conference, Beloit, Wisconsin.
1999 Late Period Phases in the Centeral Mississippi Valley: A Multivariate Approach. In *Arkansas Archaeology: Essays in Honor of Dan and Phyillis Morse,* edited by R. C. Mainfort, Jr. and M. D. Jeter, pp. 143-168. University of Arkansas Press, Fayetteville.

Marquardt, W. H.
1978 Advances in Archaeological Seriation. *Advances in Archaeological Method and Theory 1*:257-314.

Marsden, P.
1998 The Selectionist Paradigm: More Implications for Sociology. *Sociological Research Online* 3:88-108.

Mayr, E.
1959 Typological Thinking versus Population Thinking. In *Evolution and Anthropology: A Centennial Appraisal*, edited by B. J. Meggers, pp. 409-412. The Anthropological Society of Washington, Washington.
1963 *Animal Species and Evolution*. Belknap Press of Harvard University Press, Cambridge.
1976 *Evolution and the Diversity of Life: Selected Essays*. The Belknap Press of Harvard University Press, Cambridge.
1977 Darwin and Natural Selection. *American Scientist* 65:321-337.
1997 *This is Biology: The Science of the Living World*. The Belknap Press of Harvard University Press, Cambridge.

McCutcheon, P. T.
1997 *Archaeological Investigations of Stone Tool Heat-treatment Technology in Southeastern Missouri: An Experimental Approach*. Unpublished Ph.D. Dissertation, Department of Anthropology, University of Washington, Seattle.

McKern, W. C.
1939 The Mid-Western Taxonomic Method as an Aid to Archaeological Study. *American Antiquity* 4:301-313.

McNutt, C. H.
1973 On the Methodological Validity of Frequency Seriation. *American Antiquity* 38:45-60.

Meltzer, D. J.
1981 A Study of Style and Function in a Class of Tools. *Journal of Field Archaeology* 8:313-326.

Michod, R. E.
1999 *Darwinian Dynamics: Evolutionary Transitions in Fitness and Individuality*. Princeton University Press, Princeton.

Minsky, M.
1986 *The Society of Mind*. Simon and Schuster, New York.

Mooney, C. Z. and R. D. Duval
1993 *Bootstrapping: A Nonparametric Approach to Statistical Inference*. Quantitative Applications in the Social Sciences 95. Sage Publications, New York.

Moore, C. B.
1910 Antiquities of the St. Francis, White and Black River, Arkansas. *Journal of the Academy of Natural Sciences of Philadelphia* 14:255-364.
1911 Some Aboriginal Sites on the Mississippi River. *Journal of the Academy of Natural Sciences of Philadelphia* 14:367-478.

Moore, O. K.
1957 Divination – A New Perspective. *American Anthropologist* 59:69-74.

Morse, D. F.
1973 *Nodena: An Account of 75 years of Archeological Investigation in Southeast Mississippi County, Arkansas*. Arkansas Archeological Survey Research Series, No. 4, Fayetteville.
1982 Regional Overview of Northeastern Arkansas. In *Arkansas Archeology in Review*, edited by N. L. Trobowitz and M. D. Jeter, pp. 20-36. Arkansas Archeological Survey Research Series, No. 15, Fayetteville.
1989 The Nodena phase. In *Nodena: An Account of 90 Years of Aarchaeological Investigation in Southeast Mississippi County, Arkansas*, edited by D. F. Morse, pp. 97-113. Arkansas Archeological Survey Research Series, No. 30, Fayetteville.
1990 The Nodena phase. In *Towns and Temples along the Mississippi*, edited by D. H. Dye and C. A. Cox, pp. 69-97. University of Alabama Press, Tuscaloosa.

Morse, D. F. and P. A. Morse
1983 *Archaeology of the Central Mississippi Valley*. Academic Press, New York.
1990 The Spanish Exploration of Arkansas. In *Columbian Consequences 2*, edited by D. H. Thomas, pp. 197-210. Smithsonian Institution Press, Washington.
1996 Northeast Arkansas. In *Prehistory of the Central Mississippi Valley*, edited by C. H. McNutt, pp. 119-135. The University of Alabama Press, Tuscaloosa.

Morse, P.
1978 The First Excavation at Rose Mound. *Arkansas Archeological Society Field Notes* No. 163:7-9.
1981 *Parkin*. Arkansas Archeological Survey Research Series No. 13, Pine Bluff.
1990 The Parkin Site and the Parkin Phase. In *Towns and Temples Along the Mississippi*, edited by D. Dye and C. Cox, pp. 119-134. University of Alabama, Tuscaloosa.

Morwood, M. J.
1975 Analogy and the Acceptance of Theory in Archaeology. *American Antiquity* 40:111-116.

Neff, H.
1992 Ceramics and Evolution. *Archaeological Method and Theory 4:* 141-194.

Neiman, F.
1990 *An Evolutionary Approach to Archaeological Inference: Aspects of Architectural Variation in the 17th-Century Chesapeake.* Unpublished Ph.D. Dissertation, Department of Anthropology, Yale University, New Haven.
1992 Stylistic Variation in Evolutionary Perspective: Implications for Middle Woodland Ceramic Diversity. Paper presented at the 57th Annual Meeting of the Society for American Archaeology, Pittsburgh.
1993 What Makes The Seriation Clock Tick? Temporal Dynamics of Stylistic Variation Under Drift. Poster presented at the58th Annual Meeting of the Society for American Archaeology, St. Louis.
1995 Stylistic Variation in Evolutionary Perspective: Inferences from Decorative Diversity and Interassemblage Distance in Illinois Woodland Ceramic Assemblages. *American Antiquity* 60:7-36.

Nelson, N. C.
1919 The Origin and Development of Material Culture. *Sigma Xi Quarterly* 8:102-123.

Nesse, R. M. and G. C. Williams
1994 *Why We Get Sick: The New Science of Darwinian Medicine*. Vintage Books, New York.

Newell, H. P. and A. D. Krieger
1949 Analysis of Artifacts. In *The George C. Davis Site, Cherokee County, Texas,* edited by H. P. Newell and A. D. Krieger, pp. 71-74. Society for American Archaeology, Memoir 5, Menasha.

Noreen, E.
1989 *Computer Intensive Methods for Testing Hypotheses*. Wiley, New York.

O'Brien, M. J.
1994 (editor) *Cat Monsters and Head Pots: The Archaeology of Missouri's Pemiscot Bayou.* University of Missouri Press, Columbia.
1996a Archaeological Research in the Central Mississippi Valley: Culture History Gone Awry. *The Review of Archaeology* 16:23-26.
1996b Evolutionary Archaeology: An Introduction. In *Evolutionary Archaeology: Theory and Application*, edited by M. J. O'Brien, pp. 1-15. University of Utah Press, Salt Lake City.

O'Brien, M. J. and R. C. Dunnell
1998 A Brief Introduction to the Archaeology of the Central Mississippi River Valley. In *Changing Perspectives on the Archaeology of the Central Mississippi River Valley*, edited by R. C. Dunnell and M. J. O'Brien, pp. 1-31. University of Alabama Press, Tuscaloosa.

O'Brien, M. J. and G. Fox
1994 Sorting Artifacts in Space and Time. In *Cat Monsters and Head Pots: The Archaeology of Missouri's Pemiscot Bayou*, edited by M. J. O'Brien, pp. 25-60. University of Missouri Press, Columbia.

O'Brien, M. J. and T. D. Holland
1990 Variation, Selection, and the Archaeological Record. *Archaeological Method and Theory 2*: 31-80.

O'Brien, M. J. and R. L. Lyman
1998 *James A. Ford and the Growth of Americanist Archaeology*. University of Missouri Press, Columbia.
2000 *Applying Evolutionary Archaeology: A Systematic Approach.* Kluwer Academic / Plenum, New York.

Ohta, T.
1976 Role of Very Slightly Deleterious Mutations in Molecular Evolution and Polymorphism. *Nature* 252:351-354.
1996 The Current Significance and Standing of Neutral and Nearly Neutral Theories. *BioEssays* 18:673-677.

Orton, C.
1993 How Many Pots Make Five? -- An Historical Review of Pottery Quantification. *Archaeometry* 35:169-184.

Orton, C. and Tyers, P.
1990 Statistical Analysis of Ceramic Assemblages. *Archaeologia e Calcolatori* 1:81-110.
1992 Counting Broken Objects: The Statistics of Ceramic Assemblages. In *New Developments in Archaeological Science*, edited by A.M. Pollard, pp. 163-184. Oxford University Press, Oxford.

Osgood, C.
1951 Culture: Its Empirical and Non-Empirical Character. *Southwestern Journal of Anthropology* 7:202-214.

Payne, R. B.
1996 Song Traditions in Indigo Buntings: Origin, Improvisation, Dispersal, and Extinction in Cultural Evolution. In *Ecology and Evolution of Acoustic Communication in Birds*, edited by D. E. Kroodsma and E. H. Miller, pp. 198-220. Cornell University Press, Ithaca.

Payne, R. B., L. L. Payne and S. M. Doehleri
1988 Biological and Cultural Success of Song Memes in Indigo Buntings. *Ecology* 69:104-117.

Payne, R. B., W. L. Thompson, K. L. Fiala and L. L. Sweany
1981 Local Song Traditions in Indigo Buntings: Cultural Transmission of Behavior Patterns Across Generations. *Behavior* 77:199-221.

Peabody, C.
1904 *Exploration of Mounds, Coahoma County, Mississippi.* Papers of the Peabody Museum of American Archaeology and Ethnology 3, pp. 23-63. Harvard University, Cambridge.

Peregrine, P.
1991 Prehistoric Chiefdoms of the Midcontinent: A World-System Based on Prestige Goods. In *Core/Periphery Relations in Precapitalist Worlds*, edited by C. Chase-Dunn and T. Hall, pp. 193-211. Westview Press, Boulder.
1992 *Mississippian Evolution: A World-System Perspective*. Prehistory Press, Madison.
1995 Networks of Power: The Mississippian World-System. In *Native American Interactions: Multiscalar Analysis and Interpretations in the Eastern Woodlands*, edited by M. S. Nassaney and K. E. Sassaman, pp. 247-265. University of Tennessee Press, Knoxville.

Petrie, S. W. M. F.
1899 Sequences in Prehistoric Remains. *Journal of the Royal Anthropological Institute of Great Britain and Ireland* 29:295-301.

Pfeffer, M. T.
1995 Distribution and Design of Pacific Octopus Lures: The Hawaiian Octopus Lure in Regional Context. *Hawaiian Archaeology* 4:47-56.

Phillips, P.
1939 *An Introduction to the Archaeology of the Mississippi Valley.* Unpublished Ph.D. Dissertation, Department of Anthropology, Harvard University.
1940 *Archaeological Survey in the Mississippi Alluvial Valley, 1940-1947, Field Notes*. Unpublished Report. Peabody Museum, Harvard University.
1958 Application of the Wheat-Gifford-Wasley Taxonomy to Eastern Ceramics. *American Antiquity* 24:117-125.
1970 *Archaeological Survey in the Lower Yazoo Basin, Mississippi, 1949-1955.* Papers of the Peabody Museum of American Archaeology and Ethnology 60 (2 Volumes), Harvard University, Cambridge.

Phillips, P., J. A. Ford and J. B. Griffin
1951 *Archaeological Survey in the Lower Mississippi Alluvial Valley, 1940-1947.* Papers of the Peabody Museum of American Archaeology and Ethnology 25. Harvard University, Cambridge.

Phillips, P. and G. R. Willey
1953 Method and Theory in American Archaeology: An Operational Basis for Culture-Historical Integration. *American Anthropologist* 55:615-633.

Pinker, S.
1997 *How the Mind Works*. W.W. Norton, New York.

Plog, F.
1979 Prehistory: Western Anasazi. In *Handbook of North American Indians, Volume 9: Southwest*, edited by A. Ortiz, pp. 109-130. Smithsonian Institution, Washington.

Plog, S.
1976a The Inference of Prehistoric Social Organization from Ceramic design Variability. *Michigan Discussions in Anthropology* 1:1-47.
1976b Measurement of Prehistoric Interaction between Communities. In *The Early Mesoamerican Village*, edited by K. V. Flannery, pp. 255-272. Academic Press, New York.
1980 *Stylistic Variation in Prehistoric Ceramics*. Cambridge University Press, Cambridge.
1983 Analysis of Style in Artifacts. *Annual Review of Anthropology* 12:125-142.

Plog, S. and M. Hegmon
1993 The Sample Size-Richness Relation - The Relevance of Research Questions, Sampling Strategies, and Behavioral Variation. *American Antiquity* 58:489-496.

Pocklington, R. and M. L. Best
1997 Cultural Evolution and Units of Selection in Replicating Text. *Journal of Theoretical Biology* 188:79-87.

Pollnac, R. B. and R. M. Rowlett
1977 Community and Supracommunity within the Marne Culture: a Stylistic Analysis. In *Experimental Archeology*, edited by D. Ingersoll, J. E. Yellen and W. Macdonald, pp. 167-90. Columbia University, New York.

Popper, K.
1959 *The Logic of Scientific Discovery*. Basic Books, New York.

Potter, W. B.
1880 Archaeological Remains in Southeastern Missouri-- Earthworks of Missouri, Part I:

Pottery. *Archaeological Section of the St. Louis Academy of Science, Contributions to the Archaeology of Missouri* Part I:5-19.

Provine, W.
1986 *Sewall Wright and Evolutionary Biology*. University of Chicago Press, Chicago.

Pugh, T.
1990 The Distribution of Cultural Materials in the Plowed Portion of the Belle Meade Site in Crittenden County, Arkansas. Unpublished Paper in Possession of the Author.

Putnam, F. W.
1881 *Report of the Curator*. Fourteenth Annual Report of the Trustees of the Peabody Museum of American Archaeology and Ethnology, Harvard University, Cambridge.

Rakita, G. and T. Hurt (editors)
2000 *Style and Function in Archaeology*. Greenwood Press, Westport.

Ramenofsky, A.
1987 *Vectors of Death*. The University of New Mexico Press, Albuquerque.

Redman, C. L.
1977 The "Analytical Individual" and Prehistoric Style Variability. In *The Individual in Prehistory: Studies of Variability in Style and Prehistoric Technologies*, edited by J. N. Hill and J. Gunn, pp. 41-53. Academic Press, New York.

Renfrew, C.
1977 Alternative Models for Exchange and Spatial Distribution. In *Exchange Systems in Prehistory*, edited by T. K. Earle and J. E. Ericson, pp. 71-90. Academic Press, New York.

Renfrew, C. and R. M. Clark
1974 Problems of Radiocarbon Calendar and its Calibration. *Archaeometry* 16:5-18.

Rhode, D.
1988 Measurement of Archaeological Diversity and the Sample-Size Effect. *American Antiquity* 53:708-716.

Riggs, C. W.
1893 *How We Find Relics: Pleasures and Perils on the Mississippi River in a Floating Home*. W.B. Conkey, Chicago.

Rindos, D.
1984 *The Origins of Agriculture: An Evolutionary Perspective*. Academic Press, New York.
1989a Darwinianism and its Role in the Explanation of Domestication. In *Foraging and Farming: The Evolution of Plant Exploitation*, edited by D. Harris and G. Hillman, pp. 27-41. Uwin Hyman, London.
1989b Undirected Variation and the Darwinian Explanation of Cultural Change. *Advances in Archaeological Method and Theory* 12:1-46.

Rouse, I.
1939 *Prehistory in Haiti: A Study in Method*. Yale University Publications in Anthropology, No. 21, New Haven.
1955 On the Correlation of Phases of Culture. *American Anthropologist* 57:713-722.
1964 Archaeological Approaches to Cultural Evolution. In *Explorations in Cultural Anthropology*, edited by W. Goodenough, pp. 455-468. McGraw-Hill, New York.
1965 The Place of 'Peoples' in Prehistoric Research. *Journal of the Royal Anthropological Institute* 95:1-15.

Royall, P. D.
1988 *Late-Quaternary Paleoecology and Paleoenvironments of the Western Lowlands, Southeast Missouri.* Unpublished Masters of Science Thesis, Department of Geology, University of Tennessee.

Ryan, F.
1997 *Virus-X: Tracking the New Killer Plagues: Out of the Present and into the Future*. Little and Brown, Boston.

Sabloff, J. A. and R. E. Smith
1969 The Importance of Both Analytic and Taxonomic Classification in the Type-Variety System. *American Antiquity* 34:278-285.

Sackett, J. R.
1985 Style, Ethnicity and Stone Tools. In *Status, Structure and Stratification: Current Archaeological Reconstructions*, edited by M. Thompsen, M. T. Garcia and F. J. Kense, pp. 277-282. University of Calgary Archaeological Association, Calgary.
1990 Style and Ethnicity in Archaeology: The Case for Isochrestism. In *The Uses of Style in Archaeology*, edited by M. W. Conkey and C. A. Hastorf, pp. 32-43. Cambridge University Press, Cambridge.

Sagan, C.
1996 *The Demon-Haunted World: Science as a Candle in the Dark*. Ballantine Books, New York.

Sampson, C. G.
1988 *Stylistic Boundaries among Mobile Hunter-Foragers*. Smithsonian Institution, Washington.

Saucier, R. T.
1974 *Quaternary Geology of the Lower Mississippi Alluvial Valley*. Arkansas Archeological Survey Research Series No. 6, Fayetteville.

Schiffer, M. B.
1987 *Formation Processes of the Archaeological Record.* University of New Mexico Press, Albuquerque.
1995 *Behavioral Archaeology.* University of Utah Press, Salt Lake City.
1996 Some Relationships between Behavioral and Evolutionary Archaeologies. *American Antiquity* 61:643-662.

Shanks, M. and I. Hodder
1995 Interpretative Archaeologies: Some Themes and Questions. In *Interpreting Archaeology*, edited by I. Hodder, M. Shanks, A. Alexandri, V. Buchli, J. Carman, J. Last and G. Lucas, pp. 30-33. Routledge, London.

Shanks, M. and C. Tilley
1987 *Reconstructing Archaeology: Theory and Practice.* Cambridge University Press, Cambridge.

Shapiro, M.
1953 Style. In *Anthropology Today*, edited by A. L. Kroeber, pp. 287-312. Aldine, Chicago.

Shott, M. J.
1989 Diversity, Organization, and Behavior in the Material Record. *Current Anthropology* 30:283-315.

Simon, J. L.
1990 *Resampling: Probability and Statistics in a Radically Different Way.* Manuscript available from Resampling Project, College of Business, University of Maryland, College Park.

Slater, P. J. B., F. A. Clements and D. J. Goodfellow
1984 Local and Regional Variations in Chaffinch Song and the Question of Dialects. *Behavior* 88:76-97.

Slater, P. J. B., S. A. Ince and P. W. Colgan
1980 Chaffinch Song Types: Their Frequencies in the Population and Distribution Between Repertoires of Different Individuals. *Behavior* 75:207-218.

Smith, D. G.
1960 The Type-Variety Concept as a Basis for the Analysis of Maya Pottery. *American Antiquity* 25:330-340.

Smith, E. A. and J. Boone
1998 Reply to Lyman and O'Brien: The Goals of Evolutionary Archaeology: History and Explanation. *Current Anthropology* 39:615-652.

Smith, G. P.
1990 The Walls Phase and Its Neighbors. In *Towns and Temples Along the Mississippi*, edited by D. Dye and C. Cox, pp. 135-169. University of Alabama Press, Tuscaloosa.

Smith, M.
1955 The Limitations of Inference in Archaeology. *The Archaeological Newsletter* 6:1-7.

Sober, E.
1984 *The Nature of Selection: Evolutionary Theory in Philosophical Focus.* MIT Press, Cambridge.
1991 Models of Cultural Evolution. In *Trees of Life: Essays in the Philosophy of Biology*, edited by P. Griffiths, pp. 17-38. Kluwer, Brussels.

Sokal, R. R. and P. H. A. Sneath
1963 *Principles of Numerical Taxonomy.* Freeman, San Francisco.

Soltis, J., R. Boyd and P. J. Richerson
1995 Can Group-Functional Behaviors Evolve by Cultural Group Selection? An Empirical Test. *Current Anthropology* 36: 473-494.

Spaulding, A. C.
1953a Review of Measurements of Some Prehistoric Design Developments in the Southeastern States (Ford). *American Anthropologist* 55:588-591.
1953b Statistical Techniques for the Discovery of Artifact Types. *American Antiquity* 15:305-313.
1954 Reply to Ford. *American Antiquity.* 19:391-393.
1960 The Dimensions of Archaeology. In *Essays in the Science of Culture*, edited by G. E. Dole and R. L. Carneiro, pp. 437-456. T.Y. Crowell and Co., New York.

Spier, L.
1917 An Outline for a Chronology of Zuñi Ruins. *Anthropological Papers of the American Museum of Natural History* 18:207-331.

Steponaitis, V. P.
1983 *Ceramics, Chronology and Community Patterns: an Archaeological Study at Moundville.* Academic Press, New York.

Sterner, J.
1989 Who is Signaling Who? Ceramic Style, Ethnicity, and Taphonomy among Sirah Bulahay. *Antiquity* 63:451-459.

Stewart, F. M.
1977 Computer Algorithm for Obtaining a Random Set of Allele Frequencies for a Locus in an Equilibrium Population. *Theoretical Population Biology* 178:482-483.

Stiles, D.
1977 Ethnoarchaeology: A Discussion of Methods and Applications. *Man* 12:87-103.

Stuiver, M., P. J. Reimer, E. Bard, B. J. W., G. S. Burr, K. A. Hughen, B. Kromer, G. McCormac, J. Van der Plicht and M. Spurk
1998 INTCAL98 Radiocarbon Age Calibration, 24,000-0 cal BP. *Radiocarbon* 40:1041-1083.

Swallow, G. C.
1858 Indian Mounds in New Madrid County, Missouri. *Academy of Science of St. Louis, Transactions* I:36.
1875 Inventory of Objects from Mounds in New Madrid County and Brief Description of Excavations. *Peabody Museum, Harvard University, 8th Annual Report* :16-46.

Taylor, W.
1948 *A Study of Archaeology.* American Anthropological Association, Memoir Number 69. Menasha.

Teltser, P. A.
1995a Culture History, Evolutionary Theory, and Frequency Seriation. In *Evolutionary Archaeology: Methodological Issues*, edited by P. A. Teltser, pp. 51-68. University of Arizona Press, Tucson.

Teltser, P. A. (editor)
1995b *Evolutionary Archaeology: Methodological Issues.* University of Arizona Press, Tucson.

Thomas, C.
1887 Burial Mounds of the North Sections of the United States. *Bureau of American Ethnology Annual Report.* 5:9-119.
1894 Report of the Mound Explorations of the Bureau of Ethnology. *Bureau of American Ethnology Annual Report* 12:163-97.

Thompson, G.
1978 Prehistoric Settlement Changes in the Southern Northwest Coast: a Functional Approach. Unpublished Ph.D. Dissertation, Department of Anthropology, University of Washington, Seattle.

Thompson, R. H.
1956 The Subjective Element in Archaeological Inference. *Southwestern Journal of Anthropology* 12:327-332.

Tozzer, A.
1926 Chronological Aspects of American Archaeology. *Massachusetts Historical Society, Proceedings* 59: 283-292.

Trigger, B. G.
1989 *A History of Archaeological Thought.* Cambridge University Press, Cambridge.

Tringham, R.
1978 Experimentation, Ethnoarchaeology, and the Leapfrogs in Archaeological Methodology. In *Explorations in Ethnoarchaeology*, edited by R. Gould, pp. 169-199. University of New Mexico Press, Albuquerque.

Upham, S.
1982 *Polities and Power: An Economic and Political History of the Western Pueblo.* Academic Press, New York.
1983 Intensification and Exchange: An Evolutionary Model of Non-Egalitarian Socio-Political Organization for the Prehistoric Plateau Southwest. In *Ecological Models in Economic Prehistory*, edited by G. Bronitsky, pp. 219-245. Arizona State University Anthropological Research Papers, vol. 29, Tempe.

Upham, S., K. G. Lightfoot and G. M. Feinman
1981 Explaining Socially Determined Ceramic Distribution in the Prehistoric Plateau Southwest. *American Antiquity* 46:822-833.

Upham, S. and F. Plog
1986 The Interpretation of Prehistoric Political Complexity in the Central and Northern Southwest: Toward a Mending of the Models. *Journal of Field Archaeology* 13:223-238.

Waldrop, M. M.
1992 *Complexity: The Emerging Science at the Edge of Order and Chaos.* Simon and Schuster, New York.

Wallace, A. R.
1889 *Darwinism.* Macmillan, London.

Ward, J. H.
1963 Hierarchical Grouping to Optimize an Objective Function. *American Statistical Association Journal* 58:236-244.

Watson, P. J., S. A. LeBlanc and C. L. Redman
1984 *Archaeological Explanation: The Scientific Method in Archaeology.* Columbia University Press, New York.

Watson, P. J., C. Redman and S. LeBlanc
1971 *Explanation in Archaeology: An Explicitly Scientific Approach.* Columbia University Press, New York.

Watterson, G. A.
1978 The Homozygosity Test of Neutrality. *Genetics* 88:405-417.

Wayne, M. L. and K. L. Simonsen
1996 Statistical Tests of Neutrality in the Age of Weak Selection. *Trends in Ecology and Evolution* 13:236-240.

Wcislo, W. T.

1989 Behavioral Environments and Evolutionary Change. *Annual Review of Ecology and Systematics* 20:137-169.

Webb, W. S.

1938 *An Archaeological Survey of the Wheeler Basin on the Tennessee River in Northern Alabama.* Bureau of American Ethnology Bulletin 122, Washington.

Weissner, P.

1983 Style and Social Information in Kalahari San Projectile Points. *American Antiquity* 48:253-276.

1984 Reconsidering the Behavioral Basis of Style. *Journal of Anthropological Archaeology* 3:190-234.

1990 Is There a Unity to Style? In *The Uses of Style in Archaeology*, edited by M. Conkey and C. Hastorf, pp. 105-112. Cambridge University Press, Cambridge.

Wenke, R.

1981 Explaining the Evolution of Cultural Complexity. *Advances in Archaeological Method and Theory 4*: 79-119.

Whallon, R.

1968 Investigations of Late Prehistoric Social Organization in New York State. In *New Perspectives in Archeology*, edited by S. R. Binford and L. R. Binford, pp. 223-244. Aldine Press, Chicago.

1972 A New Approach to Pottery Typology. *American Antiquity* 37:13-33.

Wheat, J. B., J. C. Gifford and W. W. Wasley

1958 Ceramic Variety, Type Cluster, and Ceramic System in Southwestern Pottery Analysis. *American Antiquity* 24:34-47.

Wilhelmsen, K.

2000 Projectile Points in the Central Mississippi River Valley, in *Posing Questions for a Scientific Archaeology*, edited by T. Hunt, C.P. Lipo and S. Sterling. Greenwood Press, Westport (in press).

Willer, D. and J. Willer

1973 *Systematic Empiricism: Critique of a Pseudoscience.* Prentice Hall, Englewood Cliffs.

Willey, G. R.

1953 *Prehistoric Settlement Patterns in the Viru Valley, Peru.* Bureau of American Ethnology Bulletin 155, Washington.

Willey, G. R. and P. Phillips

1955 Method and Theory in American Archaeology. II: Historical Developmental Interpretation. *American Anthropologist* 57:723-819.

1958 *Method and Theory in American Archaeology.* University of Chicago Press, Chicago.

Williams, G. C.

1966 *Adaptation and Natural Selection.* Princeton University Press, Princeton.

Williams, S.

1954 *An Archaeological Study of the Mississippian Culture in Southeast Missouri.* Unpublished Ph.D. Dissertation, Department of Anthropology, Yale University, New Haven.

1980 The Armorel Phase: A Very Late Complex in the Lower Mississippi Valley. *Southeastern Archaeological Conference Bulletin* 22:105-110.

1999 A Discussion of Phases in Southeast Missouri. Unpublished Manuscript in Possession of Author.

Williams, S. and J. P. Brain

1983 *Excavations at Lake George, Yazoo County Mississippi, 1958-1960.* Papers of the Peabody Museum of Archaeology and Ethnology 74. Harvard University, Cambridge.

Wilson, E. O.

1994 *Naturalist.* Island Press, Washington.

1998 *Consilience: The Unity of Knowledge.* Knopf, New York.

Wissler, C.

1916a The Application of Statistical Methods to the Data on the Trenton Argillite Culture. *American Anthropologist* 18:190-197.

1916b Correlations between Archaeological and Culture Areas in the American Continents. In *Holmes Anniversary Volume: Anthropological Essays Presented to William Henry Holmes in Honor of his Seventieth Birthday, December 1, 1916*, edited by F. W. Hodge, pp. 481-490. J. W. Bryan Press, Washington, D.C.

1917 The New Archaeology. *American Museum Journal* 17:100-101.

Wright, S.

1931 Evolution in Mendelian Populations. *Genetics* 16:97-159.

1932 The Roles of Mutation, Inbreeding, Crossbreeding and Selection in Evolution. *Proceedings of the Sixth International Congress of Genetics* 1, pp. 356-366. Cornell University, Ithaca.

1940 The Statistical Consequences of Mendelian Heredity in Relation to Speciation. In *The New Systematics*, edited by J. S. Huxley, pp. 161-184. Oxford University Press, Oxford.

1948 On the Roles of Directed and Random Changes in Gene Frequency in the Genetics of Populations. *Evolution* 2:279-294.

1949 Population Structure in Evolution. *Proceedings of the American Philosophical Society* 93:471-478.

Wu, C. J. F.
1986 Jackknife, Bootstrap and Other Resampling Methods in Regression Analysis. *Annual Statistics* 14:1261-1295.

Wylie, A.
1995 An Expanded Behavioral Archaeology: Transformation and Redefinition. In *Expanding Archaeology*, edited by J. Skibo, W. Walker, A. Nielsen, pp. 198-209. University of Utah Press, Salt Lake City.

APPENDIX A: BOOTSTRAP SAMPLE SIZE TESTING PROGRAM

```
#!/usr/bin/perl
################################################################
# bootstrap_samplesize_test.pl
#
# Bootstrap Sample Size Test Program
# A simple perl program for evaluating the effect of sample
# size on type frequencies.
#
# Carl Lipo
# Department of Anthropology
# University of Washington
# Box 353100
# Seattle, WA 98195-3100
# clipo@u.washington.edu
# (206) 543-5240
#
# Version 1.0.0
# 7 April 2000
#
# Copyright 2000
# Please contact author for information on references and
# more information regarding the use of this program.
# You will need to get the statistics and math libraries for PERL from
# http://www.perl.com/cpan (any CPAN location)
# install these two modules:
#
# http://www.perl.com/CPAN/
#     modules/by-module/Statistics/Statistics-Descriptive-2.1.tar.gz
# http://www.perl.com/ CPAN/modules/
#     by-module/Math/Math-TrulyRandom-1.0.tar.gz
#
# SYNTAX:  bootstrap_samplesize_test.pl < FILE IN > < FILE OUT >
#                  < NUMBER OF RESAMPLES > < NUMBER OF INCREMENTS >
# (e.g., bootstrap_samplesize_test.pl TESTIN TESTOUT 2000 25 )
#
# SAMPLE INPUT FILE:
# <assemblage>          <counts for each type>
# Porter               64      5      1      1      83     1
# 40LA7                20      4      0      0      25     0
# Hatchie              32      6      4      4      9      0
# RichardsonsLanding   8       2      0      1      23     1
#
# SAMPLE OUTPUT (there is a section for each assemblage)
# Porter
# Number of samples per sample size 1000
# Sample Size     Mean Richness     Stdev Richness
# 3               1.987             0.610907
# 6               2.607             0.730126
# 9               2.965             0.797752
# 12              3.239             0.884681
# 15              3.488             0.93741
# …
#
##################################################################
BEGIN { $| = 1; }

use Math::TrulyRandom;
use Statistics::Descriptive;
```

```
if(!@ARGV) {
   print STDERR "\n";
   print STDERR "Usage: bootstrap_samplesize_test.pl ";
   print STDERR " <FILE IN> <FILE OUT> ";
   print STDERR " <RESAMPLE #> <INCREMENTS>\n";
   exit 1;
}

my $debug = 0;
my @assemblages = ();
my @rowtotals = ();
my @labels = ();
my $numrows = 0;
my $cols = 0;

$filein =$ARGV[0];
$fileout=$ARGV[1];
$samples = $ARGV[2];
$asseminc = $ARGV[3];
open FILEIN, $ARGV[0] or die ("Can't open $filein\n");
open FILEOUT,">$fileout" or die ("Can't open $fileout\n");

while( <FILEIN> ) {
      print FILEOUT;
      chomp;
      my @line = split /\s+/, $_;
      my $label = shift @line;
      push @labels, $label;
      $cols = scalar( @line );
      my $rowtotal = 0;
      my $richness = 0;
      my $entry;
      for ( @line ) {
            $entry = $_;
            $rowtotal += $entry;
            if ($entry>0) {
                  $richness++;
            }
      }
      push @rowtotals, $rowtotal;
      push @richness, $richness;
      my @freq = ();
      for ( @line ) {
            push @freq, ( $_ / $rowtotal );
      }
      push @assemblages, [ @freq ];
      $debug && print "---- row end ----\n";
}
$numrows = scalar( @assemblages );

srand( truly_random_value());

my $results = 0;
my $loop = $samples;
my $bigloop = $asseminc;
my $loop2 = $loop;
my $ptr1 = 0;

while( $ptr1 < $numrows ) {

      my @a =  @{ $assemblages[ $ptr1 ] } ;
      my $numa = $rowtotals[ $ptr1 ];
```

```
my ( @cum_a, $count );
$classes = scalar( @a );
my $index_a = 0;
my $total_a = 0.0;
 $count = 0;
for( $count = 0; $count < $classes; $count++ ) {
      $cum_a[ $index_a ] = $a[ $count ];
      $total_a += $a[ $count ];
      $index_a++;
}

$index_a--;

# now we loop the number times and keep track
my $cycle = $asseminc;
my $newassemsize=$numa;

print FILEOUT "\n-------------------------------------------\n";
print FILEOUT "Collection: \t ",$labels[$ptr1],"\n";
print FILEOUT "Sample Size: \t ", $rowtotals[$ptr1],"\n";
print FILEOUT "Richness: \t ", $richness[$ptr1],"\n";
print FILEOUT "Resample Size\tMean Richness\tVariance\n";
for ($xc=1;$xc<=$asseminc;$xc++ ) {

      $debug && print "(debug) cycle value: $cycle\n";

      # now we loop 1000 times and resample
      $loop = $loop2;
      $newassemsize=int($xc*$numa/$asseminc)+1;
      my $nassem=$newassemsize;
      my $stat = Statistics::Descriptive::Full->new();
      for( $xd=0;$xd<$samples; $xd++) {
            my @assem_a = ();
            my $class;
            my $total = scalar( @a );
            my $rand;
            for ($xe=0;$xe<$newassemsize;$xe++ ) {
                  $rand = rand;
                  $class = 0;
                  while(( $class < $index_a ) &&
                                          ( $rand > $cum_a[ $class ] )) {
                  $rand -= $cum_a[ $class ];
                        $class++;
                  }
                  push @assem_a, $class;
                  #print $class,"\t";
                  #print $xe,"\n";
                  $richhat[$class]++;
            }
            my ( @ahat, %aholder );
            %aholder = ();
            my $index = 0;
            my $ccount = $cols;
            for ( $x=0;$x<$ccount;$x++ ) {
                  if ($richhat[$x]>0) {
                  $rhat++;
            }
      }
      $stat->add_data($rhat);
      $rhat=0;
      @richhat = ();
      $loop--;
}
```

```
        $cycle--;
        print FILEOUT $newassemsize,"\t";
        print FILEOUT $stat->mean(),"\t";
        print FILEOUT $stat->variance(),"\n";
        undef $stat;
        $results = 0;
        }
        $ptr1++;
}
close (FILEIN);
close (FILEOUT);
```

APPENDIX B: SERIATION MACRO FOR MICROSOFT EXCEL

```
' Seriation Macro for Microsoft Excel
' c.1999 Tim Hunt, Emergent Media and Carl Lipo,
' University of Washington
' This macro produces graphical frequency bars from raw data
' that can be sorted into a
' seriation
' The program allows statistical testing of frequency orders.

Dim bar_obj(500) As Integer
Dim place_hold(500) As Double
Dim N_rem_frm_sort As Integer
Dim left_of_units As Double
Dim top_of_units As Double
Dim nrows As Integer
Dim ncols As Integer

Sub draw()
  On Error Resume Next
  Sheets("DataSheet").Delete
   Selection.Name = "sel"
   ActiveSheet.Name = "DataSheet"
   nrows = Range("sel").Rows.Count
   ncols = Range("sel").Columns.Count
   Range("sel").Range(Cells(2, 1), Cells(nrows, 1)).Name = "unit_names"
   Range("sel").Range(Cells(1, 2), Cells(1, ncols)).Name = "type_names"
   Range("sel").Range(Cells(2, 2), Cells(nrows, ncols)).Name = "data"

   Range("sel").Range(Cells(nrows + 2, 1),
      Cells(nrows + nrows + 1, ncols + 2)).Name = "sel_pcnt"
   Range("sel_pcnt").Range(Cells(nrows + 2, 1),
      Cells(nrows + nrows + 1, ncols)).Name = "sel_err"
   Range("sel_pcnt").Range(Cells(1, ncols + 3),
      Cells(nrows, ncols + ncols + 2)).Name = "hid_dat"
   Range("sel_pcnt").Range(Cells(1, ncols + ncols + 3),
      Cells(nrows, (3 * ncols) + 2)).Name = "hid_err"
   Range("sel_err").Range(Cells(nrows + 2, 1),
      Cells(nrows + nrows + 1, ncols)).Name = "sel_size"
   Range("sel_size").Range(Cells(2, 1),
      Cells(nrows, ncols)).Name = "sel_sizes"

   Range("sel_pcnt").Range(Cells(2, 1), Cells(nrows, 1)).Name = "sel_pcnt_unit_names"
   Range("sel_pcnt").Range(Cells(1, 2), Cells(1, ncols)).Name = "sel_pcnt_type_names"
   Range("sel_pcnt").Range(Cells(2, 2), Cells(nrows, ncols)).Name = "pcnts"
   Range("sel_pcnt").Range(Cells(2, ncols + 1), Cells(nrows, ncols + 1)).Name =
"pcnts_srt_index"
   Range("sel_pcnt").Range(Cells(2, ncols + 2), Cells(nrows, ncols + 2)).Name =
"snap_srt_index"


   Range("unit_names").Copy (Range("sel_pcnt_unit_names"))
   Range("type_names").Copy (Range("sel_pcnt_type_names"))

   Range("sel_err").Range(Cells(2, 1), Cells(nrows, 1)).Name = "sel_unit_names"
   Range("sel_err").Range(Cells(1, 2), Cells(1, ncols)).Name = "sel_type_names"
   Range("sel_err").Range(Cells(2, 2), Cells(nrows, ncols)).Name = "errors"

   Range("unit_names").Copy (Range("sel_unit_names"))
   Range("type_names").Copy (Range("sel_type_names"))
```

```
    Range("sel_size").Range(Cells(2, 2), Cells(nrows, ncols)).Name = "sizes"
    Range("sel_size").RowHeight = 0


    Range("sel").Cells(1, 1).Value = "Counts"
    Range("sel_pcnt").Cells(1, 1).Value = "Pcnts"
    Range("sel_err").Cells(1, 1).Value = "%Error"
    Union(Range("sel_pcnt"), Range("hid_dat"), Range("hid_err")).Select
    Selection.Name = "the_sort"
    Range("pcnts_srt_index").ColumnWidth = 0
    Range("hid_dat").ColumnWidth = 0
    Range("hid_err").ColumnWidth = 0

    Worksheets("DataSheet").DropDowns.
       Add(173.25, 99.75, 47.25, 15.75).Select
    Selection.List = Array("alpha", ".5",
       ".2", ".1", ".05", ".02", ".01", ".005", ".002", ".001", ".0005")

    Selection.ListIndex = 1
    Selection.Name = "alpha_lst"
    Selection.Left = 1
    Selection.Top = 1
    Selection.OnAction = "error_term"
    Cells(2, 2).Select
    For rw = 1 To nrows - 1
       For col = 1 To ncols - 1
          If Range("data").Cells(rw, col).Value = 0 Then
             Range("pcnts").Cells(rw, col).Value = 0
          Else
             Range("pcnts").Cells(rw, col).Value = Range("data").Cells(rw, col).Value /
Application.Sum(Range("data").Cells(rw, col).EntireRow)
          End If
          If Range("pcnts").Cells(rw, col).Value = 1 Then
             msgtxt = "Unit " & Range("unit_names").
                    Cells(rw, 1).Value & " has only one
                    type represented, remove this unit before proceeding!"
             MsgBox msgtxt, 48, "Error"
             Exit Sub
          End If

       Next col
    Next rw

End Sub

Sub error_term()
    nrows = Range("data").Rows.Count
    ncols = Range("data").Columns.Count

    Dim zvals As Variant
    Dim zval As Double
    zvals = Array(0, 0, 0.6745, 1.2816, 1.6449,
       1.96, 2.3263, 2.5758, 2.807, 3.0902, 3.2905, 3.7)
    zval = zvals(Worksheets("DataSheet").DrawingObjects("alpha_lst").Value)
    Dim N, p, q As Double
    For rw = 1 To nrows
       For col = 1 To ncols
          N = Application.Sum(Range("data").Cells(rw, col).EntireRow)
          p = Range("pcnts").Cells(rw, col).Value
          q = 1 - p
          Range("errors").Cells(rw, col).Value =
               ((zval * (Sqr((p * q) / (N - 1)))) + (1 / (2 * N)))
          If Range("pcnts").Cells(rw, col).Value = 0 Then
```

```
            Range("sizes").Cells(rw, col).Value = 0
            'Range("sizes").Cells(rw, col).Value =
             (Range("errors").Cells(rw, col).Value * 100) / 2
         Else
            Range("sizes").Cells(rw, col).Value =
              ((Range("pcnts").Cells(rw, col).Value * 100) / 2) +
              ((Range("errors").Cells(rw, col).Value * 100) / 4)
         End If
      Next col
   Next rw

  Dim h_offset() As Double
   ReDim h_offset(ncols)
   For x = 1 To (ncols)
      h_offset(x) = 0
   Next x

   Set holder1 = Range(Range("sizes").Cells(1, 1),
      Range("sizes").Cells(nrows, 1))
   h_offset(1) = 90 + (Application.Max(holder1))
   For x = 2 To (ncols)
      Set holder1 = Range(Range("sizes").Cells(1, x - 1), Range("sizes").Cells(nrows,
x - 1))
      Set holder2 = Range(Range("sizes").Cells(1, x), Range("sizes").Cells(nrows, x))
      h_offset(x) = Int(h_offset(x - 1) + 10 + (Application.Max(holder1)) +
(Application.Max(holder2)))
      'Int is an Excel function; rounds no.  down to nearest integer
      If (h_offset(x) - h_offset(x - 1)) < 50 Then
        h_offset(x) = Int(h_offset(x - 1) + 50)
      End If
   Next x
   Range("sizes").Clear
  On Error Resume Next

  Sheets("Seriation").Delete
   Sheets.Add.Select
   ActiveSheet.Name = "Seriation"
   Sheets("Seriation").Select

   Dim type_box() As Integer
   ReDim type_box(ncols) As Integer
    Dim spnnrs() As Integer
   ReDim spnnrs(ncols) As Integer

   Dim unit_box() As Integer
   ReDim unit_box(nrows) As Integer
   Dim n_box() As Integer
   ReDim n_box(nrows) As Integer
   Dim bars() As Integer
   ReDim bars(nrows, ncols) As Integer
   Dim this_row As Object
   Dim temp_grp(3) As Integer
   Dim err_wid, mn_wid As Double
   Rem ReDim serprog.bar_obj(nrows) As Integer
   v_offset = 70
   tb_start = 5
      For col = 1 To ncols
         type_box(col) = ActiveSheet.TextBoxes.
             Add(h_offset(col) - 25, 5, 50, 50).Index
         ActiveSheet.TextBoxes(type_box(col)).Text =
             Worksheets("DataSheet").Range("type_names").
             Cells(1, col).Value
         ActiveSheet.TextBoxes(type_box(col)).
```

```
            HorizontalAlignment = xlCenter
        ActiveSheet.TextBoxes(type_box(col)).
            VerticalAlignment = xlBottom
        ActiveSheet.TextBoxes(type_box(col)).Border.LineStyle = xlNone
      Next col
      Worksheets("Seriation").TextBoxes(type_box(1)).Select
      For x = 1 To (ncols)
        Worksheets("Seriation").TextBoxes(type_box(x)).Select (False)
      Next x
      Selection.Group
   For rw = 1 To nrows
      unit_box(rw) = ActiveSheet.TextBoxes.
            Add(tb_start, v_offset, 75, 11).Index
      ActiveSheet.TextBoxes(unit_box(rw)).Text =
            Worksheets("DataSheet").Range("unit_names").Cells(rw).Value
      ActiveSheet.TextBoxes(unit_box(rw)).HorizontalAlignment = xlRight
      ActiveSheet.TextBoxes(unit_box(rw)).Border.LineStyle = xlNone

      n_box(rw) = ActiveSheet.TextBoxes.
            Add(tb_start + 1030, v_offset, 35, 11).Index
      ActiveSheet.TextBoxes(n_box(rw)).Text =
            Application.Sum(Worksheets("DataSheet").
            Range("data").Cells(rw, col).EntireRow)
      ActiveSheet.TextBoxes(n_box(rw)).HorizontalAlignment = xlRight
      ActiveSheet.TextBoxes(n_box(rw)).Border.LineStyle = xlNone

      For cl = 1 To ncols

        mn_wid = (Worksheets("DataSheet").Range("pcnts").
            Cells(rw, cl).Value) * 100
        ' ignore all the non-zero entries....
        If (Worksheets("DataSheet").Range("pcnts").
            Cells(rw, cl).Value) > 0 Then
            err_wid = (Worksheets("DataSheet").
                  Range("errors").Cells(rw, cl).Value) * 100 / 2
        rec_start = h_offset(cl) - (mn_wid / 2)
        rec_end = h_offset(cl) + (mn_wid / 2)
        temp_grp(1) = Worksheets("Seriation").
            Rectangles.Add(rec_start, v_offset, mn_wid, 11).Index

        Worksheets("Seriation").Rectangles(temp_grp(1)).
            Interior.Pattern = xlGray50
        If mn_wid = 0 Then
           temp_grp(2) =
            Worksheets("Seriation").Rectangles.Add(h_offset(cl) -
            ((err_wid)), v_offset + 4, (err_wid * 2), 3).Index
            Worksheets("Seriation").Rectangles(temp_grp(1)).
                  Border.LineStyle = xlNone
            Worksheets("Seriation").Rectangles(temp_grp(2)).
                  Interior.Color = black
            bars(rw, cl) =
                  Worksheets("Seriation").Rectangles(Array(temp_grp(1),
                  temp_grp(2))).Group.Index
        Else
           temp_grp(2) =
                  Worksheets("Seriation").Rectangles.Add(rec_start -
                  (err_wid / 2), v_offset + 4, err_wid, 3).Index
                 Worksheets("Seriation").Rectangles(temp_grp(2)).
                  Interior.Color = black
                  temp_grp(3) = Worksheets("Seriation").
                  Rectangles.Add(rec_end - (err_wid / 2),
                  v_offset + 4, err_wid, 3).Index
           Worksheets("Seriation").Rectangles(temp_grp(3)).
```

```
                    Interior.Color = black
                bars(rw, cl) =
                    Worksheets("Seriation").
                    Rectangles(Array(temp_grp(1), temp_grp(2),
                    temp_grp(3))).Group.Index
            End If
            End If
        Next cl
        Worksheets("Seriation").TextBoxes(unit_box(rw)).Select
        For x = 1 To (ncols)
            Worksheets("Seriation").GroupObjects(bars(rw, x)).
                Select (False)
        Next x
        serprog.bar_obj(rw) = Selection.Group.Index
        v_offset = v_offset + 12
Next rw
   Rem ReDim serprog.place_hold(nrows)
   left_of_units = Worksheets("Seriation").GroupObjects(serprog.bar_obj(1)).Left
   top_of_units = Worksheets("Seriation").GroupObjects(serprog.bar_obj(1)).Top

   Rem find positions of objects,and order of objects
   For x = 1 To (nrows)
      serprog.place_hold(x) =
Int(Worksheets("Seriation").GroupObjects(serprog.bar_obj(x)).Top)
      Worksheets("DataSheet").Range("pcnts_srt_index").
            Cells(x, 1).Value = x
   Next x

   Range("sel").Copy (Range("hid_dat"))
   Range("sel_err").Copy (Range("hid_err"))
   Worksheets("DataSheet").Range("sel_sizes").RowHeight = 0
   Worksheets("DataSheet").Range("pcnts_srt_index").ColumnWidth = 0
   Worksheets("DataSheet").Range("snap_srt_index").ColumnWidth = 0

   N_rem_frm_sort = 0
   srt_but = Worksheets("Seriation").Buttons.Add(1, 1, 75, 20).Index
   Worksheets("Seriation").Buttons(srt_but).Text = "Sort Types"
   Worksheets("Seriation").Buttons(srt_but).OnAction = "srt_bx"
   snp_but = Worksheets("Seriation").Buttons.Add(76, 1, 75, 20).Index
   Worksheets("Seriation").Buttons(snp_but).Text = "Update Data"
   Worksheets("Seriation").Buttons(snp_but).OnAction = "snap_to"
   rem_but = Worksheets("Seriation").Buttons.Add(151, 1, 75, 20).Index
   Worksheets("Seriation").Buttons(rem_but).Text = "Remove"
   Worksheets("Seriation").Buttons(rem_but).
      OnAction = "remove_from_sort"
   add_but = Worksheets("Seriation").Buttons.Add(226, 1, 75, 20).Index
   Worksheets("Seriation").Buttons(add_but).Text = "Add"
   Worksheets("Seriation").Buttons(add_but).OnAction = "add_dlg"
   test_but = Worksheets("Seriation").Buttons.Add(301, 1, 75, 20).Index
   Worksheets("Seriation").Buttons(test_but).Text = "Test"
   Worksheets("Seriation").Buttons(test_but).OnAction = "test_dlg"
   sv_but = Worksheets("Seriation").Buttons.Add(376, 1, 75, 20).Index
   Worksheets("Seriation").Buttons(sv_but).Text = "Save"
   Worksheets("Seriation").Buttons(sv_but).OnAction = "sv_dlg"

End Sub

Sub srt_bx()
   DialogSheets("Sort_box").ListBoxes(1).RemoveAllItems
   For i = 1 To ncols
      DialogSheets("Sort_box").ListBoxes(1).
      AddItem Text:=Range("type_names").Cells(1, i).Value, Index:=i
   Next i
```

```
   DialogSheets("Sort_box").ListBoxes(1).ListIndex = 1
   DialogSheets("Sort_box").Show
 End Sub

Sub sort_cols()

   col = DialogSheets("Sort_box").ListBoxes(1).ListIndex
   If DialogSheets("Sort_box").OptionButtons(1).Value = 1 Then
      direction = xlAscending
   Else
      direction = xlDescending
   End If
   DialogSheets("Sort_box").Hide
   DialogSheets("Sort_box").ListBoxes(1).RemoveAllItems


   Worksheets("DataSheet").Range("the_sort").
      Sort key1:=Worksheets("DataSheet").Range("sel_pcnt").
      Cells(2, col + 1), order1:=direction, header:=xlYes
   y = 1
   For x = 1 To (nrows)
      If Worksheets("Seriation").GroupObjects(serprog.
            bar_obj(Worksheets("DataSheet").
            Range("pcnts_srt_index").Cells(x, 1).Value)).
            Visible = True Then
        Worksheets("Seriation").GroupObjects(serprog.
            bar_obj(Worksheets("DataSheet").
            Range("pcnts_srt_index").
            Cells(x, 1).Value)).Top = serprog.place_hold(y)
        Worksheets("Seriation").GroupObjects(serprog.
            bar_obj(Worksheets("DataSheet").
            Range("pcnts_srt_index").Cells(x, 1).
            Value)).Left = left_of_units
        y = y + 1
      End If
   Next x
   Worksheets("DataSheet").Range("hid_dat").Copy (Range("sel"))
   Worksheets("DataSheet").Range("hid_err").Copy (Range("sel_err"))

End Sub


Sub cncl_sort()
   DialogSheets("Sort_box").Hide
   DialogSheets("Sort_box").ListBoxes(1).RemoveAllItems

End Sub

Sub Button2_Click()
   sort_cols
End Sub

Sub Button3_Click()
   cncl_sort
End Sub

Sub snap_to()
   Dim snap() As Double
   ReDim snap(nrows) As Double
   For x = 1 To nrows
      snap(x) = Worksheets("Seriation").GroupObjects(serprog.bar_obj(x)).Top
   Next x
   counter = 0
```

```
    For y = 1 To nrows
       For x = 1 To nrows
         If x = Worksheets("DataSheet").Range("pcnts_srt_index").
             Cells(y, 1).Value Then
            Worksheets("DataSheet").
             Range("snap_srt_index").Cells(y, 1).Value = snap(x)
            x = nrows
         End If
       Next x
    Next y

    Worksheets("DataSheet").Range("the_sort").Sort
key1:=Worksheets("DataSheet").Range("snap_srt_index").Cells(1, 1),
order1:=xlAscending, header:=xlYes
    y = 1
    For x = 1 To (nrows)
       If Worksheets("Seriation").GroupObjects(serprog.
             bar_obj(Worksheets("DataSheet").
             Range("pcnts_srt_index").Cells(x, 1).Value)).Visible = True Then
         Worksheets("Seriation").GroupObjects(serprog.
             bar_obj(Worksheets("DataSheet").
             Range("pcnts_srt_index").
             Cells(x, 1).Value)).Top = serprog.place_hold(y)
         Worksheets("Seriation").GroupObjects(serprog.
             bar_obj(Worksheets("DataSheet").
             Range("pcnts_srt_index").
             Cells(x, 1).Value)).Left = left_of_units
         y = y + 1
       End If
    Next x
    Worksheets("DataSheet").Range("hid_dat").Copy (Range("sel"))
    Worksheets("DataSheet").Range("hid_err").Copy (Range("sel_err"))

End Sub



Sub remove_from_sort()
    Dim sel_ob As Object
    Set sel_ob = Selection
    On Error GoTo errorhandler
    Rem z = Selection.RoundedCorners

    For x = 1 To nrows
       Rem If Worksheets("Seriation").GroupObjects(serprog.
             bar_obj(x)).Selected = True Then

       If Worksheets("Seriation").GroupObjects(serprog.
             bar_obj(x)).Index = Selection.Index Then
         For y = 1 To nrows
            If Worksheets("DataSheet").Range("pcnts_srt_index").
                  Cells(y, 1).Value = x Then
               Worksheets("DataSheet").Range("sel_sizes").
             Cells(nrows, 1).Value = Range("unit_names").
             Cells(y, 1).Value
               Worksheets("DataSheet").Range("sel_sizes").
                  Cells(nrows, 2).Value = x
            y = nrows
            x = nrows
            End If
         Next y
       End If
    Next x
```

```
   Worksheets("DataSheet").Range("sel_size").Sort
key1:=Worksheets("DataSheet").Range("sel_sizes").Cells(1, 1), order1:=xlAscending,
header:=xlYes
   N_rem_frm_sort = N_rem_frm_sort + 1
   Selection.Visible = False
   snap_to
   Exit Sub
errorhandler:

   MsgBox "Must select a unit!", 48, "Error"

End Sub

Sub add_box_Button2_Click()
   add_to_sort
End Sub

Sub add_box_Button3_Click()
   cncl_add
End Sub

Sub add_dlg()
   If N_rem_frm_sort < 1 Then
        MsgBox "No Units to Add!", 48, "Error"
        Exit Sub
   End If
   DialogSheets("add_box").ListBoxes(1).RemoveAllItems
   For i = 1 To N_rem_frm_sort
      DialogSheets("add_box").ListBoxes(1).
            AddItem Text:=Worksheets("DataSheet").
            Range("sel_sizes").Cells(i, 1).Value, Index:=i
   Next i
   DialogSheets("add_box").ListBoxes(1).ListIndex = 1
   DialogSheets("add_box").Show

End Sub
Sub cncl_add()
   DialogSheets("add_box").Hide
   DialogSheets("add_box").ListBoxes(1).RemoveAllItems

End Sub
Sub add_to_sort()

   unit = DialogSheets("add_box").ListBoxes(1).ListIndex
   x = Worksheets("DataSheet").Range("sel_sizes").Cells(unit, 2).Value
   Worksheets("DataSheet").Range("sel_sizes").Cells(unit, 1).Clear
   Worksheets("DataSheet").Range("sel_sizes").Cells(unit, 2).Clear
   Worksheets("DataSheet").Range("sel_size").Sort key1:=Worksheets("DataSheet").
      Range("sel_sizes").Cells(1, 1), order1:=xlAscending, header:=xlYes
   Worksheets("Seriation").GroupObjects(serprog.bar_obj(x)).Visible = True
   N_rem_frm_sort = N_rem_frm_sort - 1
   Worksheets("Seriation").GroupObjects(serprog.
      bar_obj(x)).Top = serprog.place_hold(nrows - N_rem_frm_sort)
   Worksheets("Seriation").GroupObjects(serprog.bar_obj(x)).Left = left_of_units
End Sub

Sub test_dlg()
   For x = 1 To 5
      DialogSheets("test_box").ListBoxes(x).RemoveAllItems
   Next x
   For i = 1 To nrows
      DialogSheets("test_box").ListBoxes(1).AddItem
Text:=Worksheets("DataSheet").Range("unit_names").
```

```
        Cells(i, 1).Value, Index:=i
        DialogSheets("test_box").ListBoxes(3).
        AddItem Text:=Worksheets("DataSheet").Range("unit_names").
        Cells(i, 1).Value, Index:=i
    Next i
    For i = 1 To ncols
        DialogSheets("test_box").ListBoxes(4).
        AddItem Text:=Worksheets("DataSheet").Range("type_names").
        Cells(1, i).Value, Index:=i
    Next i
    DialogSheets("test_box").ListBoxes(5).List =
        Array(".25", ".1", ".05", ".025", ".01",
        ".005", ".0025", ".001", ".0005")
    DialogSheets("test_box").ListBoxes(2).List =
        Array("greater than", "not equal to", "less than")
     For x = 1 To 5
        DialogSheets("test_box").ListBoxes(x).ListIndex = 1
    Next x

    DialogSheets("test_box").Show
    DialogSheets("tst_reslts").Show


End Sub

Sub cncl_test()
    DialogSheets("test_box").Hide
     For x = 1 To 5
        DialogSheets("test_box").ListBoxes(x).RemoveAllItems
    Next x
End Sub

Sub test_box_Button2_2_Click()
    DialogSheets("test_box").Hide

    test_units

End Sub

Sub test_box_Button3_Click()

    cncl_test
End Sub

Sub test_box_ListBox5_Change()
    If DialogSheets("test_box").ListBoxes(2).ListIndex = 2 Then
    DialogSheets("test_box").ListBoxes(5).List =
        Array(".5", ".2", ".1", ".05", ".02", ".01",
            ".005", ".002", ".001", ".0005")
    Else
    DialogSheets("test_box").ListBoxes(5).List =
        Array(".25", ".1", ".05", ".025", ".01",
        ".005", ".0025", ".001", ".0005")
    End If
End Sub



Sub test_units()

    unit_1 = DialogSheets("test_box").ListBoxes(1).ListIndex
    unit_2 = DialogSheets("test_box").ListBoxes(3).ListIndex
    typecol = DialogSheets("test_box").ListBoxes(4).ListIndex
```

```
   tail = DialogSheets("test_box").ListBoxes(5).ListIndex
   Dim zvals As Variant
   Dim zval As Double
   zvals = Array(0, 0.6745, 1.2816, 1.6449, 1.96,
      2.3263, 2.5758, 2.807, 3.0902, 3.2905, 3.7)
   zval = zvals(tail)
   U_1 = Worksheets("DataSheet").Range("unit_names").
      Cells(unit_1, 1).Value
   U_2 = Worksheets("DataSheet").Range("unit_names").
      Cells(unit_2, 1).Value
   Ty = Worksheets("DataSheet").
      Range("type_names").Cells(1, typecol).Value
   If DialogSheets("test_box").ListBoxes(2).ListIndex = 2 Then
   al = Array("0", ".5", ".2", ".1", ".05",
      ".02", ".01", ".005", ".002", ".001", ".0005")
   Else
   al = Array("0", ".25", ".1", ".05", ".025",
      ".01", ".005", ".0025", ".001", ".0005")
   End If
   op = Array("0", "greater than", "not equal to", "less than")
   Alpha = al(tail)
   Oper = op(DialogSheets("test_box").ListBoxes(2).ListIndex)
   HO = U_1 & " is equal to " & U_2 & " for type " & Ty
   HA = U_1 & " is " & Oper & " " & U_2 & " for type " & Ty


   If DialogSheets("test_box").ListBoxes(2).ListIndex = 3 Then
      holdr = unit_1
      unit_1 = unit_2
      unit_2 = holdr
   End If
   N_one = Application.Sum(Worksheets("DataSheet").
      Range(Worksheets("DataSheet").Range("data").Cells(unit_1, 1),
Worksheets("DataSheet").Range("data").
            Cells(unit_1, ncols)))
   N_two = Application.Sum(Worksheets("DataSheet").
      Range(Worksheets("DataSheet").Range("data").Cells(unit_2, 1),
Worksheets("DataSheet").Range("data").Cells(unit_2, ncols)))
   psp = (Worksheets("DataSheet").Range("data").
      Cells(unit_1, typecol) + Range("data").Cells(unit_2, typecol)) / (N_one +
N_two)
   p_one = Worksheets("DataSheet").Range("data").
      Cells(unit_1, typecol) / N_one
   p_two = Worksheets("DataSheet").Range("data").
      Cells(unit_2, typecol) / N_two
   z = ((p_one - p_two) / ((Sqr(psp * (1 - psp))) *
       (Sqr((1 / N_one) + (1 / N_two)))))
   If z > zval Then
      res = "Ho rejected at the " & Alpha & " level"
   Else
      res = "Failed to reject Ho at the " & Alpha & " level"
   End If
   DialogSheets("tst_reslts").Labels(1).Text = HO
   DialogSheets("tst_reslts").Labels(3).Text = HA
   DialogSheets("tst_reslts").Labels(6).Text = res

End Sub

Sub tst_reslts_Button1_Click()
   cncl_res
End Sub
```

```
Sub cncl_res()
   DialogSheets("tst_reslts").Hide

End Sub


Sub sv_dlg()
      DialogSheets("sv_box").Show
End Sub

Sub save_ser()
   If DialogSheets("sv_box").EditBoxes(1).Text = "" Then
      MsgBox "Must enter a name!", 48, "Error"
      Exit Sub
   Else
      dat_name = DialogSheets("sv_box").EditBoxes(1).Text & "Data"
      ser_name = DialogSheets("sv_box").EditBoxes(1).Text & "Ser"

   End If
  On Error Resume Next
  Sheets(dat_name).Delete
  On Error Resume Next
  Sheets(ser_name).Delete

   Worksheets("Seriation").Buttons.Delete
   Worksheets("Seriation").DrawingObjects.Group
   Worksheets("Seriation").Name = ser_name

   If DialogSheets("sv_box").CheckBoxes(1).Value = xlOn And N_rem_frm_sort > 0 Then
      On Error Resume Next
      Sheets("newdata").Delete
      Sheets.Add.Select
      ActiveSheet.Name = "newdata"
      Worksheets("newdata").Range(Worksheets("newdata").
            Cells(2, 2), Worksheets("newdata").Cells(3 +
             N_rem_frm_sort, 3 + ncols)).Name = "newdat"
      Worksheets("DataSheet").Range("type_names").Copy
(Worksheets("newdata").Range(Worksheets("newdata").
            Range("newdat").Cells(1, 2), Worksheets("newdata").Range("newdat").
            Cells(1, ncols + 1)))
      z = 2
      For x = 1 To nrows
        For y = 1 To nrows
           If Worksheets("DataSheet").
                  Range("unit_names").Cells(x, 1).
                  Value = Worksheets("DataSheet").Range("sel_sizes").
                  Cells(y, 1).Value Then
              Worksheets("DataSheet").Range(Worksheets("DataSheet").
                  Range("sel").Cells(1 + x, 1), Worksheets("DataSheet").Range("sel").
                  Cells(1 + x, ncols + 1)).Copy
            (Worksheets("newdata").Range(Worksheets("newdata").
                  Range("newdat").Cells(z, 1), Worksheets("newdata").Range("newdat").
                  Cells(z, ncols + 1)))
              z = z + 1
              y = nrows
           End If
        Next y
      Next x
   End If
   If N_rem_frm_sort > 0 Then
      For x = 1 To nrows
        For y = 1 To nrows
           If Worksheets("DataSheet").Range("unit_names").
```

```
      Cells(x, 1).Value = Worksheets("DataSheet").Range("sel_sizes").
            Cells(y, 1).Value And Worksheets("DataSheet").Range("sel_sizes").
            Cells(y, 1).Value <> "" Then
              Worksheets("DataSheet").Range("data").Cells(x, 1).EntireRow.Delete
            x = x - 1
           End If
        Next y
      Next x
      For x = 1 To nrows
        For y = 1 To nrows
           If Worksheets("DataSheet").Range("sel_pcnt_unit_names").
      Cells(x,1).Value = Worksheets("DataSheet").Range("sel_sizes").
      Cells(y,1).Value And Worksheets("DataSheet").Range("sel_sizes").
      Cells(y, 1).Value <> "" Then
    Worksheets("DataSheet").Range("pcnts").
      Cells(x, 1).EntireRow.Delete
           x = x - 1
           End If
        Next y
      Next x
      For x = 1 To nrows
        For y = 1 To nrows
           If Worksheets("DataSheet").Range("sel_unit_names").
      Cells(x, 1).Value = Worksheets("DataSheet").Range("sel_sizes").
            Cells(y, 1).Value And Worksheets("DataSheet").Range("sel_sizes").
                  Cells(y, 1).Value <> "" Then
              Worksheets("DataSheet").
            Range("errors").Cells(x, 1).EntireRow.Delete
           x = x - 1
           End If
        Next y
      Next x
   End If
   Worksheets("DataSheet").Range("sel_sizes").EntireRow.Delete
   Worksheets("DataSheet").Range("pcnts_srt_index").EntireColumn.Delete
   Worksheets("DataSheet").Range("snap_srt_index").EntireColumn.Delete
   Worksheets("DataSheet").Range("hid_dat").EntireColumn.Delete
   Worksheets("DataSheet").Range("hid_err").EntireColumn.Delete

   Worksheets("DataSheet").DrawingObjects.Delete
   Worksheets("DataSheet").Name = dat_name
   DialogSheets("sv_box").EditBoxes(1).Text = ""

End Sub

Sub cncl_sv()
   DialogSheets("sv_box").EditBoxes(1).Text = ""
   DialogSheets("sv_box").Hide
End Sub
```

APPENDIX C: BOOTSTRAP SIGNIFICANCE TEST PROGRAM FOR SERIATION

```
#!/usr/bin/perl
######################################################################
# bootstrap_seriation_significance.pl
#
# Bootstrap Significance Test Program for Seriation
# A simple perl program for evaluating the significance of seriation
# orders given the effects of sample size
#
# Carl Lipo
# Department of Anthropology
# University of Washington
# Box 353100
# Seattle, WA 98195-3100
# clipo@u.washington.edu
#
# Version 1.0.0
# 7 April 1997
#
# Copyright 1997
# Please contact author for information on references and
# more information regarding the use of this program.
# You will need to get the statistics libraries for perl from
# http://www.perl.com/cpan (or any CPAN location).
# http://www.perl.com/CPAN/
#     modules/by-module/Statistics/Statistics-Descriptive-2.1.tar.gz
#
# SYNTAX:  bootstrap_seriation_significance.pl <FILENAME>
# SAMPLE DATA:
# <assemblage>  <counts for each type>
# ClayHill      190     124     1       0       166     0       2
# Soudan        123     34      2       0       55      0       0
# Davis         87      28      2       0       57      4       0
# BelleMeade    761     508     82      36      108     8       17
# SAMPLE OUTPUT:
# <ASSEM 1>     <ASSEM 2>       <p-value>       <std. dev.>
# ClayHill      Soudan          0.376           0.00128
# ClayHill      Davis           0.544           0.000480000000000001
# ClayHill      BelleMeade      0.872           0.00017
# ClayHill                      0.52            0.0011
# Soudan        Davis           0.234           0.00313
# Soudan        BelleMeade      0.634           0.00268
# Soudan                        0.774           0.00028
# Davis         BelleMeade      0.512           0.000970000000000002
# BelleMeade                    0.88            0.0008
#
######################################################################

BEGIN { $| = 1; }

use Statistics::Descriptive;

my $debug = 0;
my @assemblages = ();
my @rowtotals = ();
my @labels = ();
my $numrows = 0;
my $cols = 0;

while( <> ) {
```

```
      print;
      chomp;
      my @line = split /\s+/, $_;
      my $label = shift @line;
      push @labels, $label;
      $cols = scalar( @line );
      my $rowtotal = 0;
      for ( @line ) {
            $rowtotal += $_;
      }
      push @rowtotals, $rowtotal;
      my @freq = ();
      for ( @line ) {
            push @freq, ( $_ / $rowtotal );
      }
      push @assemblages, [ @freq ];
}
$numrows = scalar( @assemblages );

srand( truly_random_value());

my $results = 0;
my $loop = 100;
my $bigloop = 5;
my $loop2 = $loop;
my $ptr1 = 0;
my $ptr2 = 1;

# now do ALL the pairwise assemblage comparisons
# go to sleep and come back later.

while( $ptr1 < $numrows ) {
      while( $ptr2 < $numrows ) {

            my $stat = Statistics::Descriptive::Full->new();

            my @a =  @{ $assemblages[ $ptr1 ] } ;
            my @b =  @{ $assemblages[ $ptr2 ] } ;
            my $numa = $rowtotals[ $ptr1 ];
            my $numb = $rowtotals[ $ptr2 ];

            # calculate the directionality for later
            my @direction = ();
            my $num = scalar( @a );
            my $c = 0;
            for ( $c = 0; $c < $num; $c++ ) {
                  if ( $a[ $c ] < $b[ $c ] ) { push @direction, -1; }
                  elsif ( $a[ $c ] > $b[ $c ] )
                        { push @direction, +1; }
                  else  { push @direction, 0;  }
            }

            my ( @cum_a, @cum_b, $count );
            $classes = scalar( @a );
            my $index_a = 0;
            my $total_a = 0.0;
            my $count = 0;
            for( $count = 0; $count < $classes; $count++ ) {
                  $cum_a[ $index_a ] = $a[ $count ];
                  $total_a += $a[ $count ];
                  $index_a++;
            }
            $classes = scalar( @b );
```

```
my $index_b = 0;
my $total_b = 0.0;
$count = 0;
for( $count = 0; $count < $classes; $count++ ) {
      $cum_b[ $index_b ] = $b[ $count ];
      $total_b += $b[ $count ];
      $index_b++;
}

$index_a--;
$index_b--;

# now we loop 100 times and keep track
my $cycle = $bigloop;
while( $cycle ) {

      # now we loop 1000 times and resample
      $loop = $loop2;
      while( $loop ) {
            my $assemsize = $numa;
            my @assem_a = ();
            my $class;
            my $total = scalar( @a );
            my $rand;
            while( $assemsize ) {
                  $rand = rand;
                  $class = 0;
                  while(( $class < $index_a ) &&
                        ( $rand > $cum_a[ $class ] )) {
                        $rand -= $cum_a[ $class ];
                        $class++;
                  }
                  push @assem_a, $class;
                  $assemsize--;
            }
            my $assemsize = $numb;
            my @assem_b = ();
            $total = scalar( @b );
            while( $assemsize ) {
                  $rand = rand;
                  $class = 0;
                  while(( $class < $index_b )
                        && ( $rand > $cum_b[ $class ] )) {
                        $rand -= $cum_b[ $class ];
                        $class++;
                  }
                  push @assem_b, $class;
                  $assemsize--;
            }


            my ( @ahat, @bhat, %aholder, %bholder );
            %aholder = ();
            %bholder = ();
            my $index = 0;

            for ( $index = 0; $index < $cols; $index++ ) {
                  $ahat[ $index ] = 0;
                  $bhat[ $index ] = 0;
            }

            for ( @assem_a ) {
                  $aholder{ $_ }++;
```

```
            }

            for ( @assem_b ) {
                  $bholder{ $_ }++;
            }

            for ( keys %aholder ) {
                  $ahat[ $_ ] = ( $aholder{ $_ } / $numa );
            }

            for ( keys %bholder ) {
                  $bhat[ $_ ] = ( $bholder{ $_ } / $numb );
            }

            # calculate the directionality for later
            my @dir = ();
            my $num = scalar( @ahat );
            my $c = 0;
            for ( $c = 0; $c < $num; $c++ ) {
            $debug && print "loop $loop ";
            $debug && print $ahat[ $c ] - $bhat[ $c ],"\t";
            if  ( $ahat[ $c ] < $bhat[ $c ] )
                  { push @dir, -1; }
            elsif
                  ( $ahat[ $c ] > $bhat[ $c ] )
                  { push @dir, +1; }
            else  { push @dir, 0;  }
            }
            $debug && print "\n";

            # compare the two sets of results
            $num = scalar( @dir );
            $c = 0;
            my $diff = 0;
            for ( $c = 0; $c < $num; $c++ ) {
                  $debug && print "loop $loop ",
                  $direction[ $c ],"/",$dir[ $c ],"\t";
                  if ( $direction[ $c ] ==
                         $dir[ $c ] ) { next; }
                  $diff++;
            }
            $debug && print "\n";
            if ( $diff == 0 ) { $results++; }

            $loop--;
      }

      $debug && print "Results:  $results matches of";
      $debug && print " $loop2 trials\n";
      $debug && print "Probability: ";
      $debug && print $results / $loop2, "\n";

      $stat->add_data( $results/ $loop2 );
      $cycle--;

      $results = 0;

}

print $labels[ $ptr1], "\t", $labels[ $ptr2 ], "\t";
print $stat->mean(),"\t";
print $stat->variance(),"\n";
```

```
            undef $stat;
            $ptr2++;
        }
        $ptr1++;
        $ptr2 = $ptr1 + 1;

}
```

APPENDIX D: RANDOM WALK NEUTRALITY TESTING PROGRAM

```
#!/usr/bin/perl
######################################################################
# bootstrap_neutral_eval.pl
#
# Neutral Significance Test Program for Seriation
# A simple perl program for evaluating the assumptions of neutrality
#
# Carl Lipo
# Department of Anthropology
# University of Washington
# Box 353100
# Seattle, WA 98195-3100
# clipo@u.washington.edu
# (206) 543-5240
#
# Version 1.0.0
# 2 January 2000
#
# Copyright 2000
# Please contact author for information on references and
# more information regarding the use of this program.
# You will need to get the statistics and math libraries for PERL from
# http://www.perl.com/cpan (or any CPAN location).
#
# http://www.perl.com/CPAN/modules/by-module
#/Statistics/StatisticsDescriptive-1.1.tar.gz
#
# SYNTAX:  bootstrap_neutral_eval.pl <FILENAME>
# SAMPLE DATA:
#<assemblage name>     <counts for each type>
# site1                100     100     0       0
# site2                0       0       100     100
# site3                33      33      33      33
# site4                33      33      33      33
#
# SAMPLE OUTPUT:
#  (values are separated by the '|' mark for easy importation into
#     Microsoft Excel.
# Colln|Sample|I|S| MeanI|VarianceI|Exps|Theta|ExpI
# site1|200|0.5|2|0.502589900000001|1.35792922822825e-
#       05|1.9983970550615|0.179|0.848176420695505
# site2|200|0.5|2|0.502488050000001|1.26901298273275e-
#       05|1.99839705506158|0.179|0.848176420695505
# site3|132|0.25|4|0.255639118457301|1.90964582724237e-
#       05|3.99755581456441|0.636|0.611246943765281
######################################################################


use Statistics::Descriptive;
BEGIN { $| = 1; }

my $debug = 0;

my @assemblages = ();
my @rowtotals = ();
my @labels = ();
my $numrows = 0;
my $cols = 0;
```

```
while( <> ) {
       chomp;
       my @line = split /\s+/, $_;
       my $label = shift @line;
       push @labels, $label;
       $cols = scalar( @line );
       my $rowtotal = 0;
       for ( @line ) {
            $rowtotal += $_;
       }
       push @rowtotals, $rowtotal;
       $sizestat=new Statistics::Descriptive;
       my @freq = ();

       for ( @line ) {
            $i++;
            $statobj= new Statistics::Descriptive;
            push @arraystat, $statobj;
            push @freq, ( $_ / $rowtotal );
            $sizestat->AddData($rowtotal);
       }
       push @assemblages, [ @freq ];
}
$numrows = scalar( @assemblages );
$numrows = scalar( @assemblages );
$cols=scalar(@freq);

my $results = 0;
my $loop = 100;
my $bigloop = 10;
my $loop2 = $loop;
my $ptr1 = 0;
my $ptr2 = 1;


# now do ALL the pairwise assemblage comparisons
# go to sleep and come back later.

while( $ptr1 < ($numrows-1) ) {

            my @a =  @{ $assemblages[ $ptr1 ] } ;
            my @b =  @{ $assemblages[ $ptr2 ] } ;
            my $numa = $rowtotals[ $ptr1 ];
            my $numb = $rowtotals[ $ptr2 ];

            print $labels[$ptr1]," = ",$labels[$ptr2],"\n";
            # calculate the directionality for later
            my @direction = ();
            my $num = scalar( @a );
            my $c = 0;

            my ( @cum_a, @cum_b, $count );
            $classes = scalar( @a );
            my $index_a = 0;
            my $total_a = 0.0;
            my $count = 0;
            for ( $count = 0; $count < $classes; $count++ ) {
                 $cum_a[ $index_a ] = $a[ $count ];
                 $total_a += $a[ $count ];
                 $index_a++;
            }
            $classes = scalar( @b );
            my $index_b = 0;
```

```
my $total_b = 0.0;
$count = 0;
for( $count = 0; $count < $classes; $count++ ) {
     $cum_b[ $index_b ] = $b[ $count ];
     $total_b += $b[ $count ];
     $index_b++;
}

$index_a--;
$index_b--;

# now we loop 100 times and keep track
my $cycle = $bigloop;
while( $cycle ) {

     #print "(debug) cycle value: $cycle\n";

     # now we loop 1000 times and resample
     $loop = $loop2;
     while( $loop ) {
           my $assemsize = $numa;
           my @assem_a = ();
           my $class;
           my $total = scalar( @a );
      my $total = scalar( @a );
           my $rand;
           $assemsizea=$assemsize;

           while( $assemsize ) {
                $rand = rand;
                $class = 0;
                while(( $class < $index_a )
                   && ( $rand >$cum_a[ $class ] )) {
                     $rand -= $cum_a[ $class ];
                     $class++;
                }
           push @assem_a, $class;
           $assemsize--;
           }


           my $assemsize = $numb;
           my @assem_b = ();
           $total = scalar( @b );
           $assemsizeb=$assemsize;
             while( $assemsize ) {
                $rand = rand;
                $class = 0;
                while(( $class < $index_b )
                         && ( $rand >$cum_b[ $class ] )) {
                     $rand -= $cum_b[ $class ];
                     $class++;
                }
                push @assem_b, $class;
                $assemsize--;
           }

           my ( @ahat, @bhat, %aholder, %bholder );
           %aholder = ();
           %bholder = ();
           my $index = 0;
           for ( $index = 0; $index < $cols; $index++ ) {
                $ahat[ $index ] = 0;
```

```
                              $bhat[ $index ] = 0;
                         }
                    for ( @assem_a ) {
                         $aholder{ $_ }++;
                    }
                    for ( @assem_b ) {
                         $bholder{ $_ }++;
                    }
                    for ( keys %aholder ) {
                         $ahat[ $_ ] = ( $aholder{ $_ } / $numa );
                    }
                    for ( keys %bholder ) {
                         $bhat[ $_ ] = ( $bholder{ $_ } / $numb );
                    }
                    # calculate the directionality for later
                    my @dir = ();
                      my $num = scalar( @ahat );
                    my $c = 0;
                    $numclasses=$num;
                    for ( $c = 0; $c < $num; $c++ ) {

                         $differ=( $ahat[$c] -$bhat[$c]);
                         $statinstance= $arraystat[$c];
                         $statinstance->AddData($differ);
                         $times[$c]++;
                    }
                    push @difftotals, [ @diffsum ];
                    $difcount++;
                    $debug && print "\n";
                    $loop--;
               }
               $cycle--;
               $results = 0;
          }
          $ptr1++;
$ptr2=$ptr1+1;
}
print "Results\n";
print "Type: \t Mean \t Std.  Dev.\tN\t95% lower\t 95% upper \t Includes Zero?\n";
for ( $c = 0; $c < $numclasses; $c++ ) {
      print $c,"\t";
      $statinstance=$arraystat[$c];
      print $statinstance->Mean(),"\t";
      print $statinstance->StandardDeviation(),"\t";
      $mean=$statinstance->Mean();
      $stdev=$statinstance->StandardDeviation();
      $n=$times[$c];
      $lower=$mean-2*($stdev/sqrt($numrows));
      $upper=$mean+2*($stdev/sqrt($numrows));

      if ($lower <= 0.0 && $upper >= 0.0) {
            $answer="YES"; } else {
           $answer="NO";}
      print $numrows,"\t",$lower,"\t",$upper,"\t",$answer,"\n";
}
print "\n";
1;
```

APPENDIX F: LABORATORY PROCESSING AND DECORATIVE DESCRIPTIONS

Material from each location was stored in plastic bags. Each bag contained the material from a single unit within a location that was collected by an individual. For identification purposes, I labeled each of the collection bags with a sequential number within each site. Processing of the contents of the bags consisted of a series of steps:

1. The sherds in each collection bag were passed through a series of five nested geological sieves (-5.0 phi, -4.5 phi, -4.25 phi, -4.0 phi, -3.25 phi) and shaken for one minute. The sherds in the sieve were checked to be certain that minimum diameters were not less than the sieve size. Sherds that were larger than the minimum diameter of the mesh were recorded as that size class. For each bag, counts were made for each size class.

2. Each sherd was evaluated with respect to decoration. To be appropriate for analysis a sherd had to have at least one unambiguous decorative element. If the sherd was decorated, each element was described in detail (see below). If not, it was removed from analysis. In some cases, sherds were found to lack decoration after a sequential number had been assigned to them. In order to insure that sherds were not accidentally skipped, these sherds were described but their descriptive entries were left blank. In the subsequent numerical analyses, the entries for these non-decorated sherds were easily removed.

3. Each decorated sherd was given a number. Numbers given were sequential for each collection bag and written carefully on each sherd using a 0.35-mm Rotring Rapiograph pen.

4. Sherds were described and recorded on standardized forms. Workers were provided detailed instructions for filling out the forms. Entries on the forms were coded to permit fast efficient data entry. I generated all these of these forms using Teleform by Cardiff Software (see http://www.cardiffsw.com for more information), commercially available form generation software that was capable of optical character recognition (OCR) and automated data entry. I chose Teleform software (v.5.0) to build the sherd description forms as it permitted forms to contain multiple choice, constrained text, and freeform text entries. Once the forms were completed, I used a Pentium Pro 200 computer with attached scanner and sheet feeder (Epson 636 Expressionist) to scan and optically process the forms with the Teleform software. Once scanned and processed, the sherd data were easily exported in a spreadsheet format (Microsoft Excel 7.0) for tabulation and analysis.

5. The information collected about each sherd and the characteristics of the decoration was extensive. Workers were issued an identification key for describing the sherds and filling out the descriptive forms. For each sherd, the following information was recorded:

A. Collection Location: The alphanumeric characters that designate the collection location. For the purposes of these analyses, locations were coded as follows:

CL = Castile Landing	BE = Beck
BM = Belle Meade	HL = Holden Lake
CP = Cranor Place	RM = Rose Mound
NI = Nickel	

B. Sherd Type: The portion(s) of the vessel from which the sherd is derived. Sherd types include indeterminate, rim, neck, body, base, handle, and rider. The following definitions were used in the sherd type identifications (from Krause 1990):

Pottery:	Pottery is an object of the class of intentionally manufactured objects of prepared and fired clay.
Vessel:	Any member of the class of concave utensils designed to hold any substance.
Sherd:	Any fragment of a pottery vessel.
Surface:	Any two dimensional locus of points.
Exterior:	The surface of pottery vessel that is proportionally the greatest
Interior:	The surface of a pottery vessel that is proportionally the smallest.
Lip:	The intersection of the exterior and interior pottery vessel surfaces.
Rim:	The part of the upper body of a vessel that includes both the mouth and the lip.
Rim:	The part of the upper body of a vessel that includes both the mouth and the lip.
Mouth:	The minimal circumference of the upper part of the body.
Body:	The part of a vessel that does not include the lip and mouth or neck. The body is identified as having a vertically convex exterior surface.
Shoulder:	The part of a pottery vessel body with the maximal circumference.
Collar:	The inflection point joining the upper with the neck.
Bottom:	The part of a pottery vessel body with minimal circumference.
Neck:	The part of a pottery vessel uniting the rim with the shoulder including the collar. The neck is identified as having a vertically concave exterior surface.
Handle	Clay additions attached at two vertically aligned points on the rim and neck.
Rider	Clay additions attached only at one point on the rim.

C. Size Class: The graduated screen size that represents the minimum diameter of the sherd in phi size.

D. Thickness Class: The maximum thickness of the sherd in millimeters. Thicknesses were recorded as a series of mutually exclusive size classes:

Ind. =Indeterminate	6 = 4-6mm
1 = <1 mm	8 = 6-8mm
2 = 1-2mm	10 = 8-10mm
4 = 2-4mm	>10 = >10mm

E. Firing: Firing information consists of patterns of oxidation and reduction observed in terms of colors in the cross section of the sherd. Patterns that were recorded consist of fully oxidized (all buff colored), fully reduced (all dark brown colored), interior oxidized/exterior reduced (buff on the interior portion of the sherd, dark brown on the exterior), exterior oxidized/interior reduced (buff on the interior portion of the sherd, dark brown on the exterior), oxidized between reduced (a band of buff surrounded by dark brown on both sides) and reduced between oxidized (a band of buff surrounded by dark brown on both sides). A comparison gauge was used to aid the description of firing information (e=exterior, i=interior):

F. Culture-Historical Type: The culture-historical type identified using the PFG classification. The criteria for these identifications are listed in Chapter 5.

G. Sherd Orientation: Sherd orientation refers to the degree to which the sherd can be oriented to the vessel. Distingihsing vertical and horizontal axes on body sherds from spherical bodies can be difficult. The inability to do so, however, means that the axes are not measurably different. There are four possibilities for sherd orientation.

a. *Indeterminate*: No surfaces are recognizable.

b. *Interior/Exterior Surfaces*: As pots are closed curves in the horizontal dimension, all potsherds from round pots must have one axis that is convexly curved when viewed from the outside. In many cases, both axes are convexly curved. When this situation is encountered, the convex surface is the outside and the concave surface is the inside. In this case, only the interior and exterior surfaces are recognizable.

c. *Horizontal/Vertical Axes:* Most variation in curvature occurs in the vertical axis of pots. Most pots approximate a circle in horizontal section. The most nearly circular curve is the horizontal axis. In this case, sherds can recognized as to the interior/exterior surface *and* the horizontal/vertical axes.

d. *Left/Right/Up/Down:* Due to the presence of a rim or base, sherds can be oriented as to the left, right, up, and down directions (as well as interior/exterior, horizontal and vertical).

H. Interior Surface Condition: The condition of the interior surface of a sherd indicating whether the surface of the sherd is worn, eroded or fresh. Fresh surfaces are identified where the original surface is extant, incisions and punctates have sharp edges, and painted surfaces are complete. Worn surfaces are identified as those in which at least 50% of the edges of surface modifications (punctations or incisions) are rounded or smoothed and >50% of the original surface is still extant. Eroded surfaces are those with barely visible decoration, in which <50% of the original surface is extant and the >50% of the edges of surface modifications are smoothed or rounded.

I. Exterior Surface Condition: The condition of the exterior sherd surface. The same criteria for identifying interior surface conditions were used for identifying the exterior surface conditions.

J. Surface Modification: The characteristics of the surface of the sherd. Rough, smooth, polished. Polished surfaces are continuous, planar, have no irregularities, and reflect light when applied at an oblique angle to the sherd. Smooth surfaces are those that are continuous, planar, have no irregularities but are matte and do not reflect light. Rough surfaces are matte but also have irregularities and do not consist of continuous plane.

K. Temper: Temper is the non-plastic inclusions that are added to clay paste. Two kinds of temper are identified in the late prehistoric Mississippian ceramics: clay and shell. Shell-tempering was identified as any paste with visible shell in sherd cross-sections. Shell appears as white or gray-white, platy shapes. Clay tempering refers to paste tempered with clay instead of shell. This tempering is identified as blocky grains in sherd cross-sections that are the same color as the surrounding paste.

DECORATIVE ANALYSES

Along with the element analysis, I also created coded descriptions of decoration on the sherds. To allow flexibility of variability, decoration was treated as combinations of individual elements and their relationships to each other (*i.e.,* designs). Combinations of element descriptions were used to evaluate variability of PFG types since in these analyses, elements and element relationships were described in more detail than the PFG ceramic types. This descriptive system permitted me to combine element form descriptions and design variability to build a large number of potential classification units that were then evaluated as potential measurement units of homology (Chapter 8).

The design analysis started by identifying *elements*, the minimum repeated modification to the surface of a pottery vessel that comprises decoration. Elements were identified as any repeated deviation to the planar surface of a sherd that was visible to the naked eye. Elements were analyzed along the dimensions described below. At the next scale of analyses, the relationship and arrangement of elements to each other were described as *element sets*. Finally, combinations of element sets and their relationships were used to form *designs*. In the descriptions, two versions of the recording form were used (Appendix G). The 'Short Form' was used in cases where only a single element set was present. The 'Long Form' was used whenever more than one element set was described on a single sherd.

ELEMENT ANALYSES

Element Form: Element form describes the way in which the decorative element is related to the surface of the sherd. Elements can extend into the sherd surface (*e.g.,* punctates, incising), extrude out of the surface (*e.g.,* noding or modeling) or be located on the surface (*e.g.,* painting).

Element Type: Element type is the shape of the outline of the element. Element types can be linear (e.g., incising, painted lines), points (*e.g.,* nodes, punctates), surface (*e.g.,* painting) or complex (*e.g.,* cord-marking, stamping).

Element Paste: The element paste is the hardness of the paste when element was vessel. This description varies from very hard (broken chipped edges), hard (fine linear striations along application; no lip), medium (clear linear striations along application, small lip or overhanging edge) or soft (significant overhanging edges along application, broad striations along application).

Linear Elements

Linear elements are elements that are much greater in length than width. They were described in terms of:

Linearity: Linearity identifies whether the element is curvilinear, rectilinear, or complex (combination of both).

Linear Outline: Linear outline is the configuration or outline of the linear element. Modes used are line segments (unidentified portions of a linear element), single lines, intersecting lines (two or more lines intersecting each other at a 90° [± 5°] to each other), crossed lines (two or more lines intersecting each other at a <90° angle to each other), and complex figures.

Line Cross Section: Line cross section is the cross-section of the linear element. Modes used were rounded, triangular, rectangular, and complex.

Line Width: Line width is the width of the linear element. Line width categories were chosen to be consistent with PFG types (*e.g.,* Barton Incised vs. Wallace Incised [O'Brien 1994]) and consist of the following choices.

V. Narrow = <1.0 mm **Narrow** = 1.0 - 2.0 mm
Medium = 2.0 -3.0 mm **Wide** = (3.0 - 4.0 mm
V. Wide = >4.0 mm

Linear Element Surface: Linear element surface is the condition of the terminal surface of the element. Mode choices are rough (linear striations parallel to direction of line) or smooth (no linear striations).

Point Elements

Point element types such as "punctates" or "nodes" were described by a number of specific attributes:

Point Type Horizontal Outline: Point type horizontal outline is the shape of the element relative to the surface of the sherd and is recorded as triangular, rectangular, circular, semi-circular, lenticular, crescent, or complex (*i.e.,* non regular in shape).

Point Type Element Form: Point type element form describes the hole that forms the element. Element forms were described as solid (interior absence so the element consists of a simple "hole"), hollow (interior present, as when the end of a reed is used to make an indentation into a surface) or not applicable (as in the case of "nodes").

Point Type Element Lip: Point type element lip is the "burr" or extruded material caused by the insertion of the tool into the clay surface of a vessel at an oblique angle. Point type lips were described as none (no burr), little (minimal presence of burr, a simple ridge of clay less than depth of punctate), medium (ridge of clay as large as depth of punctate, no overlapping or curling), lots (large ridge of clay with a curled-over appearance) or non-applicable (as in the case of nodes).

Point Type Symmetry: Point type symmetry is the degree to which the horizontal outline is symmetric or asymmetric. Elements were described relative to a Cartesian grid using the direction of application for orientation. As shown in Figure 8.2, the direction of application was identified by the alignment of linear striations on the surface of the element and by the presence of a characteristic "lip" at the terminus of the tool mark (in the case of punctates). The Y-axis was defined as the direction of application and the x-axis was defined as perpendicular to this direction. In terms of symmetry, elements were described as having an X-axis wider than Y-axis, symmetric (X-axis equivalent to Y-axis) or having a Y-axis greater than the X-axis.

Point Type Angle: Point type angle is the angle between the surface of the sherd and the plane of greatest element depth. The angle is recorded as one of three categories: less than 45 degrees, greater than 45 degrees but less than 90 degrees, and 90 degrees (± 5°).

Point Element Surface: Point element surface describe the characteristics of the terminal surface of the element. This description can have the value of rough or smooth.

Surface Elements

Surface element types (such as those found decoration consisting of slipping or painting) are described with a unique set of attributes.

Surface Type Color: Surface type color is the color of the element added to the surface of the sherd. In late prehistoric Mississippian ceramics, surface elements are one of four colors: red, buff, black, and white.

Surface Application: Surface element application is the characteristic of application of the surface element: smooth or irregular.

ELEMENT-SET ANALYSES

Element-sets consist of groups of repeated elements (as listed above). Element-sets were described with a series of attributes about the relationships of elements to each other and to the sherd.

Element-Set Pattern: Element-set pattern is an assessment of whether or not the element set forms a linear/curvilinear pattern or is a non-linear association of elements.

Element-Set Linearity: Element-set linearity is an indication of whether elements in the set have a rectilinear or curvilinear relationship. Curvilinear element sets are those in which sequential elements form a curved pattern. Rectilinear element-sets are those that are composed of elements in a straight-line pattern.

Element-Set Type: Element-set type distinguishes whether the elements in the set form a fill pattern, form an outline around a pattern (a "boundary"), or are isolated. Boundary sets occur in a situation where one element set forms a boundary around another so its shape takes an outline form of the fill element set.

Element Set Spacing: Element-set spacing describes the spacing between element set.

Element Set Alignment: Element-set alignment is the horizontal relationship (in two dimensions) of elements in a set. Elements are arranged in parallel sets (elements placed with long axes parallel but not in a grid pattern), rectilinear sets (elements placed such that they are in a grid patterns; all axies are aligned), crossed sets (elements that overlap), concentric sets (sets that that surround one another) and sequential (elements placed one after another so that their long axes are sequentially aligned). Element-set alignment was described by whether the long axes of the elements were in aligned in one dimensions (as in the parallel or sequential alignment types) or two dimensions (as in a grid alignment pattern) and whether the elements overlap.

Element-Set Outline Shape: Element-set outline shape refers to overall shape of the element set. Outlines for element-sets include bands, circles, rectangles, trapezoids, parallelogram, triangles, and semicircles. In addition, the element-set shape category specify whether the outline shapes are symmetrical (*e.g.,* circle, square) or asymmetrical (*e.g.,* oval, rectangular).

Element Set Orientation: Orientation of elements in relationship to each other given the relationship of the elements to the vessel (see Point G). Elements can be arranged horizontally, vertically, or obliquely relative to the vessel orientation.

Element Set Application Orientation: This refers to the orientation of the application of the element in the element set relative to the sherd. Element set orientation includes none, left-to-right, right-to-left, up-to-down, down-to-up, northwest-to-southeast, southeast-to-northwest, northeast-to-southwest and southwest-to-northeast. The orientation of set can only be determined when the sherd can be oriented in relative to the vessel.

Number of Elements: Number of elements in the set.

Element-Set Completion: Set completion refers to whether the set is "complete" or not; is it possible to count all of the repeated elements in the set? This description is useful for determining if the number of the elements in the set can be compared with the number of elements in other sets.

Element-Set Panel: The sherd panel in which the element set is located. Panels are surface of pottery vessels defined by breaks in curvature (Krause 1990). Modes include the rim, plano, concave, convex, rider, and handle.

Element-Set Location: The location of the set on the sherd. Location modes were defined as exterior, interior, and top of rim.

Element-Set Boundary: Element set boundary attributes indicates whether is element set is bounded by incised or engraved lines.

DESIGN ANALYSES

The next scale of analysis consists of evaluating designs: arrangement of element sets in the same panel or contiguous panels. Sherds can have multiple designs composed of multiple combinations of element sets.

Design Set-to-Set Relations: Design set-to-set relations describe how (given sherd orientation [see Point G]) element sets are arranged in relationship to each other. Element-sets can be placed side-by-side, as boundary of one to another, overlapping one another, vertically, horizontally, angularly placed next each other or placed over, under each other, to the left, to the right, above and to the left ('northwest'), above and to the righ ('northeast'), below and to the left ('southwest') and below to the right ('southeast').

Design Rotation: If the element sets are the same but simply rotated with relationship to each other, the design rotation attribute records the degree of rotation.

Design Flipping: Design flipping is an indication of whether the element sets are the same but flipped (*i.e.,* a mirror image) with relationship to each other.

WORK EFFORT

Each of the decorated sherds was described and cataloged following the protocol described above. The process was efficient but given the volume of ceramics and the detailed and careful way in which the descriptions were made, the project consumed several months months of work. Eight people assisted in this project for three to ten hours per week during the sherd description process. Despite this assistance, the overall description process took approximately one year of nearly full-time work (2-4 individuals at any point in time). Over this year, 50,366 sherds were processed. Of these sherds, 9,131 were decorated.

ERROR CONSIDERATION

All sherds were examined at least twice to reduce the number of blunders and assignment inconsistencies. Each pair of descriptions was matched and all discrepancies between sets of descriptions were resolved by returning to the sherd and assigning the appropriate descriptive code. Describing the set of sherd twice also reduced the effects of the learning curve whereby early descriptions differed from later descriptions of the same types of sherd and of "drift" in the descriptions as decisions involved in the description process wandered during the course of the analyses (Dunnell *et al.* 1986).

Since the optical character recognition (OCR) process potentially produces data entry errors, procedures were put into place to minimize these problems. First, fill-in bubbles were used on the recording form whenever possible. Since the values for these fields depends entirely on spatial positioning of the entire form itself rather than character recognition fill-in bubbles have great precision in data entry. This is particularly true for Teleform-produced forms that use spatial guides on the corners of the form to precisely orient the form during the scanning and recognition process. Tests on these kinds of entry fields showed that misrecordings occurred with a rate of less than 0.1%. However, when space would not allow enough fill-in bubbles to fit on the form for a particular category, numeric codes were used for data entry. Although differences in handwriting potentially produce misreadings of the values, by restricting the possible set of codes that could be entered into any one field I was able to catch most errors during the scanning process. In these cases, the Teleform software simply flagged values that were not pre-defined as it converted the optical image into electronic data. For example, if the categories 1-4 were set to be allowable for a particular field, the software did not permit the entry of values of five or greater. All flagged errors were checked for handwriting problems, blunders in descriptions, or scanning software misreads and were subsequently corrected. Tests of the data entry precision were conducted by comparing multiple forms of input with each other (comparison of written and checkbox) and counting mismatches. These tests showed that error rates were much less than 0.5%.

The rather detailed manner in which the descriptions were conducted also permitted a significant amount of internal error checking. Since descriptive categories overlapped with one another, it was easy to cross-validate whether values were consistently recorded correctly. Values for a painted element, for example, must also be associated with a painted culture-historical type and only values applicable for surface elements and arrangements associated with painted decorative sets should be associated with that decription. If descriptive values for a point-element type were listed for a sherd elsewhere described as painted, these values would stand out in the spreadsheet sorting and tabluation. In this way, many blunders made during sherd description and the OCR process were easily caught.

INDEX